Power Systems

Electrical power has been the technological foundation of industrial societies for many years. Although the systems designed to provide and apply electrical energy have reached a high degree of maturity, unforeseen problems are constantly encountered, necessitating the design of more efficient and reliable systems based on novel technologies. The book series Power Systems is aimed at providing detailed, accurate and sound technical information about these new developments in electrical power engineering. It includes topics on power generation, storage and transmission as well as electrical machines. The monographs and advanced textbooks in this series address researchers, lecturers, industrial engineers and senior students in electrical engineering.

Power Systems is indexed in Scopus

S. Vijayalakshmi • J. Lekha • Lija Jacob
Savita Dahiya • R. Gunavathi
Editors

Smart Power Systems

Grid Modernization Using AI and IoT-Based Applications

Springer

Editors
S. Vijayalakshmi
School of Sciences
Christ University
Bengaluru, Karnataka, India

J. Lekha
School of Sciences
Christ University
Bengaluru, Karnataka, India

Lija Jacob
School of Sciences
Christ University
Bengaluru, Karnataka, India

Savita Dahiya
School of Engineering & Technology
Maharishi University of Information Technology
Noida, Uttar Pradesh, India

R. Gunavathi
School of Sciences
Christ University
Bengaluru, Karnataka, India

ISSN 1612-1287 ISSN 1860-4676 (electronic)
Power Systems
ISBN 978-3-031-89458-9 ISBN 978-3-031-89459-6 (eBook)
https://doi.org/10.1007/978-3-031-89459-6

This Springer imprint is published by the registered company Springer Nature Switzerland AG
The registered company address is: Gewerbestrasse 11, 6330 Cham, Switzerland

Contents

Smart Intelligence Perspective and Integrating AI with the Power System

Gaurika Chopra and Lija Jacob

1 Introduction

In the dynamic landscape of modern energy consumption, the intricate blend between technology and power systems has been nothing short of revolutionary. The evolution of power systems has been driven by an insatiable global appetite for reliable, sustainable, and efficient energy solutions. Concurrently, the rise of Artificial Intelligence (AI) has ushered in an era of unprecedented computational capabilities, enabling machines to mimic human intelligence and make decisions autonomously. The convergence of these two realms holds the potential to revolutionise how we generate, distribute, and manage electrical power.

To appreciate the profound impact of integrating AI in power systems, it is essential to delve into the historical backdrop of both these domains. Power systems, which form the backbone of modern civilisation, have evolved from humble beginnings, marked by Thomas Edison's groundbreaking inventions [1], to sprawling grids that span continents. Simultaneously, AI has emerged from its nascent stages in computer science labs to become a transformative force, influencing industries from healthcare to finance. The symbiosis between these two fields represents a pivotal juncture in our pursuit of a sustainable and intelligent energy future. Knowing the potential for synergy between these two fields became crucial as we approached the dawn of a new age in energy technology. When combined, they promised to optimise not just the power infrastructures that were already in place but also to provide the framework for a more robust and sustainable energy landscape.

In the pursuit of more intelligent and effective energy management, integrating artificial intelligence (AI) with power systems marks a major advancement. The

G. Chopra (✉) · L. Jacob
Christ University, School of Sciences, Lavasa, Pune, India

S. Vijayalakshmi et al. (eds.), *Smart Power Systems*, Power Systems,
https://doi.org/10.1007/978-3-031-89459-6_1

way power systems and smart intelligence come together might completely change how we produce, distribute, and use energy. We can achieve new heights in grid stability, predictive maintenance, and optimisation by utilising AI's analytical powers. This introduction lays the groundwork for investigating how artificial intelligence is revolutionising power systems and opening the door to a more technologically sophisticated and sustainable energy future.

2 Fundamentals of AI in Power Systems

In the realm of power systems, artificial intelligence (AI) is becoming more and more significant. It could enhance grid resilience, streamline operations, and make it possible to include renewable energy sources. AI can be used in power systems for a variety of purposes. Here are a few of them:

- **Load Forecasting.**

- Based on past data, weather predictions, and other pertinent variables, AI systems can be used to forecast future electricity demand. Energy providers may plan for effective energy generation and delivery by using accurate load forecasting.
- **Renewable Energy Integration.**
- AI can aid in the more efficient integration of renewable energy sources like solar and wind into the grid. It has the ability to estimate the production of renewable energy, estimate energy generation trends, and enhance energy storage.
- **Predictive Maintenance**
- AI is capable of predicting when components like transformers, circuit breakers, and generators are most likely to fail. This enables preventative maintenance to be carried out, minimising costs and downtime.
- **Fault Detection and Diagnostics**
- AI systems may examine sensor data to locate and assess grid issues. This aids in swiftly identifying problems and fixing them, improving grid dependability.
- **Optimal Power Flow**
- A power grid's electrical flow can be optimised by AI to reduce costs and increase efficiency. It takes into account elements like transmission restrictions, generation capacity, and demand patterns.
- **Grid Reliability**
- AI algorithms can analyse data from sensors to identify and diagnose faults in the power grid. This helps in quickly identifying and rectifying issues, thereby enhancing grid reliability.
- **Grid Continuity**
- Whether triggered by cyberattacks, natural disasters, or other emergencies, AI can assist in identifying and responding to grid disturbances. It enables services to be restored more quickly.
- **Cybersecurity**

- AI is useful for monitoring network traffic and identifying irregularities that may indicate a cyberattack. It can assist in locating and fixing security loopholes.
- **Smart Grids with Demand Response**
- AI plays a major role in enabling demand response programmes, which allow customers to adjust their power use in response to grid circumstances. Additionally, it makes smart grids more functional by enabling two-way communication between utilities and customers.
- **Optimisation of Energy Storage**
- The ideal times for charging and discharging energy storage systems can be determined using AI algorithms. By doing this, the use of stored energy is maximised.
- **Forecasting the Market and Trading**
- In order to anticipate electricity prices, AI can examine market data, weather forecasts, and other pertinent variables. For market participants like utilities and energy traders, this information is useful.
- **Regulation of Voltage and Frequency**
- Real-time voltage and frequency adjustments can be made by AI-based control systems to ensure a stable power supply within predetermined bounds.
- **Analysing the Environmental Impact**
- AI can be used to analyse the environmental effects of various energy producing techniques, assisting utilities in making more environmentally friendly choices.
- **Grid Expansion and Planning**

- By taking into account variables like demand expansion, the best location for new infrastructure, and cost-effectiveness, AI can help with planning and engineering the expansion of electricity grids.

Engineers and other professionals in the power systems sector must comprehend these foundations in order to fully realise the potential of AI for a future with more efficient, dependable, and sustainable energy sources.

3 AI-Based Algorithms for Power Systems

Machine learning and deep learning algorithms are the foundation of AI in power systems. Machine learning, a branch of artificial intelligence, makes it possible for computers to learn without explicit programming. This is accomplished by providing the computer with data so that it can recognise patterns and anticipate the future. Machine learning algorithms come in numerous forms; however, some of the more popular ones utilised in power systems are as follows:

When the intended result is known, supervised learning is employed. For instance, you can use supervised learning to train a model using historical load data in order to estimate the load on a power system. When the desired outcome is unknown, unsupervised learning is employed. Unsupervised learning, for instance, can be used to group data points based on their commonalities. This can be helpful for figuring out various load patterns.

When learning how to act in a situation to get the most out of a reward is the goal, reinforcement learning is applied. For instance, reinforcement learning can be used to train a model to regulate a power system in order to reduce power losses.

The precise problem you are seeking to tackle will determine which machine learning method is used. For instance, supervised learning is a fantastic option if you are attempting to anticipate the demand on a power system. Unsupervised learning is a fantastic option if you're seeking to group data points together. Additionally, reinforcement learning is an excellent option if you're seeking to learn how to manage a power system.

Other AI methods are also applied in power systems in addition to machine learning algorithms. These consist of:

- **Human language** is recognised and processed using natural language processing. Tasks like customer service chatbots and power system security analyses may benefit from this.
- **Images and videos** can be processed and understood using computer vision. Tasks like fault detection and asset inspection may benefit from this.

Uncertainty is reasoned about using probabilistic reasoning. Tasks like risk assessment and decision-making can benefit from this. While the use of AI in power systems is still in its early stages, it has the potential to fundamentally change the way these systems are operated and maintained. AI can assist us in enhancing the security, dependability, and efficiency of our power systems. It lessens our dependency on fossil fuels by enabling the smooth integration of renewable energy sources.

The framework and organisation that make it possible to include artificial intelligence methods into the functioning of power systems are referred to as the architecture of AI-based power systems. These systems seek to improve the processes involved in power generation, transmission, and distribution in terms of dependability, efficiency, and sustainability.

The design of AI-based power systems is built on the analysis of massive volumes of data gathered from diverse sources inside the power grid using sophisticated algorithms, machine learning models, and data analytics techniques. This data contains details on the production of electricity, load demand, weather patterns, the status of the equipment, and other pertinent information.

AI-based power systems often have an architecture made up of several essential elements. These comprise a monitoring and diagnostics module, an AI algorithm and model, a control and optimisation module, and a data collecting and preprocessing module. Together, these elements gather, examine, and take action on data to improve the performance of the power system.

Advanced algorithms, machine learning models, and data analytics methods are combined in the architecture of AI-based power systems to produce intelligent and adaptable power grids. These systems have the potential to transform the way power systems are managed by utilising AI capabilities, resulting in increased effectiveness, dependability, and sustainability in the production and delivery of electrical energy.

The components of an AI-based power system can be categorised into three main areas: data collection and preprocessing, AI algorithms and models, and decision-making and control (Fig. 1).

3.1 Data Collection and Processing

Sensors and Measurement Devices: These devices collect data on electricity production, load demand, weather conditions, equipment status, and other relevant parameters.

Data Acquisition System: This system gathers data from sensors and measurement devices and prepares it for further processing.

Data Preprocessing: Raw data is cleaned, filtered, and transformed to remove noise, outliers, and inconsistencies. It may involve data normalisation, feature extraction, and data integration from multiple sources.

3.2 AI Models

Machine Learning Algorithms: These algorithms are used to train models on historical data and learn patterns, correlations, and trends. Common machine learning algorithms include regression, classification, clustering, and deep learning.

Feature Selection and Engineering: This process involves selecting relevant features from the data and creating new features that capture important information for the AI models.

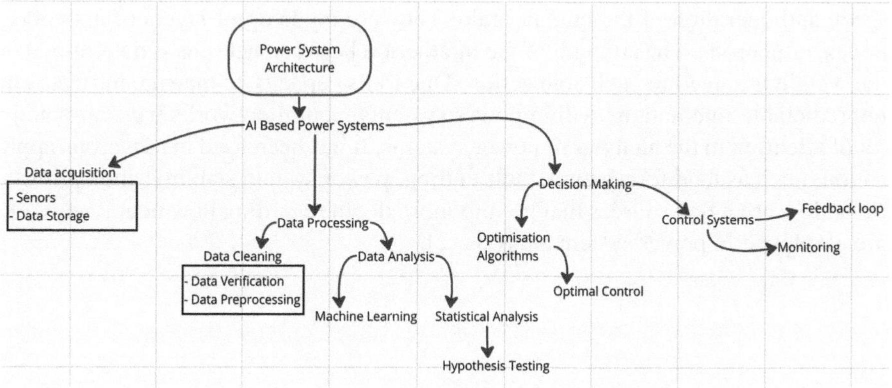

Fig. 1 Architecture of AI power systems

Model Training and Validation: The selected AI models are trained using labelled data and validated to ensure their accuracy and performance.

Model Optimisation: The trained models are optimised to improve their efficiency and effectiveness in making predictions and decisions.

3.3 Decision-Making and Control

Predictive Analytics: The AI models use the collected and preprocessed data to make predictions about power demand, equipment failures, and other relevant factors.

Optimisation and Control: Based on the predictions and other inputs, optimisation algorithms are used to optimise power generation schedules, load balancing, and other operational decisions.

Anomaly Detection: AI algorithms can detect anomalies in the power system, such as equipment malfunctions or cybersecurity threats, and trigger appropriate actions.

Real-Time Monitoring and Visualisation: The system provides real-time monitoring of power system parameters and visualises the data and insights for operators and decision-makers.

3.4 Neural Networks and their Role in Power System Analysis

Neural networks, whose design was modelled after that of the human brain, have proven to offer amazing modelling and prediction abilities. Neural networks have become highly effective instruments in the domain of power system analysis, offering a versatile and adaptive approach to solving complex problems. These computer models, known as artificial neural networks (ANNs), are modelled after the composition and operation of the human brain. They are made up of layers of networked nodes, or neurons. The strength of the interaction between neurons is determined by the weight assigned to each connection. Due to its capacity to manage intricate and unpredictable interactions within power systems, neural networks have attracted a lot of attention in the analysis of power systems. It has been used in numerous applications such as load forecasting, fault finding, power system stability, and optimisation. Here are a few sources that go into more depth regarding how neural networks are employed in power system analysis [2].

3.5 Load Prediction/Load Forecasting

Based on past data and a variety of influencing factors, neural networks can forecast future electricity consumption. For utilities to plan their generation and distribution strategies and guarantee a steady supply of electricity, this is essential. For the design and operation of electricity systems, load forecasting is a crucial task. Particularly, long short-term memory networks (LSTMs) and recurrent neural networks (RNNs) have demonstrated greater performance in both short- and long-term load forecasting. [3].

3.6 Fault Detection

In power systems, neural networks have been used successfully for defect detection and categorisation. Short circuits and equipment breakdowns are only two examples of the various issues in power systems that neural networks can identify and categorise. This feature helps to minimise downtime and potential damage by rapidly recognising and resolving issues. In terms of identifying and categorising various defect kinds, methods such as feedforward neural networks (FNNs) and convolutional neural networks (CNNs) have shown to be highly accurate. [4].

3.7 State Estimation

State estimate is essential for keeping a watchful eye on and controlling power systems. Radial basis function networks (RBFNs) and multilayer perceptrons (MLPs) are examples of neural networks that have been used to improve the reliability and accuracy of state estimation algorithms. [5].

3.8 Optimisation of Power Systems Via Optimal Power Flow (OPF)

The objective of OPF, a crucial role in power system operation, is to decrease generation costs while meeting a variety of operational requirements. Finding optimal operating points can be aided by using neural networks to approximate the nonlinear relationship between system parameters. In order to improve the effectiveness and performance of power systems, neural networks have been incorporated into optimisation frameworks. Applications include voltage regulation, unit commitment, and economic dispatch.

3.9 Evaluation of Voltage Suitability

A dependable power source depends on maintaining stable voltage levels. In order to anticipate possible problems with voltage instability, neural networks are able to simulate the intricate interactions that occur between various power system components.

4 Applications of AI in Power Systems

AI's integration with power systems is bringing in a new era of sustainability, dependability, and efficiency. These situations show how AI applications have a real-world influence. Unquestionably, AI will play a significant role in creating a smarter, more resilient world as the electricity sector develops.

4.1 Fault Detection and Diagnostics

Fault Detection and Diagnostics play a pivotal role in maintaining the reliability and stability of power systems. This application harnesses the capabilities of Artificial Intelligence (AI) to swiftly identify and rectify issues within the power grid. AI algorithms are employed to continuously monitor the performance of various components in the power system, such as transformers, circuit breakers, and transmission lines.

When a fault or anomaly is detected, the AI system immediately alerts operators or initiates automated responses to address the issue. This proactive approach is crucial in preventing potential disruptions in the energy supply. By swiftly identifying and mitigating faults, the power system can maintain uninterrupted service, even in the face of unexpected events or equipment failures.

Fault Detection and Diagnostics systems do not rely solely on pre-defined thresholds for triggering alerts. Instead, they leverage advanced machine learning techniques to analyse complex data patterns and trends. This enables them to detect subtle deviations from normal operating conditions that may go unnoticed by traditional monitoring methods.

One example is the use of SmartMeters to notify PG&E's Control Centre of voltage situations that may raise the danger of an ignition incident. This technology enables PG&E to remotely de-energise a line from the Control Centre in minutes, allowing for faster mitigation and a reduction in the amount of time a line is energised while down. It is used to identify wire-down conditions for lower-current fault conditions, which may not be reliably mitigated by EPSS. [6].

The San Francisco-based startup Pano AI claims to have developed a technology that helps utilities identify, evaluate, and locate new fires sooner in order to extinguish them while they are still minuscule. The platform continuously rotates a network of linked, ultra-high definition cameras atop high observation points to capture a panoramic 360-degree view of the terrain once per minute. Pano AI's artificial intelligence uses camera data in conjunction with satellite images, emergency call details, and other data feeds to monitor for risks and deliver actionable intelligence through a unified, user-friendly interface. [7].

In general, the application of AI in Fault Detection and Diagnostics represents a significant advancement in the field of power systems. It not only enhances the reliability of energy supply but also contributes to the overall resilience of the grid in the face of unforeseen challenges. This technology-driven approach ensures that power systems can adapt and respond swiftly to maintain uninterrupted service, ultimately benefiting consumers, industries, and economies as a whole.

4.2 Load Forecasting and Demand Prediction

Load Forecasting and Demand Prediction stands as a critical application in the realm of power systems, providing utilities with invaluable insights into future energy consumption patterns. This application leverages the computational capabilities of Artificial Intelligence (AI) to meticulously analyse a diverse range of data sources.

Historical consumption patterns, which encapsulate how energy usage fluctuates over time, serve as a foundational dataset. This information is supplemented by weather data, as meteorological conditions profoundly influence energy consumption. For instance, extreme temperatures can lead to surges in heating or cooling needs, affecting overall demand. Additionally, economic indicators, such as industrial activity, population growth, and economic development, are factored in. These indicators offer crucial context for understanding the underlying drivers of energy consumption trends.

By assimilating and processing this multifaceted data, AI algorithms can generate highly accurate load forecasts and demand predictions. These forecasts provide utilities with a glimpse into anticipated energy requirements over varying timeframes, spanning from short-term periods to long-range projections.

The benefits of precise load forecasting and demand prediction are profound. Utilities can proactively allocate resources, ensuring that the right amount of energy is generated and distributed to meet the expected demand. This optimisation is essential for preventing overproduction, reducing energy wastage, and ultimately minimising costs for both the utility and consumers.

As India's population continues to rise and more consumers adopt air conditioners and electric vehicles, the demand for electricity, particularly for cooling, is poised for a significant surge. [8]. This is evident in TATA's venture. A strategic partnership has been formed between California-based AutoGrid, a global leader in AI-based energy management solutions, and Tata Power Delhi Distribution Limited (TPDDL), a leading power utility that provides electricity to over 7 million people in North Delhi. The partners will work together on joint projects in the areas of demand response (DR), distributed energy resource management (DERMS), electric vehicles (EVs), virtual power plants (VPPs), energy storage optimisation, flexibility management, forecasting, and renewable integration. Modern energy management technologies will be developed and implemented by both parties in an effort to maximise usage and benefit Tata Power-DDL customers overall. In an effort to lower peak demand and network capital costs, Tata Power-DDL and AutoGrid are collaborating to introduce a behavioural demand response programme for the company's residential customers. By assisting consumers in comprehending their purchasing habits and assessing the efficacy of demand response (DR) programmes, this pilot initiative seeks to empower consumers. [9].

Also, accurate demand predictions allow utilities to strategically plan for peak periods of usage, avoiding potential grid congestion or overload situations. Additionally, this application supports the integration of renewable energy sources by aligning generation with expected demand patterns. [10].

Basically, load forecasting and demand prediction using AI represents a paradigm shift in how power networks adjust to shifting customer demands. Utilities can enhance their operational efficiency, mitigate their environmental footprint, and guarantee dependable energy accessibility for both consumers and industry by leveraging artificial intelligence (AI) and data analytics.

4.3 Renewable Energy Integration and Grid Balancing

Renewable Energy Integration and Grid Balancing stands at the forefront of transforming the energy landscape, and Artificial Intelligence (AI) plays a pivotal role in making this transition seamless and efficient. This application harnesses the power of AI to facilitate the harmonious integration of renewable energy sources with conventional ones, ensuring a stable and reliable energy supply.

AI algorithms are employed for real-time monitoring and management of renewable energy generation. These algorithms continuously analyse a plethora of data points, including weather conditions, solar and wind patterns, and energy production levels. By processing this information, AI systems can accurately forecast the availability of renewable energy at any given moment.

One of the key challenges in renewable energy integration is its intermittent nature. Unlike conventional sources, such as coal or natural gas, renewable energy production is subject to fluctuations based on weather conditions. AI steps in to mitigate this challenge by dynamically adjusting energy generation and distribution strategies in real time. For example, if a cloud covers a solar panel array, AI algorithms can prompt the grid to compensate with alternative sources or stored energy reserves.

Moreover, AI-driven grid balancing optimises the allocation of renewable energy within the broader energy mix. By intelligently distributing energy resources, utilities can minimise waste and maximise the utilisation of clean, sustainable power sources. This not only reduces carbon emissions but also contributes to a more resilient and adaptable energy infrastructure. A stark example of this application is Denmark. Approximately half of the country's electricity is sourced from renewables, predominantly wind power. The national objective is to achieve full reliance on renewable energy, reaching 100% by the year 2030. [11].

This endeavour presents distinct challenges for Energinet, Denmark's electric transmission systems operator, given the inherent variability of renewable energy sources. The organisation must meticulously oversee the grid to ensure the continued security of the energy supply. This is being achieved through the Big data and AI analysis tools of IBM. [12].

Furthermore, the integration of AI in renewable energy management fosters a more responsive and adaptive grid. It enables utilities to anticipate and manage fluctuations in energy supply and demand, ultimately leading to a more stable and reliable energy supply for consumers.

Thus, the use of AI in grid balancing and renewable energy integration is a critical step towards the development of sustainable energy systems. AI is contributing significantly to the creation of a cleaner, more dependable energy future through efficiently overseeing the integration of renewable sources and streamlining grid operations. This is in line with environmental objectives and creates the foundation for an energy infrastructure that is more flexible and robust. [13].

4.4 Energy Storage Optimisation

Energy Storage Optimisation is a crucial facet of modern energy management, and the integration of Artificial Intelligence (AI) revolutionises the way stored energy is utilised. This application harnesses AI algorithms to fine-tune the operation of energy storage systems, ensuring that stored energy is used in the most efficient and effective manner possible.

AI algorithms continually analyse a multitude of variables, including real-time demand patterns, grid conditions, and energy production levels. By processing this data, the AI system can make rapid, data-driven decisions on when to charge, discharge, or hold onto stored energy. This dynamic approach ensures that energy is available precisely when it is needed most, optimising resource allocation and reducing wastage. [14].

During periods of peak demand, when electricity costs are typically higher, AI algorithms strategically release stored energy to meet the heightened requirements. This not only helps stabilise the grid but also minimises the reliance on conventional, often more polluting, energy sources. Conversely, during off-peak hours when demand is lower, excess energy can be stored for later use, thereby avoiding wastage and contributing to a more sustainable energy system.

In addition to being sustainable, this grid-scale energy storage project—which is being closely coordinated with the South Australian Government and Neoen—will help manage summertime peak load, lessen intermittencies, and address power shortages, all of which will increase the dependability of South Australia's electrical infrastructure [15]. Residential installations of Tesla's Powerwall are currently underway and progressing swiftly throughout Australia. Homeowners can harvest energy throughout the day to store and make it available day and night, providing uninterrupted electricity even in the event of a grid outage, using the same technology that can assist stabilise the South Australian system. [16].

Furthermore, Energy Storage Optimisation enhances the overall resilience of the energy grid. By strategically managing stored energy resources, utilities can better navigate unforeseen events, such as sudden spikes in demand or unexpected disruptions in supply. This adaptability is crucial for maintaining a reliable energy supply in the face of dynamic and sometimes unpredictable conditions. [17].

In summary, the application of AI in Energy Storage Optimisation marks a significant leap forward in the pursuit of a more efficient, reliable, and sustainable energy infrastructure. By leveraging advanced algorithms, utilities can ensure that stored energy is deployed with precision, maximising its impact and contributing to a more resilient and adaptable energy grid. This not only leads to cost savings but also advances the transition towards a greener, more sustainable energy future. [18].

5 Mind Map of Benefits and Challenges

See Tables 1 and 2.

Table 1 Benefits of AI in power systems

Benefits	Justification
Enhanced reliability and stability	AI-powered systems enable real-time monitoring and swift response to faults, ensuring uninterrupted power supply, which is crucial for maintaining grid reliability and stability
Improved energy efficiency	AI-driven load forecasting and demand prediction optimise resource allocation, reducing wastage and leading to more efficient energy utilisation
Reduced operational costs	Predictive maintenance and fault detection algorithms prevent costly breakdowns, leading to decreased maintenance expenses and improved operational efficiency
Proactive response to power outages	AI-based fault detection allows for immediate identification and resolution of issues, minimising downtime and enhancing the resilience of the power system
Facilitating a sustainable transition	AI assists in seamless integration of renewable energy sources, contributing to a more sustainable energy landscape by minimising reliance on fossil fuels

Table 2 Challenges of AI in Power Systems

Challenges	Justification
Data quality and quantity	Ensuring high-quality, comprehensive data is essential for accurate AI modelling and predictions. Inadequate or erroneous data can lead to inaccurate results
Model interpretability and trust	Understanding and trusting AI models' decisions is crucial, especially in critical systems like power grids. Ensuring transparency in AI algorithms is imperative
Security concerns in AI-driven power systems	Implementing robust cybersecurity measures is essential to safeguard against potential vulnerabilities in AI-powered systems, protecting against cyber threats
Technical and computational challenges	Developing and deploying AI systems in complex power grid environments may require substantial technical expertise and computational resources
Economic and policy challenges	The initial investment in AI implementation and ensuring compliance with regulations and policies can pose economic and administrative challenges

6 Future of AI in Power Systems

- **The Role of Quantum Computing and Edge Computing**
- As power systems become increasingly complex, the integration of Quantum Computing and Edge Computing emerges as a transformative force. Quantum computing's immense processing power holds the potential to revolutionise grid optimisation and complex simulations, enabling utilities to make more accurate forecasts and dynamic adjustments in real time [19].
- Concurrently, Edge Computing empowers localised data processing, allowing for quicker response times and reduced latency in critical grid operations. Google's Quantum AI Lab has made significant strides in harnessing the power of quantum computing. Their research has demonstrated the potential to solve complex optimisation problems that could revolutionise energy grid management.
- **Decentralised AI-Driven Microgrids**
- The decentralisation of power generation through microgrids, coupled with AI-driven intelligence, heralds a new era in energy resilience. By leveraging AI algorithms, microgrids can autonomously manage energy production, storage, and distribution within localised communities. [20].
- This approach not only enhances grid resilience against disruptions but also fosters greater integration of renewable energy sources. [21].
- The Brooklyn Microgrid project in New York utilises blockchain technology and AI to enable localised energy trading among neighbouring households. This decentralised approach enhances energy efficiency and reduces reliance on centralised grids [22]
- **Integration of IoT and AI for Advanced Grid Management.**

- The convergence of the Internet of Things (IoT) and AI introduces a new paradigm in grid management. IoT sensors gather real-time data from various grid components, enabling AI algorithms to make precise decisions for optimal grid performance. [23].
- This synergy leads to improved fault detection, load balancing, and predictive maintenance. PG&E, a major utility in California, has implemented an IoT-driven grid management system. This system integrates with AI algorithms to enhance real-time monitoring and decision-making, resulting in a more reliable and resilient grid [24].
- The integration of Quantum Computing, Edge Computing, decentralised AI-driven microgrids, and IoT in power systems is shaping a future where energy management is intelligent, adaptive, and efficient. These technological advancements not only enhance grid reliability but also pave the way for a more sustainable and resilient energy landscape.

7 Conclusion

In reflecting upon the current state of AI in power systems, it is evident that we stand at the precipice of a transformative era. The integration of Artificial Intelligence has ushered in a new paradigm, revolutionising the way we generate, distribute, and manage electrical power. From fault detection to renewable energy integration, AI applications have demonstrated remarkable efficacy in enhancing grid reliability, optimising resource allocation, and driving the transition towards sustainable energy sources. [11].

However, it is imperative that we recognise that the journey is far from over. As the global demand for energy continues to surge, along with the imperative to reduce carbon emissions, the role of AI in power systems will only become more critical. This calls for a resounding call to action for researchers, policy makers, and industry leaders alike.

7.1 Reflection on the Current State of AI in Power Systems

The current state of AI in power systems is marked by remarkable progress and untapped potential. We have witnessed tangible benefits in fault detection, load forecasting, and renewable energy integration. The fusion of AI with quantum and edge computing, coupled with decentralised microgrids, has opened up new avenues for grid resilience and sustainability.

Also, the convergence of IoT with AI is revolutionising grid management, offering real-time insights and enabling more informed decision-making. However, it is crucial to acknowledge that challenges such as data quality, model interpretability, and security concerns persist and warrant further attention. [25].

7.2 Call to Action for Researchers, Policy Makers, and Industry Leaders

The journey towards an intelligent, resilient, and sustainable energy future requires concerted effort from various stakeholders. Researchers must continue to push the boundaries of AI applications in power systems, developing innovative solutions to address existing challenges and uncover new opportunities.

Policy makers play a pivotal role in creating an enabling environment for the widespread adoption of AI in the energy sector. This includes fostering regulations that promote data sharing, privacy, and cybersecurity, while also incentivising investment in AI-powered infrastructure.

Industry leaders have a responsibility to embrace these technologies and pioneer their integration into existing power systems. Collaboration between public and private sectors is essential in driving innovation and ensuring that AI is effectively deployed to meet the growing energy demands of the future.

In uniting our efforts, we have the potential to revolutionise the energy landscape, ushering in an era where power systems are not only reliable and efficient, but also sustainable and environmentally conscious. The time for action is now, and together, we can forge a brighter and more sustainable future for generations to come.

References

1. P. J. Kiger, *6 Inventions by Thomas Edison That Took the World by Storm*. HISTORY, Mar. 06, 2020. https://www.history.com/news/thomas-edison-inventions
2. Aggarwal, R., & Song, Y. (1998). Artificial neural networks in power systems. III. Examples of applications in power systems. *Power Engineering Journal, 12*(6), 279–287. https://doi.org/10.1049/pe:19980609
3. Chen, K., Chen, K., Wang, Q., He, Z., Hu, J., & He, J. (2019). Short-term load forecasting with deep residual networks. *IEEE Transactions on Smart Grid, 10*(4), 3943–3952. https://doi.org/10.1109/TSG.2018.2844307
4. Guo, M.-F., Yang, N.-C., & Chen, W.-F. (2019). Deep-learning-based fault classification using Hilbert–Huang transform and convolutional neural network in power distribution systems. *IEEE Sensors Journal.*, 1. https://doi.org/10.1109/JSEN.2019.2913006
5. A. Kumar and S. Chakrabarti, *ANN-based hybrid state estimation and enhanced visualisation of power systems*. ISGT2011-India, Kollam, India, 2011, pp. 78–83, doi: https://doi.org/10.1109/ISET-India.2011.6145359.
6. P. G. and E. Company, *As Peak wildfire season nears, PG&E deploys innovative technologies to enable layers of protection to mitigate wildfires*. PR Newswire: press release distribution, targeting, monitoring and marketing. Retrieved October 27, 2023, from https://www.prnews-

wire.com/news-releases/as-peak-wildfire-season-nears-pge-deploys-innovative-technologies-to-enable-layers-of-protection-to-mitigate-wildfires-301858237.html#:~:text=This%20technology%20helps%20PG%26E,line%20is%20energized%20while%20down
7. C. E. C. Directors, *PG&E and PGE test AI-based wildfire detection technology.* POWERGRID International, Nov. 02, 2021. Retrieved October 31, 2023, from https://www.power-grid.com/td/pge-and-pge-test-ai-based-wildfire-detection-technology/#gref.
8. *Press Release Details*, Tatapower-ddl.com. [Online]. Retrieved October 31, 2023, from https://tatapower-ddl.com/pr-details/199/2153286/in-an-industry-first,-tata-power-delhi-distribution-limited-partners-with-anyline-to-deploy-ai-based-ocr-application-for-meter-reading.
9. *Press Release Details.* www.tatapower-ddl.com. Retrieved October 31, 2023, from https://www.tatapower-ddl.com/pr-details/199/1753686/tata-power-ddl-joins-hands-with-autogrid-to-deploy-ai-enabled-smart-energy-management-system.
10. *Press Release Details.* www.tatapower-ddl.com. Retrieved October 31, 2023, from https://www.tatapower-ddl.com/pr-details/199/2513286/tata-power-ddl-collaborates-with-us-based-utiltyx-to-ensure-cyber-security-and-seamless-power-supply-to-consumers.
11. B. P. Bhattarai et al., "Big data analytics in smart grids: state-of-the-art, challenges, opportunities, and future directions," IET Smart Grid, vol. 2, no. 2, pp. 141–154, 2019, doi: https://doi.org/10.1049/iet-stg.2018.0261.
12. *Frontpage | Energinet.* en.energinet.dk. https://en.energinet.dk
13. *PATHWAYS TOWARDS A ROBUST FUTURE ENERGY SYSTEM Analysis focusing on cost-effective and robust system solutions for climate neutrality, large-scale offshore wind power and the green energy industry.* Energinet.dk, 2022. [Online]. Retrieved October 31, 2023, from https://energinet.dk/media/y5rhoqjy/pathways-towards-a-robust-future-energy-system_energinet-2023-01-23.pdf.
14. P. Chakraborty, *GE Digital's GridOS software is focused on helping electric utilities transition to a clean energy grid: Parag Marathe.* DATAQUEST, Apr. 25, 2023. Retrieved October 31, 2023, from https://www.dqindia.com/ge-digitals-gridos-software-is-focused-on-helping-electric-utilities-transition-to-a-clean-energy-grid-parag-marathe/.
15. S. Hanley, *Tesla to install world's largest grid-scale battery in South Australia*, CleanTechnica, 07-Jul-2017. [Online]. Retrieved October 31, 2023, from https://cleantechnica.com/2017/07/07/tesla-install-worlds-largest-grid-scale-battery-south-australia/.
16. *Tesla Powerpack to Enable Large Scale Sustainable Energy to South Australia*, www.tesla.com, Jul. 06, 2017.https://www.tesla.com/is_IS/blog/tesla-powerpack-enable-large-scale-sustainable-energy-south-australia
17. *GE Digital | Accelerating Digital Transformation* | GE.com, 2018. https://www.ge.com/digital/
18. *GridOS® - Modern Grid & Digital Utility Software.* Retrieved October 31, 2023, from www.ge.com. https://www.ge.com/digital/industry/grid.
19. C. Zoufal, A. Lucchi, and S. Woerner, Quantum Generative Adversarial Networks for learning and loading random distributions NPJ Quantum Information, vol. 5, no. 1, Nov. 2019, doi: https://doi.org/10.1038/s41534-019-0223-2.
20. *Design and Implementation of Adaptive Path Finding Based on CPS for Evacuation Guidance System in Disaster Situation.* Retrieved October 31, 2023, from ieeexplore.ieee.org. https://ieeexplore.ieee.org/document/6847381 (accessed Oct. 31, 2023).
21. *Brooklyn microgrid*, Brooklyn Microgrid. [Online]. Retrieved October 31, 2023, from https://www.brooklyn.energy.
22. Silicon, *New York neighbours power up blockchain-based Brooklyn Microgrid*, Silicon Republic, Sep. 27, 2017. Retrieved October 31, 2023, from https://www.siliconrepublic.com/machines/brooklyn-microgrid-blockchain-energy-networks.
23. *Brooklyn Microgrid: A Blockchain-based Platform for Locally Traded Electricity*, The New Stack, Apr. 03, 2016. https://thenewstack.io/brooklyn-microgrid-blockchain-based-platform-locally-traded-electricity/

24. *Smart Grid*, Retrieved October 31, 2023, from www.pge.com.https://www.pge.com/en_US/safety/how-the-system-works/electric-systems/smart-grid/smart-grid.page?WT.mc_id=Vanity_smartgrid.
25. A. Ramamurthy and P. Jain, *The Internet of Things in the Power Sector Opportunities in Asia and the Pacific and Sustainable Development Working Paper Series No. 48*, 2017. Retrived from : https://www.adb.org/sites/default/files/publication/350011/sdwp-48.pdf

Boosting Power Grid Efficiency: Meta-RL Approaches

Surya Batchu, Cherindranath Reddy Vanguru, Gurram Yashwanth, and Satyesh Das

1 Introduction

1.1 Motivation and Importance of the Problem

Electrical power grids are an essential, yet extremely complicated piece of the infrastructure of city life. Responsible for the generating electricity and delivering it to the final consumers, the sheer sizes of some of these power grids are a testimony to the difficulty of managing them. The Synchronous Grid of Continental Europe [1], for example, provides electricity to over 400 million people in 24 countries. In North America, two power grids are responsible for providing energy to the entire region, with the exception of Texas, Quebec, and Alaska [2].

Managing such power grids is as complex as it is relevant. Unexpected fluctuations in supply and demand may lead to power shortages (such as the 2003 Northeast Blackout, that affected an estimated 55 million people in the USA and Canada [3], or to other adverse effects (such as when clocks in Europe ran slow because Kosovo used more electricity than anticipated as a result of its conflict with Serbia in 2018 [4]. Human tampering is also another cause for concern, as seen in the California energy crisis, when electricity companies such as Enron deliberately created a

S. Batchu
National Institute of Technology, Durgapur, India

C. R. Vanguru (✉)
Bangalore Institute of Technology, Bengaluru, Karnataka, India

G. Yashwanth
Indian Institute of Technology, Dhanbad, India

S. Das
Rajiv Gandhi Institute Of Petroleum Technology, Jais, Amethi, India
e-mail: 21cs3042@rgipt.ac.in

S. Vijayalakshmi et al. (eds.), *Smart Power Systems*, Power Systems,
https://doi.org/10.1007/978-3-031-89459-6_2

shortage in electricity supply in order to inflate prices [5]. Even when power grids are operating normally, there is still a loss due to the complexities of managing such behemoths. In the USA, for instance, as much as 5% of all generated energy is lost before reaching the customers [6], leading states like New York to plan spending US$30 billion over the next decade in power grid maintenance alone [7].

Smart grids have been proposed as a mean to manage power grids more efficiently [8]. By implementing telemetry across the entire grid and employing algorithms that can seamlessly adjust the generation, transmission, and distribution of energy, experts argue we could reduce losses and environmental impact without significant impact to current electricity prices. To be precise, as far as residential and commercial buildings are concerned, estimates are that the CO_2 emission of buildings can be reduced by 50–90% [9].While plans have been proposed to substitute both the North American [10] and the European [11] current power systems by smart grids, actually implementing smart grids, however, has caveats and challenges of its own. Among the challenges, one that is of particular concern to Artificial Intelligence (AI) scholars is how to train the algorithm that will manage such grids. Within this study, we employ Reinforcement Learning (RL) to educate an agent in overseeing a compact energy network generated by the CityLearn simulation toolkit [12]. We use this as an instance to illustrate how training times may be too long to be practically feasible, thereby compromising the potential values of such agents in real-world applications. We further discuss alternatives to speed up the training process.

2 Literature Review

2.1 General Statement of the RL Problem

RL, an essential aspect of Machine Learning (ML) alongside Supervised and Unsupervised Learning, revolves around the interplay between an agent and its environment. The agent discerns environment features, takes actions, obtains returns, and induces changes in the environment. This cyclic process involves the agent continually adapting its actions based on its perception of the environment's state, with the ultimate goal of maximizing the cumulative return. The essence of RL lies in guiding the agent to optimize its decision-making while seeking to enhance the expected cumulative return.

By "teach the agent how to act," we mean deciding which action to take for each observed state of the environment. This is called a *policy* and do not need to be deterministic: one may perfectly well have a policy that states that, when faced with an environment in a certain state, the agent should randomly pick an action from a given probability distribution defined over the set of admissible actions. Ultimately, the "answer" for a Reinforcement Learning problem is a certain policy, i.e., a function that connects each state in the environment, either to an action to be taken (if the policy is deterministic) or to a probability distribution from which the action is to be drawn (if the policy is stochastic).

By "expected value," we mean the exact same thing that is meant in finance by the homonymous expression. Namely, those imminent rewards are perceived as more valuable than similar rewards in a distant future, so we apply a *discount factor* for every period we have to wait in order to perceive a reward. Imminent rewards are more valuable because rewards in a distant future have a lot of risk as there might unexpected obstacles in future that decrease the reward. As a result, the values of rewards decrease exponentially with the time we have to wait to perceive them.

By amalgamating these concepts into a unified mathematical representation, we can formulate the essence of RL as the pursuit of solving the subsequent challenge: discerning a solution for the following quandary, presented with a collection of context-specific initial circumstances and limitations.

$$\pi^* = \arg\max_{\pi} \mathbb{E}_{\pi}\left[\sum_{t=0}^{+\infty} \gamma^t r(s_t, a_t)\right] \tag{1}$$

Here, γ is the discount factor and $r(s_t, a_t)$ is the reward obtained at time t as a result of taking the action (a_t) prescribed by policy π given the state (s_t) the agent finds itself at that time. The reader may recognize the term within the square brackets as the present value of a cash flow of rewards. But since the policy is arguably stochastic, there is uncertainty on the exact course of action to be followed and therefore on the rewards to be perceived. Thus we cannot maximize the present value of expected rewards, but only its expected value given the policy the agent undertakes. Our objective is to ascertain the strategy that optimizes this anticipated value to its maximum potential.

In particular, if $\gamma \to 0$, our policy consists on choosing what is the immediate best solution, with no regards for what might happen in future time steps. It's a greedy policy. Conversely, with $\gamma \to 1$, no discounting is in place, and similar rewards are perceived as having equal values regardless of how much the agent has to wait in order to perceive them.

$\rightarrow$

Furthermore, note that no mention is made as to how the environment moves from one state to another, or the probabilities with which states succeed each other. All that is necessary is to observe the rewards that result from following the prescribed policy. In this sense, RL is *model free*. This is an advantage of RL for real-life problems, where we are often be ignorant on how the real-world works, and are only conscious of our actions and their consequences. This is, in a way, quite poetic. There is no blueprint to life. We learn by doing, acting, and facing the consequences of our actions. That's how RL works, and that's a feature that makes it very applicable to so many real-life problems.

In addition, states need not be perfectly observed. Thus, in the first paragraphs of this section, we were careful in our choice of words, when we said that "agent observes certain features of the environment and acts on what he observes." The agent does not need to observe *all* features of the environment. But whatever it does observe, that's what it acts upon. Our agent will try to find the best actions it can, given the information it has, however partial or noisy they may be. The fact that information may not be neither complete nor perfect is very realistic and accounts for another advantage of RL over other ML approaches such as Supervised or Unsupervised Learning, which are much more silent about these issues.

2.2 Overview of Heuristics for Solving the Reinforcement Learning Problem

So how does one go about solving Eq. (1)?

There are essentially two ways to do this: one can either follow a *value-based approach* or a *policy-based approach*.

In a *value-based approach*, one tries to determine the value of taking each possible action on each possible state. By value we mean the expected value in Eq. (1), i.e., the thing which we want to maximize. If an agent knows the value of taking each possible action on each possible state, it can easily figure out the best action to take in whichever state it happens to be in. All it has to do is pick whichever action moves it to the next state of highest value.

In more mathematical terms, in the value-based approach, one has to discover a function $Q: \mathcal{S} \times \mathcal{A} \rightarrow \mathbb{R}$ that associates the value of taking a given action $a \in \mathcal{A}$ when in state $s \in \mathcal{S}$. By value we mean the expected value of all future rewards gained by the agent.

The optimal value of this function will typically be improved at every iteration according to an Eq. (2) as given below and known as the Bellman equation:

$$Q^{new}(s_t, a_t) \leftarrow \underbrace{Q(s_t, a_t)}_{\text{old value}} + \underbrace{\alpha}_{\text{learning rate}} \cdot \overbrace{\left(\underbrace{\underbrace{r_t}_{\text{reward}} + \underbrace{\gamma}_{\text{discount factor}} \cdot \underbrace{\max_a Q(s_{t+1}, a)}_{\text{estimate of optimal future value}}}_{\text{new value (temporal difference target)}} - \underbrace{Q(s_t, a_t)}_{\text{old value}} \right)}^{\text{temporal difference}}$$

One challenge to this approach arises when the number of states is very large, and in particular, when it is infinite. In this scenario, some states may only seldom be visited. One way to deal with this is by encoding each state by a finite vector of essential features ϕ. It's as if, rather than having a list of infinite possible states, we had a small set of features that matter for decision-making. Mathematically, this means we move from having an infinite number of possible states, $\{s_0, s_1, s_2 \ldots\}$, to having a function $s(\phi)$, which enables us to estimate which state we are in based on a small set of features ϕ. Since it is impossible to visit an infinite number of states, one must learn how to infer the rewards of each state based on the values of ϕ observed for the set of states that actually were visited. This task is often accomplished with Deep Learning techniques.

In a *policy-based approach*, one tries to determine the optimal policy directly, without resorting to the intermediate step of calculating the values of taking each action on each possible state. This is done by expressing the policy as a function $\pi_\theta(a, s)$ that assigns a probability of taking an action a when on a state s, and then seeking to find the optimal values of the parameter vector θ.

2.3 The Soft Actor-Critic (SAC) Algorithm

Each of these approaches has its strengths and weaknesses. One approach to combine both approaches and thus leverage their strengths is the *actor-critic* algorithm. The "actor" acts following a certain policy. The "critic" takes note of the rewards it detects and employs this information to compute the worth of every choice within each circumstance. These values are then used by the actor to estimate the optimal policy. It then enacts this policy, providing more information for the critic to update its values estimated. And thus, both actor and critic improve their respective performances, until the actor's estimate of the optimal policy converges.

Two neural networks, the actor and the critic, fulfill these roles in practice. The critic aims to discover the function connecting ϕ with the rewards observed in a specific state, whereas the actor aims to determine the θ variables that result in the optimal performance of π_θ. Both neural networks interact and evolve together. This approach of combining neural networks is no novelty to AI literature and has been successfully implemented in applications such as Generative Adversarial Neural Networks or Variational Autoencoders.

The actor-critic algorithm can be improved by fostering exploration. In all RL algorithms, there is an inherent concern over how to balance exploration and exploitation. The latter (exploitation) consists of taking those actions that are already known to perform the best, while the former (exploration) consists of attempting actions which may seem senseless on the hopes that they may help us discover a better policy than the one we currently have. No learning is possible without some degree of exploration. Yet, since exploration leads to failure much more often than it leads to success, it's easy to overlook it. This is especially striking in continuous spaces of action, where the probability of exploring any given specific action tends to zero, as does the probability of that specific action being the jackpot. So how can we encourage an agent to explore new courses of action on a continuous-space of possible actions?

The *soft-actor-critic* (SAC) algorithm does that by adding an additional term to the traditional RL objective function. This function is (proportional to) the entropy of the policy. Hence, it praises policies which acknowledge doubt on to what is the best course of action to pursue and punishes policies that are overly confident that only one specific action can be taken in any given case. Under this framework, Eq. (3) becomes:

$$\pi^* = \arg\max_{\theta} \mathbb{E}_{\pi(\theta)} \left[\sum_{t=0}^{+\infty} \gamma^t r\left(s_t(\phi), a_t\right) + \alpha \mathcal{H}_{\pi} \right]$$

where H_π is the entropy of policy π, now parametrized in terms of θ, and α is a measure of how much one wishes to penalize a policy for failing to explore. The critic will try to arrive at the value of ϕ best estimates $Q_\pi(s,a) = \sum_{t=0}^{+\infty} r\left(s_t(\phi), a_t\right)$, while the actor will attempt to find the value of θ that produces the optimal policy. Note that $Q_\pi(s, a)$, the function estimated by the critic, is dependent on π, the policy provided by the actor. Conversely, π^*, the policy that the actor believes to be optimal, is dependent on $Q_\pi(s, a)$, the function estimated by the critic.

2.4 *Imitation Learning*

There is yet another way to solve the Reinforcement Learning that evades solving Eq. (1) altogether. Rather than letting a computer figure out the best policy by acting and suffering the consequences of its actions, why not teach the computer simply to copy the behavior of someone who already knows what to do?

The task therefore becomes to find out what would an "expert" do in *every* situation given that we have a dataset of what he said should be done in *some* situations. The "expert" behavior thus becomes a target, and our goal is to find a predictor for it. Therefore, our challenge in reinforcement learning transforms into a conventional supervised learning scenario [13].

Granted, the performance of the trained algorithm will be upper bounded by the performance of the expert. Such caveat notwithstanding, this approach is very convenient when an "expert" is easily available, and when quantifying the rewards of each action is difficult [14]. Put differently, imitation learning proves valuable when an expert can more easily demonstrate the necessary actions than deduce the rewards associated with every potential move.

Such is the case, for example, on autonomous cars [15]. One can easily find a driver and observe that he signals a left turn before actually turning left. Quantifying the reward obtained by such signaling, however, is much harder.

Another advantage of imitation learning is that it draws upon algorithms from classical supervised learning, which tend to be much more simpler and faster to run than those of RL.

A downside of using classical supervised learning algorithms, however, is that these algorithms often suppose that training samples are independent. This is seldom the case. In the self-driving car example, what the driver sees in one moment cannot be said to be independent from what he will see in the following moment.

Quite on the contrary, the latter is a direct consequence of the actions taken by him on the former. By treating observations as independent, errors committed in one circumstance add up, leading the algorithm to learn a behavior that may ultimately be very different than what the expert would actually do [13].

2.5 Inference

Suppose we want to create an agent that learns how to manage a power grid in New York. Suppose we already have such an agent, but it has actually learnt how to manage the power grid in London, not in New York. Anyone faced with this situation will be tempted to get the same algorithm that has been working in London and apply it in New York. Yes, the climates are different, so we would expect the agent's performance in New York to be inferior than the agent's

performance in London. Still, given the great time it takes to train an agent from scratch, copying the agent from London seems like an economically sensible solution. This is called *inference*.

If the agent were copied from London to New York and then continued to be trained on New York, this would be called *transfer learning*.

If, however, the agent did all its learning in London and then were merely applied to New York (with no further learning there), this would be called inference.

It is expected (but not mathematically necessary) that the agent's performance on the second location will be inferior to the performance on the first location though this may change with further learning. Moreover, it is reasonable that the change in performances observed in both contexts should increase with the dissimilarity between such contexts. Thus, copying an agent from London to New York shouldn't entail a dramatic drop in performance as copying it from London to Nairobi.

In this paper, as we shall further explain, we consider four different climate zones and employ agents trained in one zone to act on another.

2.6 Rule-Based Control

Rule-based control (RBC) consists on having an agent act according to a set of rules that he is given. Technically, it doesn't qualify as machine learning, let alone as reinforcement learning. Yet, precisely for not qualifying as machine learning, it serves as a noteworthy benchmark as to the gains that can be achieved by proper Machine Learning algorithms.

Whereas in ML a decision rule is derived from the data, in RBC the decision rule is explicitly given to the machine. Such rule may be derived from expert knowledge, that is, from human knowledge. Or it can be even simpler than that. In the case of a power grid management, for example, an RBC approach might be to never stock any energy for future use, and just use energy from the grid, however much it costs.

Alternatively, the rule could be to stock energy for future use when its price is less than a certain threshold value. The key point is that these rules are given, not learned. Thus, for being the quintessential opposite of ML, RBC serves as an interesting benchmark as to whether the machine has learned anything at all.

Having reviewed the field of RL from a general point of view, we now present the simulator we will be using for this paper. Subsequently, we will delve into how extant literature has employed RL to tackle the challenges of power grid management and, in particular, to address the issue of demand response.

2.7 The CityLearn Power Grid Simulator

CityLearn [12] is a simulation environment for energy demand response. In its simplest form, it considers a generator (that supplies energy) and a building (that consumes energy). Demand for energy varies according to time zone, time of day, and types of buildings. The structure has the capacity to store energy during periods of minimal usage and subsequently utilize it during times of heightened demand. However, there exists a constraint on the maximum energy the building can store. The building seeks to minimize a cost function which considers a series of factors: amount spent on energy, risk of shortage, risk of rampage, and others. In a not-so-simple scenario, there are multiple generators and multiple buildings. An agent decides how much energy should be used by or stocked at each building, so as to minimize the cost function for the *network* (not any individual building). Diverse forms of energy can be taken into account, including thermal energy, battery storage, and air-to-water heat pumps. Although these energy sources can be stored, their costs vary.

To be more specific, CityLearn simulates an environment where a set of N buildings are, at any given time t, in a state that is completely defined by 28 variables. These variables are of the following types:

- Temporal variables as “month,” “day,” or “hour”
- Temperature variables
- Humidity variables
- Diffuse solar radiation variables
- Storage variables(How much energy is stored)

We find it important to highlight that all but the last type of variable are related to the demand of energy. Indeed, the demand for energy varies through time depending on temperature, humidity, or the speed with which energy dissipates. The last kind of variable, however, refers to storage. And storage is ultimately a “managerial” decision: it’s how the agent chooses to manage the network. Such choice has implications that deserve our attention: If a building stocks energy today, it is betting that energy today will be cheaper than in the future. Moreover, if it chooses to use energy it has previously stocked, it is limited by the choice of how much it decided to stock in the past. Hence, the storage variables insert a path dependency into our problem.As for the actions, there are only two actions possible for the agent to take

- Increasing or decreasing the energy in cooling storage
- Increasing or decreasing the energy in domestic hot water storage

Since each building has its own storage, two such actions are, in reality, two actions *per building*. In other words, the agent must decide how much energy will be put in storage (or taken out of storage) for each building.

Such agent may be a single agent, managing the entire network, or a set of agents, one responsible for each building, which act independently but can communicate with each other.

Having presented both the field of RL and the simulator we will employ, we now proceed to discuss how RL scholars have previously addressed the issue of power grid management. We shall be particularly concerned with a particular problem in power grid management (namely, demand response) for which we will give a precise definition in the upcoming section. Finally, we will show how this paper articulates itself with extant literature, and how it relates to the ongoing debate on how can RL be used to improve current practices of power grid management.

2.8 *Reinforcement Learning for Demand Response*

The task of managing a power grid conforms well to the basic tenants of RL. For each action one takes on a power grid, one has a corresponding consequence to either enjoy or suffer. The grid can have a myriad of different states and unknown transition probabilities. In short, one must teach an agent which actions to take given only the rewards of such actions. That's the quintessential example of RL [16]. Indeed, there is a vast literature on employing RL to managing power grids (e.g., [17]).

Glavic ([18]) performed a literature review where he analyzed extant challenges and concerns about the employment of RL to power grid management concluding the overall sentiment in the scientific community is that RL is a very promising approach to deal with the task at hand.

Indeed, RL has been employed to multiple problems, all of which fall under the general category of power grid management. Thus, some authors focus on optimizing the generation of electricity (e.g., [19]), others focus on its distribution [20], while others focus on managing the consumption of energy itself [21]. Likewise, some authors focus on controlling the power grid [22], others in stabilizing it [23], while others attempt to modulate demand by setting energy prices dynamically [24].

This latter stream of research, in particular, has been receiving a significant share of scholarly attention, and even has a name of its own: demand response. Demand response consists of managing the power grid by modulating the demand for energy. This will typically be done by some sort of price incentive, lowering energy prices when one wishes to increase demand, and raising energy prices when one wishes to decrease it [25, ?]. Our paper falls into this category.

We consider the problem of a small power grid to which buildings of different kinds are connected. The price of electricity varies with time and, in particular, it increases as a response to an unexpected ramping in demand. Buildings use electricity to either heat up or cool down, thereby maintaining a pleasant temperature. The need to heat up or cool down is determined by the outside temperature which varies throughout the year in a way that is specific to the climate zone where the power grid is located. Buildings have the option to store either hot or cold water that can be used to either heat up or cool down the building at some later time. Hence, if a building is anticipating hot weather on the upcoming days and the price of electricity is cheap at present, it may wish to store cold water, so it can cool itself down when the temperature rises and energy prices are no longer cheap. All data is simulated, and we use the CityLearn package to this end.

Unlike supervised and unsupervised learning, RL often makes use of simulated data (with some exceptions, e.g., [21]). Kang [26] notes that one reason for the ubiquity of simulated data on RL research is because RL requires copious amounts of data, to the extent where getting such data might be very costly, if at all feasible. Beattie et al. [27] make an interesting remark on this point. They argue that, in some real-world application domains, data may be generated slowly. This is, indeed, the case with our paper. In order to learn how to efficiently manage energy throughout yearly cycles of succeeding seasons, one would need many years of data. By using simulations, such data can be obtained with no need to wait.

There is, of course, a trade-off here. While simulated data is much faster and cheaper to obtain in large amounts, it may fail to accurately replicate what real-data would look like. In plain terms, one can always cast doubt on whether a simulator is indeed realistic [28]. This has led some researchers to endeavor to combine real-data and simulated data [26].

In spite of such promising developments in the field, there has been a remarkable development in simulators targeted at RL research (e.g., [27]). At present, most research on RL relies on simulated data. Our paper follows this tradition, and uses the CityLearn simulator [12].

The CityLearn simulator simulates a small power grid network. As concerns its size, our grid is an intermediate between the one modeled by Kuznetsova et al. [29] (which consists of a consumer, a generator, and a battery for storage), and the one modeled by Yousefian and Kamalasadan [22] (where dispersion over a large area is a feature that must be considered). In our power grid, there are multiple buildings, but they are sufficiently close together for delays between generation and consumption to be neglected.

In addition, we have multiple buildings and pay no attention to how is energy distributed inside a building. In this respect, our paper is opposite to that of Tsui and Chan [24], where one considers a single home but worries about the multiple appliances it contains.

It is important to highlight, however, that nine different types of buildings are considered. This should arguably enhance the adherence of our simulated power grid to reality since in the real world there exist buildings with different demands for energy. A commercial building has high demand for electricity during work hours, but a smaller demand during the weekend. The opposite is true for residential buildings.

Likewise, we consider five different climate zones, which means there are five distinct ways to simulate how temperature (and hence energy demand) evolves throughout the year.

In our simulator, buildings can use energy to store either warm or cold water that will subsequently be used for cooling or heating. Scholars such as Tsui and Chan [24] and Xu et al. [21] allow for many more uses of electricity within a building, but their interest lies on a single household. Conversely, scholars such as Ademoye and Feliachi [17] and Mocanu et al. [30] deal with the energy demand as a black box, with no distinction whatsoever as regards its use. Our paper strikes a balance between both approaches: while not delving into all the possible uses of electricity within a building, by allowing two possible uses, it acknowledges that this utility may be used in different ways.

In addition, it's worthy to note that buildings can only store or use energy. They cannot generate energy. In this regard, our approach differs from Wang et al. [19], where buildings are equipped with photovoltaic generating systems. This should arguably increase the need for energy storage. While data from the International Energy Agency [31] suggests global photovoltaic capacity has been doubling every 2.4 years, the same report shows that such capacity is capable of meeting only 3% of the global demand for energy. Hence, at present, it is more realistic to suppose buildings do not have the ability to generate their own energy than to suppose otherwise.

When all such buildings are brought together and integrated into a power grid, one must make the decision as to how such grid is to be managed. One possibility, for example, is to have a single agent decide how each building should act, so as to minimize the cost of managing the entire grid. Such an agent would need to have information on each building's energy needs and storage levels. In its own turn, the agent would combine all this information into a single objective function that considered the costs of all buildings, thus achieving what Economic literature calls the social optimum. Yousefian and Kamalasadan [22] is an example of such an implementation.

One practical problem of the single agent approach is that granting its information on the energy consumption of each individual building may be perceived as a violation of privacy rights. Moreover, granting control over the power grid to a single agent raises concerns over malpractices such as observed during the California Energy Crisis [5]. Finally, minimizing the collective cost of energy doesn't necessarily guarantee that each individual building will have its own cost minimized, which may also pose an issue in more individualistic societies.

Another possibility as to how to manage a power grid is to do precisely the opposite: have each building decide on its own how much energy to use or to store. In plain terms, rather than having a single agent decide what everyone should do, each building shall decide for itself. In this multi-agent approach, each building optimizes its own objective function. In principle, nothing precludes buildings from sharing information among themselves, so it is still possible to incorporate information from multiple buildings within each objective function. Still, since each building optimizes its own objective function (possibly conditioned on its expected behavior of the remaining buildings), there is no guarantee that a social optimum will be achieved. Nonetheless, concerns over privacy, excessive market power, and the limitation of the individual's freedom to choose have been greatly reduced. Li et al. [32], Dhamankar et al. [33], Xu et al. [21], and Ademoye and Feliachi [17] are examples of this approach.

In our paper, we use a multi-agent approach with no information sharing. This means each building maximizes its own objective function, with no regards whatsoever for other buildings. We chose a multi-agent approach because, in our understanding, it is more realistic, as it evades the caveats cited previously. In addition, by doing so, we align ourselves with the majority of previous research. A third, but equally important reason, results from a meeting we had with the developer of the CityLearn simulator, in which he suggested we used the multi-agent approach, since its code is more stable, and has been more tested, than the single agent.

In reality, a hybrid model is often observed, where a large network is broken into regions and, within each region, the generation, transmission, and distribution of energy are managed by a single-agent fashion, whereas the buildings are granted autonomy to decide for themselves how much to consume or store. In other words, within a certain region, the decisions pertaining generation, transmission, and distribution are made by centralized agents, whereas the actual consumption of energy is made by decentralized agents. The Synchronous grid of Continental Europe [1], for example, is managed by the European Network of Transmission System Operators for Electricity (ENTSO-E), an organization that is, in turn, comprised by 42 other organizations (called Transmission System Operators, or TSOs) that manage the generation, transmission, and distribution of electricity within specific regions of the grid. Since the entry costs of generation, transmission, and distribution are formidably high, these activities are natural monopolies, so concerns over market power of TSOs cannot be solved by competition. Thus, a single agent approach lends itself naturally to this scenario. The same cannot be said, however, for the demand side of the grid. Hence, as far as buildings are concerned, they are autonomous to make their own decisions regarding energy consumption. Here, what we have is a multiple-agent approach. We have been unable to find any papers that employ such a hybrid model of power grid management. While we acknowledge that such vacuity may be a result of the increased complexity of a hybrid model, we should also point out that this is (to the best of our knowledge) an open field for future research.

3 Data

3.1 Simulator

CityLearn is an OpenAI Gym environment, facilitating RL agent implementation in multi-agent demand response scenarios. It reshapes electrical demand by managing energy storage across diverse buildings [12]. Its main goal is to standardize RL

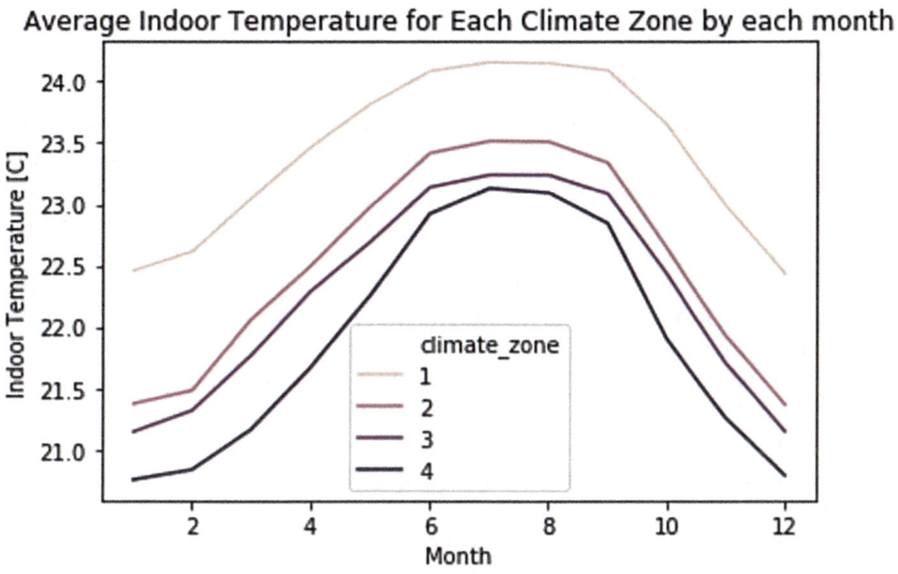

Fig. 1 Average indoor temperature setting by climate zone

agent evaluation for benchmarking algorithms. CityLearn includes energy models for air-to-water heat pumps, electric heaters, and pre-computed energy loads, covering functions like space cooling, dehumidification, appliances, domestic hot water (DHW), and solar generation. We're introducing a novel RL algorithm to CityLearn for managing DHW and chilled water storage. RL agents provide hourly control actions and receive states and rewards in response. Building indoor temperatures remain constant, as the environment constrains controller actions and ensures energy supply sufficiency. RL agents control cooling and heating energy storage and release. A backup controller in CityLearn prioritizes meeting building energy demand over additional energy storage. The simulator allows us to configure the building properties and their number to be controlled by the agent(s). In addition, it also provides four climate zone setting to evaluate if the agent can works generally in several environments. As Fig. 1 shows the climate properties, such as temperature inside/outside the buildings differ by the climate zone.

3.2 *State, Reward, and Action for Reinforcement Learning Agent*

CityLearn simulates an environment where a set of N buildings are, at any given time t, in a state that is completely defined by 28 variables. These variables are of the following types:

- Temporal variables as "month," "day," or "hour"
- Temperature variables

- Humidity variables
- Diffuse solar radiation variables
- Storage variables (How much energy is stored)

We find it important to highlight that all but the last type of variables are related to the demand for energy. Indeed, the demand for energy varies through time, with temperature, humidity, or with the speed with which energy dissipates. The last kind of variable, however, refers to storage. And storage is ultimately a "managerial" decision: it's how the agent chooses to manage the network. Such choice has implications that deserve our attention: If a building stocks energy today, it is betting that energy today will be cheaper than in the future. Moreover, if it chooses to use the energy it has previously stocked, it is limited by the choice of how much it decided to stock in the past. Hence, the storage variables insert a path dependency on our problem. As for the actions, there are only two actions possible for the agent to take—increasing or decreasing the energy in cooling storage—increasing or decreasing the energy in domestic hot water storage since each building has its own storage, two such actions are, in reality, two actions per building. In other words, the agent must decide how much energy will be put in storage (or taken out of storage) for each building. Such a central agent may be a single agent, managing the entire network, or a set of agents, one responsible for each building, which acts independently but can communicate with each other. We have considered a set of one agent per building acting independently. For simplicity, however, we often refer to these agents in a singular form.

4 Methods and Experimental Results

As we mentioned in Section "Motivation and Importance of the Problem", our objectives are reducing the training time while having an appropriate performance for actual real-world use. The baseline performance is calculated by RBC, and our evaluation metric, *cost*, is calculated by comparing with RBC result. Figure 2 shows a training curve of agents trained in climate zone 1 and 4. As the figure shows, after 4-year and 7-year simulations for climate zone 1 and 4, respectively, the agents can produce better operations than RBC's one. Therefore, if energy management by this reinforcement learning technique is required in another climate zone, it might take another at least 4 year to train its agent, which is not an efficient approach. In this section, we introduce two methods we considered to reduce the training time. One is inference approach and another one is behavior cloning. In those subsections, we mention how they work and show the experimental results. Then, we discuss the strengths and weaknesses of each approach.

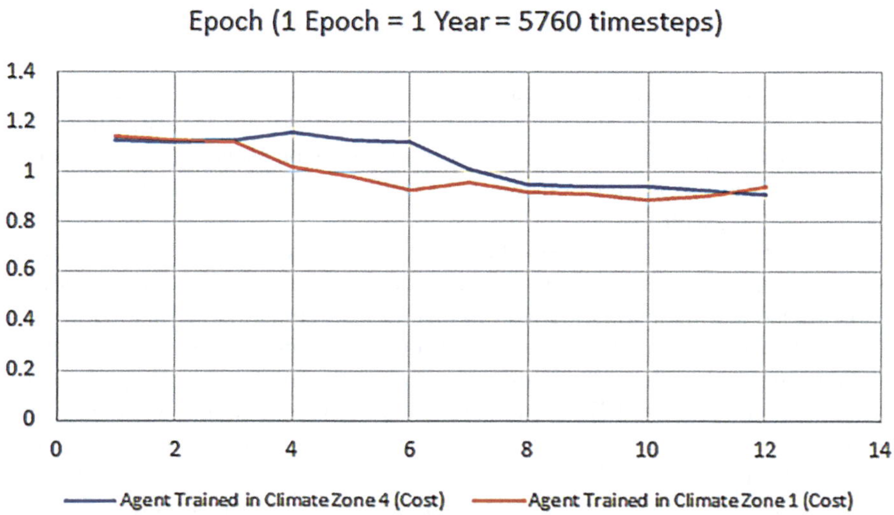

Fig. 2 Training curves of agents trained in climate zone 1 and 4

4.1 Inference

Reinforcement learning made remarkable advancement in the past few years. However, reinforcement learning still faces few difficulties such as the exploration and exploitation dilemma. [34] The exploitation of knowledge about the environment to improve an agent's performance can start after the agent explored sufficiently enough. For example, one of the reasons the training time is too long to adopt as a real-world application is SAC agent needs to explore about 3 years to gather enough states and actions for its replay buffer. After exploration, it starts to update weights of its policy and value function networks. As a result, inference, which is an approach to take advantage of external expertise from other tasks to apply into target task, is extremely efficient as it does not require any training or exploration.

In our research, we trained a policy through minimizing *cost* in a particular climate zone. Also, we assume environments, CityLearn simulator, has the same number and types of buildings in all climate zone. Let $\{Z_s, Z_t\} \in Z$ be environment of the source domain and target domain, respectively and shown in Eq. (4)

$$\pi^* =_\pi \mathrm{E}_{s\sim\mu_0, a\sim\pi}\left[Q_Z^\pi(s,a)\right] \tag{2}$$

where μ_0 is the set of initial states. $a \in A$, $s \in S$ denotes an action the agent can take and a state from CityLearn simulate in a particular climate, Z, respectively. The policy π can be described by Eq. (5):

$$\pi = \phi\left(D_s \sim Z_s, D_t \sim Z_t\right) : S_t \rightarrow A_t \tag{3}$$

In terms of ϕ, we use a deep neural network as a policy network. Regarding our objective with policy, we aim to find $\pi*$ for the target climate zone, Z_t minimizing $|D_t|$ that is the training time in the target climate zone. Concisely saying, we use pre-trained policy and value function network (critic) trained on Z_s and evaluate on Z_t to see how well it performs.

We evaluate how an agent trained in a specific climate zone for 12 years simulation works in another climate zone. Figure 2 shows experimental results that how an agent trained on climate zone 1 performs on the target climate zone 4 and vice versa. Costs obtained by policy from zone that it has trained on and from zone that it has evaluated on are less than 1.0 in both case of climate zone 1 and 4. They are appropriate costs compared to RBC's ones. Hence, we can mention that while originally it takes 12 years to train the agents, inference is an efficient method as it still provide appropriate costs compared to RBC. On the other hand, we can confirm the difference of costs between ordinary learning and inference, which arise from generalization problem. All climate zones CityLearn simulator provides are in the USA. Hence, there is a possibility that inference makes a larger difference if the source climate zone has no similarity to the target climate zone, especially in actual use. Therefore, as an improvement of this approach, generalization of reinforcement learning is required. It means generating an agent that provides an averagely appropriate cost on any other climate zone and transferring its policy and value function networks (Fig. 3).

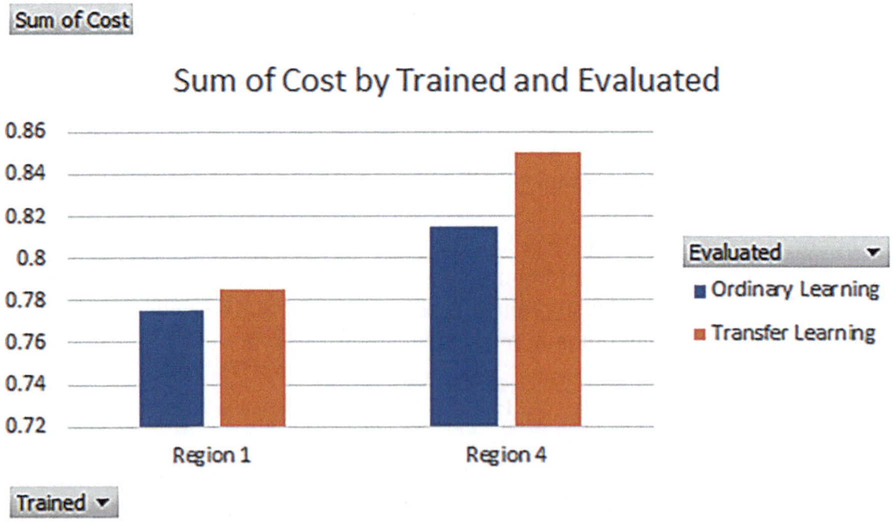

Fig. 3 Comparison of costs obtained after 12-year training and transfer learning

4.2 Behavior Cloning

In the preceding Section "Inference", we optimized training time with inference techniques. We also recognized the importance of extending the inference approach's applicability, particularly in dissimilar climate zones from the agent's original training environment. In this section, similarly to the inference, we consider using a policy network that mimics RBC. As we introduced in Section "State, Reward, and Action for Reinforcement Learning Agent", it calculates policy deterministically based on given state variables. Our idea is if an agent can imitate RBC using deep reinforcement learning algorithm, our agent can generate a cost that is close to 1.0 without any training. Then, we can train additionally to adapt to each climate zone. It is noteworthy that RBC is one of the generalized agents, and it can produce a cost that is 1.0 on every climate zone because it is a baseline to calculate cost itself.

In Algorithm 1, the algorithm of behavior cloning is written. First, we gather RBC's demonstrations by letting RBC interact with the target environment. In this process, a pair of state and RBC's actions is appended into the demonstration dataset. Using the demonstration dataset, the policy network extracted from the deep reinforce learning agent is trained. As indicated by the algorithm, it undergoes training using a supervised learning approach, and once convergence is achieved, the network is fully trained.

We experiment that a reinforcement learning agent is trained by behavior cloning and is also additionally trained by its own interaction with the environment and policy updating. We use climate zone 4 as an environment and compare its training curve with one generated from SAC agent. As Fig. 4 illustrates, the agent pre-trained by RBC demonstration dataset produces a lower cost (close to 1.0) than one SAC provides at the beginning of the training duration. However, during the training, the additional training does not improve the agent's performance.

Algorithm 1: Behavior Cloning
Require: Z: environment of the target climate zone
Require: f, θ: policy network and its weights of deep reinforcement learning agent
Require: L, α: loss function and step size parameter for update
1: interact RBC with Z and store (s, a) into expert data D
2: **while** stopping criterion not met **do**
3: Sample a minibatch of b examples from D, $s_1, \cdots, s_b$ with corresponding actions a_i
4: Compute gradient estimate: $\hat{g} = +\frac{1}{b}\nabla_\theta \sum_i L\left(f\left(s_i;\theta\right),a_i\right)$
5: Update: $\theta = \theta - \alpha\hat{g}$
6: **end while**

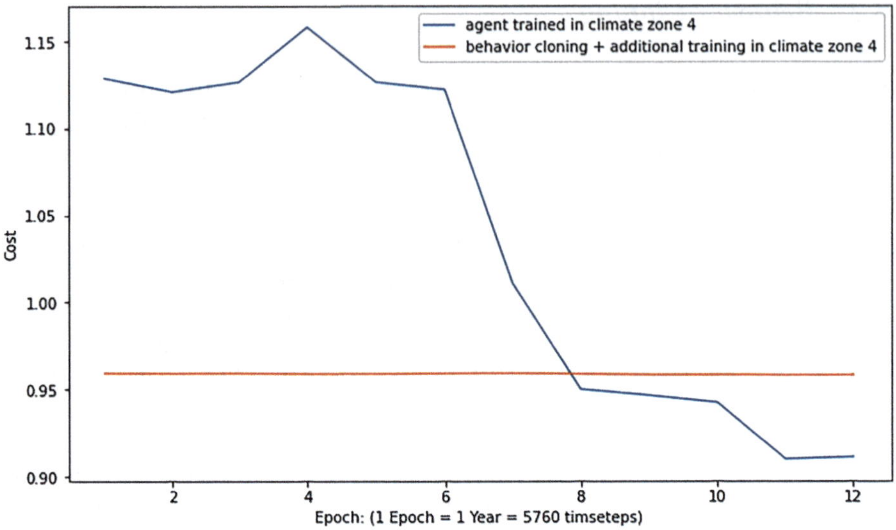

Fig. 4 Comparison of training curves between SAC and BC

The agent has actor (policy) and critic (value function) networks like SAC agent. In the pertaining process by behavior cloning, only the policy network is updated and value function network still has initial weights. Actor-Critic reinforcement algorithm updates their network weights by the loss calculated from its policy and value function networks. [35] Hence, when calculating the loss, the larger coefficient of value function network's loss causes that the policy network forgets its pre-trained weights drastically. [36] To prevent this catastrophic interference from the value network, we tuned the coefficient in a smaller value. Figure 5 describes only the training curve of the agent pre-trained by behavior cloning for 100-year simulation period. Because of the smaller coefficient of the value network, we can confirm the cost decreasing slightly and gradually.As an evaluation of this approach, it can generate better cost compared to SAC agent. However, it is fatal that the additional training is not efficient, and it takes a longer time to adapt to the environment. Behavior cloning can be used if we do not need to adapt to the environment and just imitate an external expert. In addition to the previous Section "Inference", we can conclude that both policy and value networks should be pre-trained on certain region, as it outperforms behavior cloning method.

5 Conclusion/Discussion

Model-Agnostic Meta-LearningThe limitations of our method are in generalization and the policy's long training period. We have shown that policy does not do well on generalization over other regions, especially compared to the trained regions in the cost. Due to a lack of generalization, we might need to train the policy for each

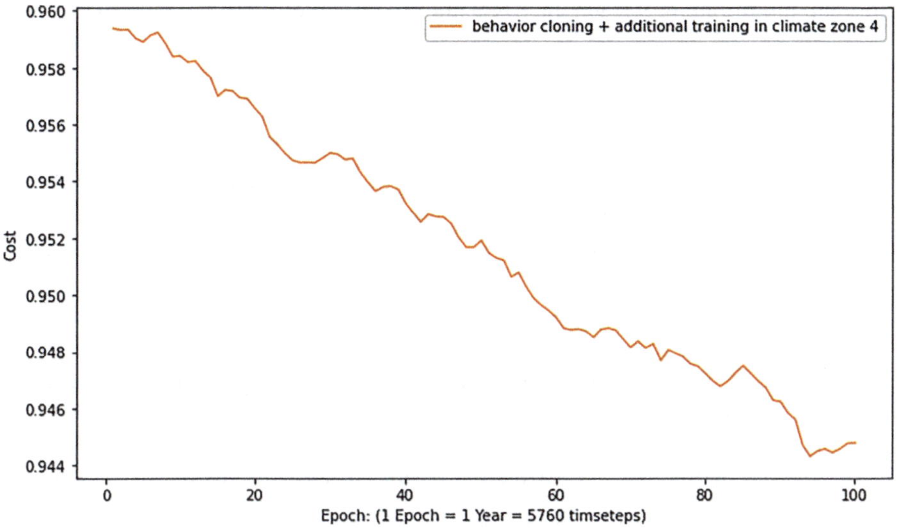

Fig. 5 Enlarged view of training curve of behavior cloning

region separately. However, it takes at least 10 years to train an adequate policy. To address this problem, we plan to incorporate Model-Agnostic Meta-Learning(MAML) algorithm, which will deal with over-fitting problems and decrease the policy's training period. [37] There are numerous interpretation of MAML. One of them is that the algorithm is designed to make the policy to learn shared components between different regions. It learns shared components by averaging different gradient descents updates from each region. In more detail, it will sample trajectories from each n sampled regions. It will update its parameters for n policies, copies of the original policy, using gradient descent. Using those n updated policies, it will sample another h trajectories. It will update the original policy by averaging those different gradient descents updates from h trajectories. In theory, the learning of unique components will decrease from this average process. After learning shared components, we would have fine-tuning process, in which policy learns its unique components. Since the policy only has to learn the unique components after using MAML, the required training period will decrease, especially compared to learning from scratch.We did face some obstacles from implementing MAML into our environment. Most of the packages for MAML are for a single agent. However, our CityLearn environment has multiple agents. Furthermore, we are interested in optimizing the policy in a competitive environment, so we also need to optimize other remaining policies to establish this environment. We thought our problem was similar to Gibbs sampling problem, a sampling problem for distributions that has a dependent relationship, as our optimizing agent also depends on other agents' policy. In this case, we cannot directly optimize or sample from a single agent or single parameter. Inspired by Gibbs sampling algorithm, we would iteratively update each policy, so each iteration the policy would be, in theory, optimized in a more competitive environment than previous iterations. The pseudocode for our algorithm is written in Algorithm 2. We are currently working on this problem.

Algorithm 2: Multi-agent MAML
Require: $P(Z)$: distribution of zones
Require: θ_i: policy for each building in the environment
Require: α, β: step size parameter for update
1: **while** not done **do**
2: **for all** θ_i **do**
3: sample batch of zones $Z_j \sim P(Z)$
4: **for all** Z_j **do**
5: Sample K trajectories $D = \{(x_1, a_1, \ldots x_H)\}$ using f_{θ_i} in the environment with zone Z_j
6: Evaluate $\nabla_{\theta_i} L_{Z_j}\left(f_{\theta_i}\right)$ using D and L_{Zj}
7: compute adapted parameters using gradient descent $\theta'_{i,j} = \theta - \alpha \nabla_{\theta_i} L_{Z_j}\left(f_{\theta_i}\right)$
8: Sample trajectories $D'_j = \left\{\left(x_1, a_1, \ldots x_H\right)\right\}$ using $f_{\theta'_{i,j}}$ in zone Z_j
9: **end for**
10: Update $\theta_i = \theta_i - \beta \nabla_{\theta_i} \sum_{Z_j \sim P(Z)} L_{Z_j}\left(f_{\theta_{i,j}}\right)$ using each D'_j and L_{Z_j}
11: **end for**
12: **end while**

5.1 Applicability

There are some concerns about whether the trained policy could be applicable in real life. One of the concerns is whether the simulation from the CityLearn environment will be similar to what is actually happening in the real-life environment. For example, CityLearn environment has only nine types of buildings. It is unlikely that those nine types of buildings will represent all different building types in real life. Furthermore, the CityLearn environment assumes that nine types of buildings will be the same for different zones. However, each zone's building types should be different as they would be built differently in material and structure. For example, an architect from warmer zones will use other materials compared to an architect from colder zones to account for climate and temperature. If there is such a difference, then buildings should generate different states if they have the same weather. Yet, such circumstances are not accounted in the CityLearn environment. Therefore, we need to study further how climate affects architecture and apply it to CityLearn to overcome this problem.

Another concern is whether policy that had been optimized in the environment with nine buildings will generalize into the environment with at least more than 1000 buildings. Original scholars and we used an environment with only nine buildings due to computational cost. Optimization could be more complicated and different when the number of buildings increases in the environment. If this is the case, we might impose a bias into our policy, which might lead to worse outcome than training from scratch. To overcome this problem, we should also look into how we could generalize the policy when there are a different number of buildings in the environment.

References

1. *Synchronous grid of Continental Europe*. Wikipedia. Retrieved from https://en.wikipedia.org/wiki/Synchronous_grid_of_Continental_Europe
2. *North American power transmission grid*. Wikipedia. Retrieved from https://en.wikipedia.org/wiki/North_American_power_transmission_grid
3. *Northeast blackout of 2003*. Wikipedia. Retrieved from https://en.wikipedia.org/wiki/Northeast_blackout_of_2003
4. *Serbia, Kosovo power grid row delays European clocks*. Reuters. Retrieved from https://www.reuters.com/article/serbia-kosovo-energy/serbia-kosovo-power-grid-row-delays-european-clocks-idUSL5N1QP2FF
5. *2000–01 California electricity crisis*. Wikipedia. Retrieved from https://en.wikipedia.org/wiki/200001_California_electricity_crisis
6. *Frequently Asked Questions (FAQs)*. U.S. Energy Information Administration (EIA). Retrieved from https://www.eia.gov/tools/faqs/faq.php?id=105&t=3
7. *Smart Grid*. NYSERDA. Retrieved from https://www.nyserda.ny.gov/Researchers-and-Policymakers/Smart-Grid
8. *What is a Smart Grid?* Live Science. Retrieved from https://www.livescience.com/41920-smart-grid.html
9. *First Int'l Workshop on Reinforcement Learning for Energy Management in Buildings Cities (RLEM)*. Retrieved from https://rlem-workshop.net/#program
10. *Unified National Smart Grid*. Repower America. Retrieved from https://web.archive.org/web/20081112091929/http://www.repoweramerica.org/elements/unified-national-smart-grid/
11. *SuperSmart Grid*. Potsdam Institute for Climate Impact Research. Retrieved from https://www.pik-potsdam.de/en/output/projects/projects-archive/supersmart-grid-1
12. V'azquez-Canteli, J.R., Kämpf, J., Henze, G., Nagy, Z.: CityLearn v1.0: An Open AI gym environment for demand response with deep reinforcement learning. In: *BuildSys 2019 - Proceedings of the 6th ACM International Conference on Systems for Energy-Efficient Buildings, Cities, and Transportation*. pp. 356–357. Association for Computing Machinery, Inc., New York, NY, USA (2019). https://doi.org/10.1145/3360322.3360998., https://dl.acm.org/doi/10.1145/3360322.3360998
13. Rashidinejad, P., Zhu, B., Ma, C., Jiao, J., Russell, S.: Bridging offline reinforcement learning and imitation learning: A tale of pessimism (Mar 2021), http://arxiv.org/abs/2103.12021
14. *A brief overview of imitation learning*. Smart-Lab AI – Medium. Retrieved from https://smartlabai.medium.com/a-brief-overview-of-imitation-learning-8a8a75c44a9c
15. Pomerleau, D.A. *ALVINN: An autonomous land vehicle in a neural network*. Tech. Rep.
16. Lambert, N.: Constructing axes for reinforcement learning policy - Democratizing automation (2020), https://robotic.substack.com/p/rl-policy
17. Ademoye, T., & Feliachi, A. (2012). Reinforcement learning tuned decentralized synergetic control of power systems. *Electric Power Systems Research, 86*, 34–40. https://doi.org/10.1016/j.epsr.2011.11.024
18. Glavic, M., Fonteneau, R., & Ernst, D. (2017). Reinforcement learning for electric power system decision and control: Past considerations and perspectives. *IFAC-PapersOnLine, 50*(1), 6918–6927. https://doi.org/10.1016/j.ifacol.2017.08.1217

19. Wang, Y., Lin, X., & Pedram, M. (2016). A near-optimal model-based control algorithm for households equipped with residential photovoltaic power generation and energy storage systems. *IEEE Transactions on Sustainable Energy, 7*(1), 77–86. https://doi.org/10.1109/TSTE.2015.2467190
20. Gemine, Q., Ernst, D., & Corn'elusse, B. (2017). Active network management for electrical distribution systems: problem formulation, benchmark, and approximate solution. *Optimization and Engineering, 18*(3), 587–629. https://doi.org/10.1007/s11081-016-9339-9
21. Xu, X., Jia, Y., Xu, Y., Xu, Z., Chai, S., & Lai, C. S. (2020). A multi-agent reinforcement learning-based data-driven method for home energy management. *IEEE Transactions on Smart Grid, 11*(4), 3201–3211. https://doi.org/10.1109/TSG.2020.2971427
22. Yousefian, R., & Kamalasadan, S. (2016). Design and real-time implementation of optimal power system wide-area system-centric controller based on temporal difference learning. *IEEE Transactions on Industry Applications, 52*(1), 395–406. https://doi.org/10.1109/TIA.2015.2466622
23. Hadidi, R., Jeyasurya, B.: Reinforcement learning based real-time wide-area stabilizing control agents to enhance power system stability. IEEE Transactions on Smart Grid 4(1), 489–497 (2013). https://doi.org/10.1109/TSG.2012.2235864
24. Tsui, K. M., & Chan, S. C. (2012). Demand response optimization for smart home scheduling under real-time pricing. *IEEE Transactions on Smart Grid, 3*(4), 1812–1821. https://doi.org/10.1109/TSG.2012.2218835
25. Christensen, M., Ernewein, C.: *Demand response through price-setting multi-agent reinforcement learning*. Retrieved from https://www.youtube.com/watch?v=_f0L313_4xE
26. Kang, K., Belkhale, S., Kahn, G., Abbeel, P., Levine, S.: Generalization through Simulation: Integrating simulated and real data into deep reinforcement learning for vision-based autonomous flight. *Proceedings - IEEE International Conference on Robotics and Automation 2019-May*, 6008–6014 (Feb 2019), http://arxiv.org/abs/1902.03701
27. Beattie, C., Leibo, J.Z., Teplyashin, D., Ward, T., Wainwright, M., Kuttler, H., Lefrancq, A., Green, S., Vald'es, V., Sadik, A., Schrittwieser, J., Anderson, K., York, S., Cant, M., Cain, A., Bolton, A., Gaffney, S., King, H., Hassabis, D., Legg, S., Petersen, S.: *DeepMind lab*. Tech. Rep. (2016)
28. Dulac-Arnold, G., Mankowitz, D., & Hester, T. (2019). Challenges of real-world reinforcement learning. *arXiv*. http://arxiv.org/abs/1904.12901
29. Kuznetsova, E., Li, Y. F., Ruiz, C., Zio, E., Ault, G., & Bell, K. (2013). Reinforcement learning for microgrid energy management. *Energy, 59*, 133–146. https://doi.org/10.1016/j.energy.2013.05.060
30. Mocanu, E., Mocanu, D. C., Nguyen, P. H., Liotta, A., Webber, M. E., Gibescu, M., & Slootweg, J. G. (2019). On-line building energy optimization using deep reinforcement learning. *IEEE Transactions on Smart Grid, 10*(4), 3698–3708. https://doi.org/10.1109/TSG.2018.2834219
31. *Snapshot 2020. IEA-PVPS*. https://iea-pvps.org/snapshot-reports/ snapshot-2020/
32. Li, F. D., Wu, M., He, Y., & Chen, X. (2012). Optimal control in microgrid using multi-agent reinforcement learning. *ISA Transactions, 51*(6), 743–751. https://doi.org/10.1016/j.isatra.2012.06.010
33. Dhamankar, G., Vazquez-Canteli, J., Nagy, Z.: *Benchmarking multi-agent deep RL algorithms on building energy demand coordination task*. Retrieved from https://www.youtube. com/watch?v=eWlVPKIB23o
34. Zhu, Z., Lin, K., Zhou, J.: Transfer learning in deep reinforcement learning: A survey (2021).
35. Sutton, R. S., Mcallester, D., Singh, S., & Mansour, Y. (2000). Policy gradient methods for reinforcement learning with function approximation. In *Advances in neural information processing systems* (Vol. 12, pp. 1057–1063). MIT Press.
36. Jena, R., Liu, C., Sycara, K. Augmenting gail with bc for sample efficient imitation learning (2020).
37. Finn, C., Abbeel, P., Levine, S.: Model-agnostic meta-learning for fast adaptation of deep networks (2017).

Enhancing Power Distribution Efficiency Through AI-IoT and ML-Based Methodologies

Suresh Kumar Natarajan, Shajahan Basheer, S. Anthoniraj, S. Jayanthi, Hoshiyar Singh Kanyal, and Rajeshwar Voleti

1 Introduction

1.1 Importance of Efficient Power Distribution

Excellent power distribution system is the main stay of modern economy, which is further critical for ensuring that the constant natural flowing power reaches to homes, commercial businesses, and industries without any perturbation [1]. The significance of efficient power distribution can be understood through various dimensions:

Reliability and Accessibility: An exceptional power distribution system forms the bedrock of a modern economy, guaranteeing the seamless transmission of electricity to residences, businesses, and industries. This reliability and accessibility are pivotal, ensuring a continuous supply of energy that is indispensable for critical infrastructures such as healthcare facilities, communication networks, and emergency response systems. Hospitals, for instance, rely on uninterrupted

S. K. Natarajan (✉) · S. Basheer · S. Anthoniraj
School of Computer Science and Engineering, Jain (Deemed-to-be University), Ramanagara, Karnataka, India

S. Jayanthi
Department of DS & AI, Faculty of Science and Technology, ICFAI Foundation of Higher Education (IFHE), Hyderabad, Telangana, India

H. S. Kanyal
Department of Computer Science and Engineering, ABES Institute of Technology, Ghaziabad, India

R. Voleti
School of Computer Science and Engineering, Jain (Deemed-to-be University), Ramanagara, Karnataka, India

S. Vijayalakshmi et al. (eds.), *Smart Power Systems*, Power Systems,
https://doi.org/10.1007/978-3-031-89459-6_3

power to sustain life-saving equipment and patient care, while communication networks must remain operational for personal and national security reasons. In times of crisis, such as natural disasters, a dependable power grid ensures the resilience of essential services, underscoring the significance of reliable and accessible electricity distribution.

Economic Growth and Development: Efficient power distribution systems serve as catalysts for economic growth and development by fostering a conducive environment for businesses and industries to thrive. Industries heavily depend on a stable and reliable power supply to sustain production processes, thereby minimizing downtime and operational losses [2]. Moreover, a reliable power infrastructure enhances the competitiveness of economies by reducing operational risks associated with power outages, thereby attracting investors and spurring business expansions. Commercial entities also benefit from consistent electricity, ensuring uninterrupted operations and contributing to overall economic activity. Thus, the role of power distribution in driving economic growth and development cannot be overstated.

Quality of Life: The quality of life is profoundly influenced by a dependable power distribution system, which underpins various aspects of daily living. Electricity is integral to modern conveniences such as lighting, heating, cooling, cooking, and entertainment, enhancing comfort and productivity for individuals and communities. Reliable power supply ensures that homes and workplaces are well equipped and functional, supporting activities ranging from work and study to leisure and recreation. Furthermore, community services such as schools, libraries, and recreational facilities rely on a steady power supply to serve residents effectively. Thus, the importance of power distribution in enhancing the quality of life is paramount.

Environmental Sustainability: Efficient power distribution systems play a crucial role in promoting environmental sustainability by optimizing energy use and reducing waste [3]. Modern technologies improve energy efficiency, minimizing losses during transmission and distribution and resulting in lower overall energy consumption. Moreover, the integration of renewable energy sources, facilitated by efficient power distribution, reduces reliance on fossil fuels and mitigates greenhouse gas emissions. By adopting sustainable practices and reducing environmental degradation associated with energy generation and consumption, power distribution systems contribute to global efforts to combat climate change and preserve natural resources.

Resilience and Disaster Management: Resilient power distribution systems are essential for effective disaster management and recovery, ensuring the continuity of critical services during emergencies [4]. By safeguarding critical facilities such as hospitals, emergency shelters, and communication centers, resilient power infrastructure supports emergency response efforts and enhances community resilience. Advanced grid technologies and smart distribution systems enable rapid restoration of power after disruptions, minimizing downtime and accelerating recovery processes. Furthermore, by providing reliable power to systems and technologies that monitor and respond to crises, resilient power infrastructure enhances emergency preparedness and fosters community safety.

Technological Innovation: The integration of advanced technologies such as the Internet of Things (IoT), artificial intelligence (AI), and machine learning (ML) is driving technological innovation in power distribution. These technologies optimize grid operations by automating processes, providing real-time insights into grid performance, and offering predictive capabilities for maintenance. Automation and optimization enhance efficiency and reliability, while predictive maintenance reduces downtime and maintenance costs. Moreover, advanced monitoring technologies offer real-time visibility into the state of the power grid, enabling better decision-making and more effective management of power distribution. Thus, technological innovation is reshaping power distribution, paving the way for a more efficient, reliable, and sustainable energy future (Fig. 1).

1.2 Evolution of AI-IoT and ML in Power Distribution

The technology novelty in power system control and cybersecurity features attributable to the application of AI, IoT, and ML has made fundamental improvements in energy saving, reliability, and sustainability [5]. The evolution of AI-IoT and ML in power distribution can be traced through several key stages:

Early Adoption and Experimentation: In the initial stages, power distribution utilities began testing AI, IoT, and ML technologies with the goal of enhancing grid operations, improving asset management, and strengthening failure detection capabilities [6]. The earliest applications entailed the implementation of smart meters for immediate energy consumption observation, the creation of sensor networks for monitoring conditions of the assets, as well as data analytics generated for preventive maintenance purposes.

Advancements in Data Analytics: In the era that volume, velocity, and variety of the power distribution systems data produced are growing, there is a rising trend of the importance of advanced techniques of data analytics [7]. The use of ML algorithms was applied to the historical data to uncover trends, patterns, and forecasts on certain factors like load demand, equipment health, and grid performance.

Integration of IoT Devices: The evolution of the IoT and the sensors system also accelerated the influx of AI and ML engineering in power distribution [8]. IoT devices were used throughout distribution networks to monitor real-time voltage levels, current flows, heat, humidity, and other conditions that provide good situational awareness that helped in decision-making.

Emergence of Smart Grids: The idea of smart grid gained currency, suggesting a new concept of intelligent power supply that would incorporate modern digital technologies, clean energy sources, and energy storage systems into traditional power grid [9]. The AI technology proved to be invaluable in terms of efficiency

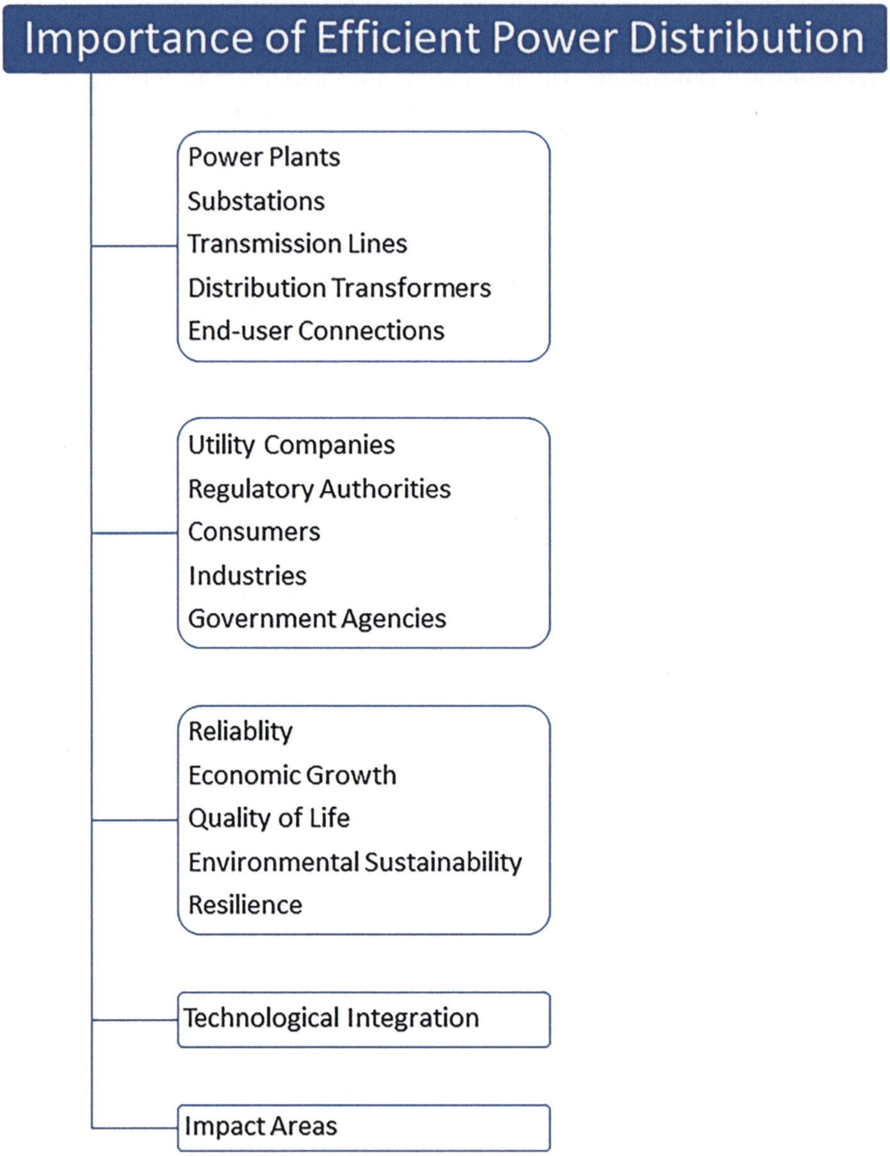

Fig. 1 The importance of efficient power distribution

of managing smart grids, as well as addressing the issues of demand response, distributed energy resources, and grid stability and resilience.

AI-Driven Automation and Control: AI development brought in a lot of modern features. This includes AI-controlled systems and automation to get a real-time optimized distribution flow. Through data science, ML models were deployed for the predictive and preventive decrease of the voltage fluctuations, provision of help in the current optimization, balancing supply and demand, as well as detection and response of the grid disruptions [10].

Holistic Approach to Grid Optimization: AI-IoT and ML in the energy production have brought a new way to compose grid optimization which includes predictive maintenance, fault localization, energy management, and consumer interaction. Combining AI platforms into the mix creates out an avenue for utilities to gain vital analytics based on the data analytics. In turn, utilities generate decisions that can reduce operational costs and offer better service to consumers.

Continued Innovation and Collaboration: The path from AI-IoT and ML in power distribution activity includes ever-faster development in collaboration between power utilities, the technology suppliers, research establishments, and regulators. The continuous evolutions and upgrades of AI algorithms, capability advancements of IoT devices, issues related to cybersecurity, and interoperability and scalability are the focal points of the ongoing research and development work.

The use of AI-IoT and ML in power distribution is being seen as a pivot whereby more calculative, resilient, and greener power systems with smart infrastructure will be developed. They can also tackle the issues that traditional energy setup is facing in the emergence of more complicated and dynamic energy scenario via the adoption of digital technologies, thus advancing a more sustainable and efficient way of operation.

1.3 Essentials of Power Distribution

This chapter is organized to provide a detailed discovery of how Artificial Intelligence (AI) does a very good job of making distribution of power more efficient, helping IoT, and also ensuring that Machine Learning (ML) is secured through reliable data transfer. The chapter is organized as follows:

Introduction: The chapter starts with a short note that signifies the emergence of AI-IoT and ML technologies as well the progress of these technologies in the field of efficient power distribution. Setting the grounds for further debate is one of the main tasks for this introduction. It is always recommended to point out the relevance and scope of the chosen topic.

Fundamentals of Power Distribution: This segment talks about power distribution systems, explains about the installation and function of its components, and outlines the difficulties in distribution networks. It seeks to make sure that the development of an alternative approach for power distribution will be done in order to address the complications that will emerge constantly and the market demands that will keep on increasing.

Role of AI-IoT in Power Distribution: This section further talks the headway AI as well IoT applications embarks in power distribution systems into consideration. It investigates how AI algorithms and IoT devices can be applied within grid framework for realizing grid operations scheduling, equipment maintenance, and not forgetting general efficiency. The spotlight is on particular scenarios including: predictive maintenance, fault detection, energy optimization, and load forecasting. These are provided in detail.

ML-Based Approaches in Power Distribution: Herein, the discussion moves on to ML algorithms and their utility in system wide reliance in power distribution optimization. This part starts with the description of machine learning (ML) methods, like supervised learning, unsupervised learning, as well as reinforcement learning, and explains how they are applied to the fields of load forecasting, fault detection, grid optimization, and so on.

Case Studies and Applications: This part gives the light on implementation scenario of AI-IoT and ML in distribution of power systems. Cases of smart grid architectures, control and distribution management systems, and demand side approaches are discussed presenting their successful results, effects, and influence on cost effectiveness, but also conservation of natural resources.

Challenges and Future Directions: The chapter closes by means of a dialogue on the challenges and directions for power distribution of AI-IoT and ML next. Constraints in present-day approaches, problems with scalability, and human issues along with research opportunities are taken into consideration, showing us that there is a need for more research and collaboration in this field.

2 Fundamentals of Power Distribution

2.1 Overview of Power Distribution Systems

The power distribution systems establish the key connection between electricity and industrial sites, providing the required power flow control and customer peak power shaping. The basis of function and maintenance of power distribution systems is critical to their performance [11]. Proactive companies will profit from the resultant growth in efficiency and reliability. This segment of the analysis gives an explanation of the power distribution systems and the working, components, and principles of those; it also delves deeper into that latter.

2.2 Components of Power Distribution Systems

Generation Stations: The power distribution commences at generation stations which are setup to generate kinetic energy using various methods including thermal, hydroelectric, nuclear, and renewal sources.

Transmission Network: The electricity produced is transmitted through high-voltage transmission lines that run to distribution substations located in proximity to consumers.

Substations: Substations therefore hold the crucial position in voltage transformation from the transmission levels to the distribution levels suitable to the consumers [12]. They also might as well include machine sets for regulation of voltage, change, and protection functions. Use our AI to write for you about in the nonfiction given sentence, the author states the possibility of the technology being valued even as a collector's item.

Distribution Lines: Conduits like overhead lines and underground cables are responsible for delivering electricity from substations to homes, businesses, and factories which act as the consumers.

Transformers: Distribution transformers are positioned at multiple places from the beginning of the line and down to the last in order to help reduce the voltage level and the voltage's compatibility with specific loads.

Switchgear and Protection Devices: Switchgear and sectionalizing equipment are incorporated into the system to control the amount of electricity, isolate the parts of the network that have problems (the faulty areas), and ensure safety and continuity.

Metering and Monitoring Equipment: The use of metering equipment in the form of meters and smart meters has been instrumental in measuring the electricity consumption at different consumer premises. Today monitoring devices allow collection of all types of data in real time using various tools, such as sensors, and telemetry systems.

2.3 Functions of Power Distribution Systems

Voltage Regulation: The power distribution systems keep the range of voltage within limits to enable the compatibility of the connected equipment and appliances. It concerns this issue when the range of voltage is not suitable for the appliances.

Load Balancing: Load balancing is achieved throughout system distribution networks considering various feeders and phases, to avoid overloading and to best utilize network resources.

Fault Management: Protection structures that allow to identify and insulate faults, thus, decreasing the global interruption and time without the supply are inherent to distribution systems and are among the major advantages of the new technological achievements (Fig. 2).

Energy Loss Reduction: Overall, the focus is on reducing the losses of energy which are achieved by means of the following: equipment design that is adequate; running the equipment well; and adjusting the operating parameters properly.

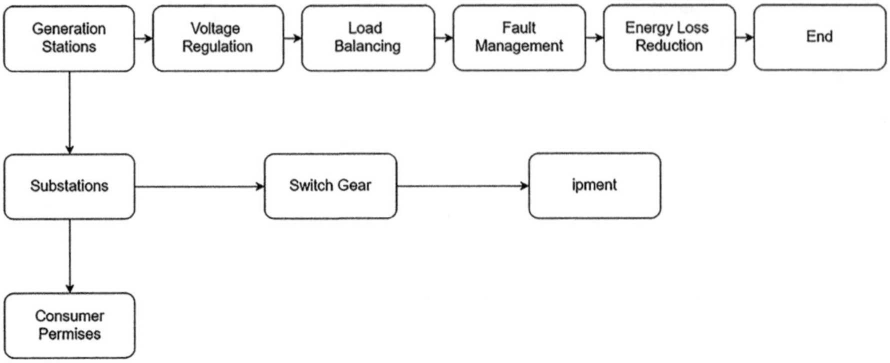

Fig. 2 The overview of power distribution systems

Resilience and Reliability: Normally, the distribution network of power is designed to cope with diverse environmental and operational challenges, thus guaranteeing the on-working supply of power to customers.

2.4 Challenges in Traditional Power Distribution

Central networks of old power system models experience a number of different obstacles, such as the quality of end product, reliability, and stability. Both challenges are essential for a reliable and modern electrical grid. They help to develop good grid technologies [13]. This section highlights some of the key challenges in traditional power distribution:

Aging Infrastructure: About the aged infrastructure, many power distribution systems worldwide have well-known problem areas like substations, step-down transformers, and power distribution lines. The machinery aging results in more frequent technical faults and becomes less reliable, thus causing the costs of maintenance to further increase.

Limited Visibility and Monitoring: Regular government electricity distribution networks that is unable to be operated and controlled fully. Being unable to see the quality of the networks makes the sensor's operator difficult to pin the problem spots, inspect the systems, and optimize the usage of frameworks instantaneously.

Voltage Instability and Fluctuations: Power quality issues might be caused by issues with line strength and level stability, which result from factors like rapidly changing loads, equipment breakdowns, and insufficient voltage control mechanisms. As a result, these problems limit the efficiency of interlinked equipment and ultimately result in their breaking down or inoperability.

Overloading and Congestion: In the case when distribution pedestal gets overloading or congestion due to times of peak demand or unbalanced load distribution, the networks may experience unstable consumption [14]. An overloaded circuit will definitely lead to issues like voltage drops, potential power outages, and low system capacity.

Limited Integration of Renewable Energy: With the progress of renewable energy sources such as solar and wind, the traditional power distribution system becomes a victim of this problem because this system is required to integrate large amounts of fluctuating and intermittent generation of renewables. The system's requirements of superior control and grid management are more sophisticated ones.

Cybersecurity Risks: Main centers of the energy distribution systems are the potential targets for the cybercrime, and the attack to control systems, communication networks and data infrastructures represent a big threat for security. Cybersecurity breaches may result in denial or delay of services, leak of confidential information, and undermining of the stability and effectiveness of systems.

Lack of Flexibility and Resilience: Systems based on traditional power distribution have generally been ones that hold on to rigid topologies and centralized control systems. Flexibility becoming elusive is a relevant barrier to overcoming issues such as incorporation of the distributed energy resources, managing unexpected evolutions of operational conditions, and growing the resilience against disasters.

Environmental Impact: Ordinary power circulation systems to a great extent improvised from fossil fuels and as a consequence, air pollution, greenhouse gases, and resources were depleted. Ensuring the environment sustainability implies changing to renewable energy source, energy saving, and dropping of greenhouse emissions (Fig. 3).

2.5 Need for Innovation in Power Distribution Methodologies

The custom channels through which power distribution takes place are experiencing added demands as a result of the shifting energy environment, emerging technologies and changing landscape of the consumer market. This section highlights the pressing need for innovation in power distribution methodologies to address emerging challenges and capitalize on new opportunities:

Rising Energy Demand: Along with the increase in global human population, urbanization, and economic development, the need for energy also rises accordingly. With the steadily increasing demand for energy utilities cannot remain traditional, their system must be modernized because it will be more effective and environmentally friendly.

Integration of Renewable Energy: Just the move to renewable energies like solar, wind, and hydro-power depicts the reality of an aspect and challenge to distribution of power [15]. The grid needs new methodologies to be able add intermittent

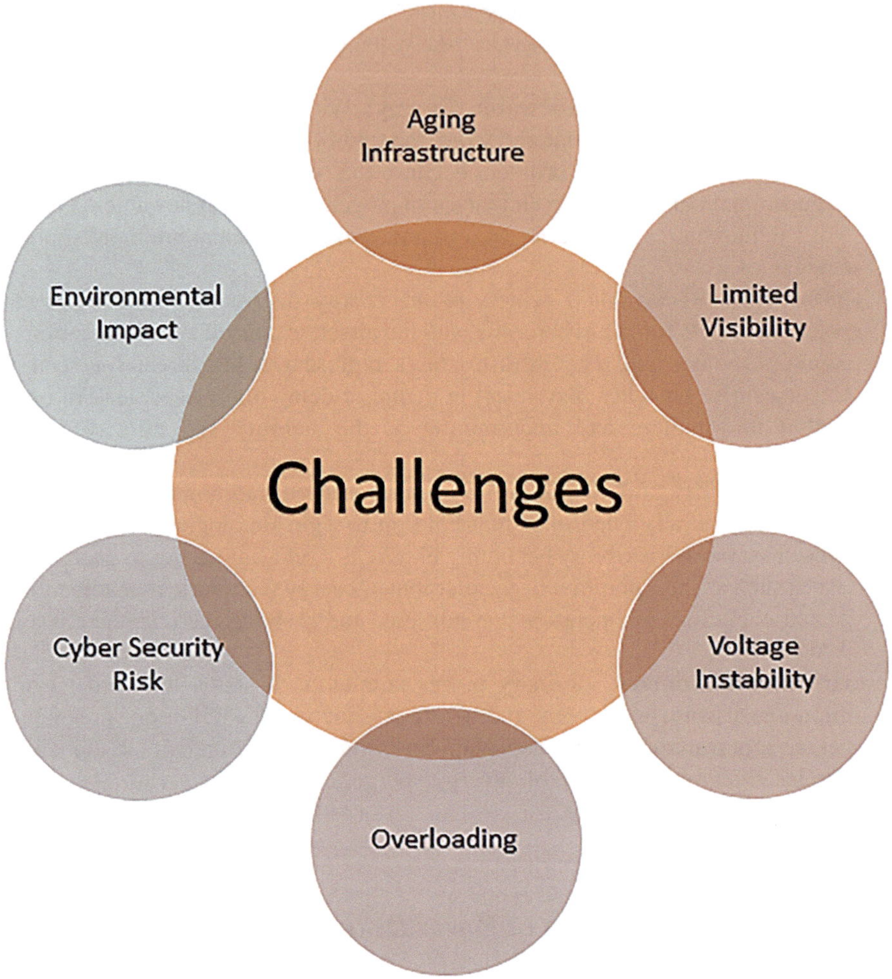

Fig. 3 The challenges in traditional power distribution

renewables, to fully understand how energy is stored and to also maintain the stability the grid.

Emergence of Distributed Energy Resources (DERs): Distributed energy resources, such as the solar panels where some households have installed on their roofs, various storage systems including batteries, and electric cars are redefining the energy scheme. New distribution channels adapt to DERs's inverting power in both directions to small-scale transactions among consumers.

Advancements in Digital Technologies: The digital technology diffusion of AI, IoT, ML, and BC provides incredible chances to significantly improve distribution operations, resilience of power grids, and engagement with utility customers. Smart grid innovation is in the center of these approaches, which use these technologies in order to develop advanced real-time load monitoring, proactive load management, and dynamic control of energy usage.

Demand for Grid Resilience and Reliability: In recent years, the increasing occurrence of extreme weather events and sophisticated cyber threats has placed significant pressure on power distribution networks. This has created an urgent need for electric grids that are not only dependable but also resilient to disruptions. To meet this demand, modern utilities are adopting advanced technologies that support automation, promote self-recovery during faults, and enable more centralized control of grid infrastructure. These innovations are key to strengthening the grid's ability to withstand and quickly recover from unexpected disturbances.

Customer-Centric Approaches: Today's customers are increasingly looking for ways to remain involved in their energy system as a party having full freedom in energy cost issues and environmental issues. Besides, novel distribution models that are customer-oriented tend to utilize centered on customer services that are personalized, energy efficiency project, as well as response to the customers.

Environmental Sustainability: Sustainable distribution in commercial order fulfillment needs process innovations like energy efficiency, eco-friendly, and carbon emission reduction. This translates to integration of smart grid technologies and also implementation of energy-efficient equipment, in addition to green infrastructure investments.

Regulatory and Policy Drivers: Regulatory environments and policies that promote this change are some of the major factors that drive the development of voltage regulation innovation technologies. To this end, governments and regulators through various regulations are slowly edging toward smart grids, improvisations to renewable deployment initiatives, and investment promotion of cutting-edge distributed solutions.

3 Role of AI-IoT in Power Distribution

3.1 Introduction to AI and IoT Technologies

AI and the IoT technologies are transforming the power distribution systems with the loyal-words of developing technologies that assist the power grids on optimizing operations and ensure asset management and efficiency. This section provides an introduction to AI and IoT technologies and their relevance to power distribution.

3.1.1 Artificial Intelligence (AI)

AI can be understood as the ability of machines and computers to mimic the learning, deductive reasoning, and problem-solving processes which are predominantly associated with human intelligence.

Key Concepts: AI is built on different areas including machine learning, natural language processing, computer vision, and robotics that are grouped as its core.

Applications in Power Distribution: AI technologies which their owners apply to grid management allow to analyze large data pools, extract meaningful information, and then use it to construct an automated decision-making process [9]. The AI algorithms can be used to do the predictive maintenance, fault detection, load forecasting, energy optimization, and grid management.

3.1.2 Internet of Things (IoT)

IoT implies the network of interrelated devices embedded with sensing, control, and communication units that allow to convey, receive, and interact with information in this way.

Key Concepts: Talking about IoT devices, we have sensors, smart meters, actuators, and embedded systems which open up the possibilities for direct monitoring and management control over physical assets and environments.

Applications in Power Distribution: Smart technologies help to doing the networks of sensors throughout the distribution systems which allows utilities to supervise equipment operation, witness energy consumption, detect faults and/or anomalies, and optimize the system performance [16]. Smart meters may be called-upon to facilitate the demand response programs and more accurate metering and as well as offering the necessary consumer engagement.

Relevance to Power Distribution: Data-driven decision-making: Utilities of power distribution can take the benefit from AI and IoT technologies for making data-driven decisions by means of real-time data from IoT sensors and previously collected data via AI-enabled analytics.

Predictive Maintenance: Algorithms based on AI can process sensor data to foresee equipment malfunctions before they happen, allowing utilities to carry out routine maintenance as an incidence of equipment failure before it occurs will contribute to the decreasing of costs of utility associated with downtime ensuring a smooth productivity.

Fault Detection and Diagnostics: Using sensors intelligently (IoT) and (AI) analytics to detect faults and anomalies at an early stage in the distribution networks, utilities get to respond and fix the problems promptly, resulting in minimization of service disruptions.

Energy Optimization: AI algorithms are going to process power utilization by an analysis of consumption patterns, forecasting demand, and regulation of grid systems in real time in order to provide the very optimum efficiency as well as the least losses.

Grid Management and Automation: AI interfaced automation system through IoT enabled to keep an eye on and to dynamically adjust grid attributes. As a result, AI improves the management, stability, and resilience of grids (Fig. 4).

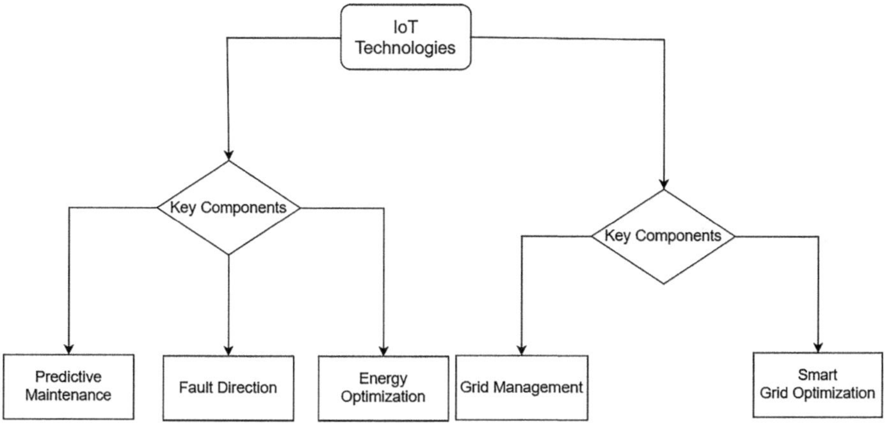

Fig. 4 The introduction to AI and IoT technologies in the context of power distribution

3.2 Integration of AI and IoT in Power Distribution Systems

The use of AI and IoT technologies brings high potential for the next generation power distribution systems to be highly adaptive, smart which can handle the excess supply and varying demand. This section explores how AI and IoT are integrated into power distribution systems:

Data Acquisition and Sensor Networks: Through deployed IoT devices, which are provided with sensors, the real-time data is collected about voltage levels, current flows, temperature, and humidity, and other parameters are recorded. The interconnected sensor networks offer detailed information on the status of distribution assets and how they are functioning to enable utilities to control the system and notice abnormalities.

Data Analytics and Predictive Maintenance: AI algorithms examine the overwhelming amount of data generated from the IoT sensory inputs to detect the patterns, trends, and the unusual behavior. AI-powered maintenance models obtain a predictive function to anticipate equipment failures and deterioration, this way allowing utilities to schedule the maintenance activities prior to the downtime occurrence.

Fault Detection and Diagnostics: IoT sensors along with AI feed-forward systems enable real-time tracking of any anomaly that may be taking place within the distribution process. AI algorithms apply data from sensors to give pinpoint diagnoses of faults which, by being timely and accurate, allow for the implementation of immediate corrective measures and, as a result, prevent service outages.

Energy Optimization and Demand Response: AI-powered analytics focus on energy distribution efficiency by learning about consumption patterns, anticipating demand trends, and adapting grid operations real time. IoT-enabled smart meters are pivotal in establishing demand response programs, and utilities can propose energy conservation during peak time to consumers by giving them due to incentives.

Grid Automation and Control: The AI-powered dynamics-controlling systems enable automatic operation and power flow optimization increasing grid stability and resiliency. IoT devices be used for both remotely accessed monitoring and control of grids assets and dynamically changing operation processes while the machines self-adapt to apply reasonable response.

Customer Engagement and Service Enhancement: IoT-ready smart meters and AI-run analytics offer for households data on the moment. These are two ways that allow consumers to peek into their energy usage and the costs. AI-driven chatbots and virtual helpdesk would help the customer by giving personalized energy management advice and providing 24/7 support for customer questions and complaints.

Security and Privacy Considerations: AI and IoT security is the matter of concern as well as bringing up the privacy issues, and therefore, it requires reliable defense mechanisms and recommended actions. Utilities should therefore provide utmost of confidentiality, integrity, and availability of the data used by AI-IoT systems in developing reliable data pathways and platforms which will guarantee stakeholders trust and assurance at all times.

The infusion of AI and IoT as tools of power distribution systems assists utilities to improve efficiency, reliability, and resiliency which are at another level. Utilities can realize that power by way of application of analytics of data, automation, and real-time monitoring which in turn facilitates them to enhance operation of the grid, increases customer satisfaction, and lays foundation for a sophisticated energy tomorrow.

3.3 Applications of AI-IoT in Improving Power Distribution Efficiency

AI with IoT technology provides an array of the purpose consists in increasing the reliability and efficiency of power distribution companies' operations. This section explores four key applications of AI-IoT in power distribution:

Predictive Maintenance: AI-IoT systems have the capacity to do predictive maintenance by essentially running in a constant mode to correctly determine the health and performance of distribution assets like transformers, switchgears, and cables. Using the IoT sensors, data that shows whether the equipment is properly under operating conditions, temperature, vibration, and other parameters is then examined by AI algorithms to identify abnormalities and impending breakdowns. Predictive maintenance models demand machine learning methodologies through which the number of useful life for certain assets is considered and the maintenance tasks are highlighted in this way. They also underscore the period through which interventions take place prior to occurrence of critical failures. The pre-emption of malfunctioning maintenance will help to bring the interruption of service, prolong the life cycle of assets and control maintenance cost at a minimum level.

Fault Detection and Diagnosis: AI-IoT systems use machine learning and data analytics which quickly spot and characterize errors and malfunctions in the distribution network. IoT sensors placed in different parts of the grid for continuous monitoring of the voltage, current flow, frequency, and and repeated other parameters transmit this data in real time, which helps a lot in knowing on the network conditions. AI algorithms use this data for identifying the fault patterns that may be presented with the voltage sag, asymmetrical phases, or abnormal load profiles; these are among the examples. The problems diagnosis puts into application of machine learning can recognize the cause of fault mainly from the equipment malfunction, surroundings' conditions, or external disturbances. It is possible to monitor and discover failures early with prompt detection and diagnosis of the faults ensuring that the utility companies take minimal service disruptions, effective outage responses and hence, make the grid more reliable.

Energy Optimization: AI-IoT systems fundamentally transform the norms of energy distribution and consumption by their capabilities to analyze the patterns, forecasting needs, and design of smart grid operations in a time-to time mindfully manner. The Internet of Things (IoT)-based smart meters provide meters with intricate data of the energy use at customer premises, which the utilities can therefore observe and learn the consumption trends and patterns. AI algorithms detect such loading patterns for the optimization of grid operations, regulation of supply and demand, controlling energy losses via load shedding, voltage regulation, and demand response programs. Through machine learning, energy optimization models are trained to accurately predict energy demand, which is subsequently used for the planning of the generation and distribution schedules with a focus on the usage of renewable energy sources based on supply fluctuations. Via delivering and distributing energy optimally, utilities tend to increase the efficiency, lessen the costs, and can be more environmental friendly.

Load Forecasting: For instance, IoT sensors send us weather information, load profiles, historical on peak load, economic indicators, and other factors which determine the energy mentioned consumption rates. AI algorithms review this data in real time and use it as an informant to produce the anticipated energy consumption reports with various time ranges, from hours to days or even weeks ahead. This is where the models that predict the load of power plants utilize machine learning techniques like neural nets, time series analysis, and regression that capture the complex patterns and relationships of energy consumption features. Through successful capacity planning, therefore, utilities can arrange the optimal utilization resources, prepare the infrastructure developments, and hence, maintain the overall grid security. The boundaries of AI-IoT applications within the power distribution domain go beyond predictive maintenance, fault detection, energy efficiency, and load forecasting to extend to an array of functionalities driven by the desire to improve operational efficiency, enhance grid resilience, and provide unmatched services to customers. Via strict instruments of data analytics, automation, and continuous monitoring, utilities completely open a long way of development and start revolutions on the scene of energy systems.

References

1. Miao, G., Himayat, N., Li, G. Y., & Talwar, S. (2011). Distributed interference-aware energy-efficient power optimization. *IEEE Transactions on Wireless Communications, 10*(4), 1323–1333.
2. Ayres, R. U., Turton, H., & Casten, T. (2007). Energy efficiency, sustainability and economic growth. *Energy, 32*(5), 634–648.
3. Chen, W., Alharthi, M., Zhang, J., & Khan, I. (2024). The need for energy efficiency and economic prosperity in a sustainable environment. *Gondwana Research, 127*, 22–35.
4. Yang, Z., Dehghanian, P., & Nazemi, M. (2020). Seismic-resilient electric power distribution systems: Harnessing the mobility of power sources. *IEEE Transactions on Industry Applications, 56*(3), 2304–2313.
5. Debauche, O., Mahmoudi, S., Mahmoudi, S. A., Manneback, P., & Lebeau, F. (2020). A new edge architecture for AI-IoT services deployment. *Procedia Computer Science, 175*, 10–19.
6. Ahmad, T., Madonski, R., Zhang, D., Huang, C., & Mujeeb, A. (2022). Data-driven probabilistic machine learning in sustainable smart energy/smart energy systems: Key developments, challenges, and future research opportunities in the context of smart grid paradigm. *Renewable and Sustainable Energy Reviews, 160*, 112128.
7. Adi, E., Anwar, A., Baig, Z., & Zeadally, S. (2020). Machine learning and data analytics for the IoT. *Neural Computing and Applications, 32*, 16205–16233.
8. Zhang, X., Manogaran, G., & Muthu, B. (2021). IoT enabled integrated system for green energy into smart cities. *Sustainable Energy Technologies and Assessments, 46*, 101208.
9. Khan, A. A., Laghari, A. A., Rashid, M., Li, H., Javed, A. R., & Gadekallu, T. R. (2023). Artificial intelligence and blockchain technology for secure smart grid and power distribution automation: A state-of-the-art review. *Sustainable Energy Technologies and Assessments, 57*, 103282.
10. Aldoseri, A., Al-Khalifa, K., & Hamouda, A. (2023). *A roadmap for integrating automation with process optimization for AI-powered digital transformation*. https://doi.org/10.20944/preprints202310.1055.v1
11. Mahzarnia, M., Moghaddam, M. P., Baboli, P. T., & Siano, P. (2020). A review of the measures to enhance power systems resilience. *IEEE Systems Journal, 14*(3), 4059–4070.
12. Ma, W., Wang, W., Chen, Z., Wu, X., Hu, R., Tang, F., & Zhang, W. (2021). Voltage regulation methods for active distribution networks considering the reactive power optimization of substations. *Applied Energy, 284*, 116347.
13. Saeed, M. H., Fangzong, W., Kalwar, B. A., & Iqbal, S. (2021). A review on microgrids' challenges & perspectives. *IEEE Access, 9*, 166502–166517.
14. Haider, Z. M., Mehmood, K. K., Khan, S. U., Khan, M. O., Wadood, A., & Rhee, S. B. (2021). Optimal management of a distribution feeder during contingency and overload conditions by harnessing the flexibility of smart loads. *IEEE Access, 9*, 40124–40139.
15. Jin, B. (2023). Impact of renewable energy penetration in power systems on the optimization and operation of regional distributed energy systems. *Energy, 273*, 127201.
16. Kuthadi, V. M., Selvaraj, R., Baskar, S., Shakeel, P. M., & Ranjan, A. (2022). Optimized energy management model on data distributing framework of wireless sensor network in IoT system. *Wireless Personal Communications, 127*(2), 1377–1403.

Power Consumption Forecasting with AI and IOT

G. Murugesan, B. Radha, A. S. Nithya, and R. Gunavathi

1 Introduction

An essential component of energy management and grid operations is energy performance forecasting, often known as load forecasting or power demand forecasting [1]. Predicting future patterns of electricity output or consumption is one way to help the energy sector plan, allocate resources, and make decisions more effectively. Smart city energy power enhances the efficiency, sustainability, dependability, and accessibility of urban energy systems by combining cutting-edge technologies and data-driven methodologies. The goal of smart city programs is to replace outdated energy infrastructure with intelligent, networked systems that can minimize environmental impact and meet the increasing energy needs of urban populations. Here, we present the idea of energy power in smart cities [2]. An overview of power consumption prediction is as follows:

- The main purpose of energy forecasting is to provide accurate estimates of future energy demand that can be used for various purposes, such as:

    Grid Operations: Performance prediction is a tool used by grid operators to guarantee a steady and dependable power supply. Precise projections aid in the management of electricity production, transmission, and distribution, enhance grid stability, and avert overloads and blackouts.

G. Murugesan (✉) · A. S. Nithya
Department of Artificial Intelligence and Machine Learning, Sree Saraswathi Thyagaraja College, Pollachi, Tamil Nadu, India

B. Radha
Department of Computer Technology and Data Science, Sri Krishna Arts and Science College, Coimbatore, Tamil Nadu, India

R. Gunavathi
Data Science Department, Christ (Deemed to be University), Pune, Maharashtra, India

S. Vijayalakshmi et al. (eds.), *Smart Power Systems*, Power Systems,
https://doi.org/10.1007/978-3-031-89459-6_4

Resource Planning: Forecasts are used by utilities and power generation firms to plan investment strategies and resource allocation. The construction of new power plants, the updating of infrastructure, and the effective use of resources are all guided by these forecasts.
Energy Trading: Accurate power forecasting is essential for energy traders and other market players in deregulated energy markets. These aid in risk management, trading strategy optimization, and the purchase and sale of power at competitive pricing [1, 3].
Demand Response: Forecasting is a tool that utilities employ to run demand response programs. In order to successfully balance supply and demand, consumers can be encouraged to cut back on or alter their electricity consumption during peak demand hours.
Renewable Energy Integration: Power forecasting aids in the prediction of energy production patterns for renewable energy sources like wind and solar power. Power grid operators need this information in order to handle intermittent renewable energy efficiently.
Energy Efficiency: Energy forecasting allows energy consumers such as commercial buildings and industries to optimize their energy consumption behavior. This saves energy and reduces costs.
Infrastructure Maintenance: Reducing downtime and increasing reliability can be achieved by using accurate projections to help plan maintenance schedules for power generating and distribution equipment. Predicting performance accurately is essential to maintaining grid stability, optimizing resource use, and managing energy costs.

Energy performance forecasting will play a more significant role in determining the direction of the energy sector as the integration of renewable energy sources, electrification of transportation, and development of smart grids continue to transform the energy landscape. Energy performance is predicted using a range of methods, including statistical modeling, machine learning, time series analysis, and sophisticated algorithms. It is based on historical statistics on production and consumption as well as outside variables including the state of the economy, population growth, and weather.

2 Evolution of Energy Usage

In the past, the average person's access to power was limited. The Industrial Revolution allowed nations to boost output and effectively automate processes. Many industrial machinery, technologies, and inventions that are essential to human life have been made possible by the good source of electrical energy. Great inventions have altered people's lives since the nineteenth century. Through the production and use of electricity, civilizations throughout the industrial age legitimized national economic power. Because large industries, mines, and industrial projects need more power, there is an increase in the demand for electricity. Inventors come up with ways to find various resources to produce a lot of electricity. Heat sources are excellent for generating a lot of energy. Natural resources, however, were decimated by industrial growth. On the other hand, global warming is a serious issue that the industrial revolution has brought forth. People's attitudes have altered as a result of the CO_2 emissions issue. Let's also consider renewable energy. Sadly, along the course of our industrial revolution, we have learned to generate electricity utilizing nuclear and thermal energy, which hastens the occurrence of natural calamities. As a result, the creation of renewable energy could not quickly replace available resources. Furthermore, this technology offers ways to use electrical energy more efficiently [4]. In the past, light bulbs and straight tube lamps in every home used a lot of electrical energy in this technological age. Product and appliance consumption is significantly lower than that of conventional electrical consumables. Humanity has adapted to the difficulties of the times by utilizing the new aspects of globalization in the areas of resource allocation, economic development, and environmental protection to provide the framework for the safe, efficient, and sustainable growth of the planet. Energy needs to be understood.

Long-term electricity consumption estimates are important for planning and development at the national and regional levels each year. When it comes to making decisions, businesses and organizations that design the energy system need to know this information. Several methodologies, including statistics, computational intelligence, and a mix of techniques, have been employed by models to provide such predictions [5]. In certain models, the primary variable is broken down into its component parts, examined in isolation, and then combined back into the primary variable. For instance, the energy demands of each sub-region within a region that can be subdivided can be used to calculate the overall energy demand of the area [6]. After that, the energy demand of this sub-region is examined, and the region's energy demand projection is produced by adding together each sub-region's separate demand forecasts. Different levels of this variable breakdown may take place. Two treatment approaches, referred known as "bottom-up" and "top-down," can be distinguished by looking at whether variables can be broken down into their constituent parts. Top-down modeling gives the direct outcomes of the variables of interest by treating them as a whole. The variable of interest is separated into components in bottom-up models, to which treatments are applied. The variable of interest is then formed by adding the results of these components. When these two modeling approaches are compared, it becomes evident that while top-down modeling produces findings with less data, it is less clear how certain external factors influenced those conclusions. Not at all. Bottom-up requires a lot of data, but it provides a deeper understanding of how outside influences affect outcomes. Bottom-up modeling thus necessitates both huge and high-quality data sets. There is more complication the more detail there is.

3 Energy in Smart Building

Creating sustainable, effective, and comfortable interior spaces requires careful consideration of energy management in smart buildings Modern technology and clever data-driven solutions are used in smart buildings to maximize energy utilization, eliminate waste, lower expenses, and raise system efficiency [7]. The following are important facets of energy management in smart buildings: energy consumption in smart buildings is a dynamic, integrated process that strives to minimize energy expenditures and environmental effect while balancing operational efficiency and occupant comfort [8]. These initiatives support the development of a more ecologically conscious and sustainable method of building management and construction. In addition to lowering energy expenses and consumption, intelligent energy management in buildings improves occupant comfort, the built environment's overall productivity, and sustainability. This is an essential element in creating smart, resilient, and environmentally friendly buildings.

4 Building Automation Systems (BAS)

Building Automation Systems (BAS), also referred to as Building Management Systems (BMS), are centralized systems used to monitor and control various building operations. A building control system or BCS is a centralized control system that uses advanced technology and software to monitor, manage, and optimize various building systems and services. Building automation systems are essential for lowering operating costs and having a positive environmental impact while increasing user comfort, energy efficiency, and overall building performance. These form the basis of contemporary building management in the business and private domains [9]. Building automation systems (BAS) are used in residential, commercial, and industrial settings to enhance operational performance, comfort, safety, and energy efficiency. Control devices and sensors offer several benefits, including improved energy efficiency, enhanced comfort, and real-time monitoring of building conditions. Sensors are devices that are positioned all over a structure to keep an eye on different characteristics like CO_2 levels, light levels, temperature, humidity, and occupancy. Actuators use data from sensors to control appliances like blinds, lights, and HVAC systems. The building is equipped with sensors that are able to record several environmental factors, including temperature, humidity, air quality, and occupancy levels. This data is used to adjust lighting, HVAC, and other systems in real time to optimize energy consumption and occupant comfort.

The controller is the brain of BAS. The controller receives data from sensors, makes decisions based on pre-programmed logic and user-defined parameters, sends commands to actuators, and adjusts building systems accordingly. A user interface allows building operators and facility managers to monitor and control the system. This is a graphical user interface (GUI) on your computer or mobile device.

A communication network connects all components of the BAS and enables the exchange of data and commands. Common protocols used in BAS networks include BACnet, Lon Works, and Modbus. AS systems often have data logging and storage capabilities that allow users to analyze historical data for trends, performance optimization, and troubleshooting. Energy optimization smart buildings often use energy-efficient LED lighting systems. These systems can be integrated with occupancy sensors and dimming controls to turn on lights only when and where needed. Energy-efficient building envelopes: Insulation: Smart buildings are often designed with high-quality insulation to minimize winter heat loss and summer heat gain, thereby improving HVAC. This reduces the need for windows: energy-efficient windows with low-emissivity coatings and multiple panes improve insulation and reduce heat transfer. Energy storage: some smart buildings are equipped with energy storage systems, such as batteries, that store excess energy during off-peak hours and release it during periods of high demand or power outages. Renewable energy integration: solar panels and wind turbines can be integrated into building designs to generate renewable energy on-site. Excess energy can be stored or fed back to the power grid.

Heating and cooling typically account for a significant portion of a building's energy consumption. Smart buildings utilize advanced HVAC systems that can adapt to changing conditions. Predictive algorithms, temperature and humidity sensors, and zone-based controls can be used to maintain comfort while minimizing energy consumption [10]. Window automation: smart windows with auto-darkening and glazing features regulate indoor temperatures and reduce the need for heating and cooling. Energy management software platform tools and applications collect, analyze, and visualize data from a variety of building systems. This enables facility managers to find potential for energy savings, monitor performance, and make informed decisions. Based on past data and current circumstances, predictive analytics can forecast energy demand and optimize energy consumption [9]. Demand response: programs that entail temporarily lowering energy use during times of high demand are available to smart buildings. In response to signals from the utility or grid operator, automated systems can modify heating, cooling, lighting, and other services.

5 Advanced Building Architecture and Construction

In addition to other techniques, smart buildings are made to use natural light, orientation, and ventilation in order to reduce their energy use [11]. Because of this, routine maintenance is necessary to ensure that all of a building's systems are operating at peak efficiency. On the other side, predictive maintenance makes use of sensors and data analysis technologies to identify equipment problems early on, allowing for their repair prior to energy waste or system failures [12, 13]. One feature connected with some smart homes, for example, is the ability to charge electric vehicles. As a result, energy use in these complexes is a complex process that incorporates occupant pleasure with operational efficiency goals and aims to minimize both the environmental impact and power consumption expenses. In such like ways do these undertakings contribute toward sustainable development through environmentally conscious building management and construction practices.

Workplace Culture and Ethics: Feedback loops, incentives, awareness programs, etc. could be employed to engage consumers, thereby creating a user base who should adopt energy-saving habits. Indoor quality largely depends on how much power is consumed within intelligent structures since it contributes greatly toward making them more efficient cleaner healthier places to live in; also these kind of houses utilize sophisticated technology together with data-driven approaches aimed at cutting down wastage curbing expenses lowering overall consumption levels while improving performance of different energy types altogether.

6 Energy in Smart City

Energy consumption in smart cities is a complex and interconnected system that spans different sectors and services. Smart cities use advanced technology and data-driven approaches to optimize energy consumption, reduce waste, and improve overall energy efficiency. Residential and commercial buildings: smart homes utilize energy-efficient appliances, lighting, and HVAC systems. Automation and sensor technology help optimize energy consumption depending on usage and environmental conditions. Smart buildings include energy management systems that monitor and control HVAC, lighting, and other systems. These systems adjust settings in real time to reduce energy waste [14].

Electromobility: Smart cities facilitate the adoption of electric vehicles (EVs) and provide charging infrastructure. Electric buses and public transport help reduce emissions and dependence on fossil fuels.

Intelligent Transportation Systems (ITS): ITS technology optimizes traffic flow, reduces congestion, and minimizes fuel consumption through real-time traffic data, intelligent traffic lights, and route optimization. Smart cities are converting street lighting to energy-efficient LED technology. Automatic lighting control adjusts brightness based on pedestrian and vehicle traffic to save energy. Smart city data centers and IT infrastructure are designed with energy efficiency in mind and integrate technologies such as virtualization and energy-efficient servers.

Renewable Energy Integration: Solar energy: Smart cities often integrate solar panels on buildings and public infrastructure to generate renewable energy. Excess energy can be stored or fed back to the power grid.

Wind Power: If properly placed, wind turbines can generate electricity to power city services or contribute to the electrical grid.

Waste-to-Energy: Waste conversion: Some smart cities use waste-to-energy technology to convert organic waste into biogas or electricity, reducing the environmental impact of landfilling and generating renewable energy.

Advanced grid management systems are used in smart cities to provide two-way communication between electricity providers and customers. These networks can facilitate demand response initiatives, incorporate renewable energy sources, and maintain supply and demand equilibrium. Demand response programs can help buildings and industries in smart cities use less energy during peak hours. Sensors and monitoring systems are used in smart cities to gather information on temperature, air quality, and other environmental factors. The information in this report is helpful for both emergency response and energy efficiency initiatives [15, 16]. Hospitals and other vital infrastructure, like emergency services, are frequently outfitted with backup power systems in smart cities to guarantee that vital services can continue in the event of a power loss or disaster.

7 Social Awareness in Smart Cities

Smart cities use incentives, campaigns to raise awareness, and real-time data on energy consumption to get businesses and communities involved in energy conservation. The energy environment of smart cities is significantly shaped by laws and regulations from the government. Smart city programs frequently include incentives for the deployment of renewable energy, building rules, emission reduction targets, and energy efficiency standards [17]. Diverse parties, including governmental organizations, commercial businesses, academic institutions, and the general public, must work together successfully to achieve smart city energy projects. Project implementation and financing can be aided by public–private partnerships. By tackling energy-related possibilities and constraints, smart cities seek to build more livable, sustainable, and efficient urban settings. These initiatives have a positive long-term economic impact, less negative effects on the environment, and better living standards for locals.

8 Importance of Energy Efficiency

Energy efficiency is critical for a number of strong reasons. Environmental impact: The environmental benefits of energy efficiency may be its greatest justification. One of the main causes of greenhouse gas emissions that contribute to climate change is the burning of fossil fuels for energy. Energy efficiency initiatives lower the amount of energy used for certain jobs, as well as the corresponding CO_2 emissions and environmental harm. Climate change mitigation: using energy efficiently can help reduce the effects of climate change. We can drastically cut greenhouse gas emissions and lessen the effects of rising global temperatures, as well as the accompanying repercussions of extreme weather, sea level rise, and ecosystem loss, by consuming less energy and switching to cleaner energy sources.

Energy Security and Conservation: One type of energy conservation is energy efficiency. It allows you to use less energy to get the same or better results as before, helping you to make the most of your limited energy resources. By conserving, we lessen the depletion of fossil fuels—a limited supply of non-renewable energy. Energy bills for individuals, companies, and governments are frequently decreased by energy-efficient devices and practices. Reduced energy use results in significant savings over time, freeing up funds for other crucial expenditures. Improving energy efficiency helps guarantee a steady supply of energy and lessen reliance on imported energy. Countries can increase their energy security and lessen their susceptibility to supply disruptions by adopting more efficient energy use practices [18].

Conservation of Resources: In order to produce energy, additional natural resources like water, minerals, and land are frequently extracted and used. In addition to lowering the need for resources, energy efficiency helps save ecosystems and slow down environmental deterioration.

Global Energy Access: By increasing energy efficiency, we can make better use of the resources at our disposal and provide electricity to more people in areas where access to energy is scarce. Technologies that use less energy are very good at reducing energy poverty.

9 Improvement of Social Wings

Jobs are generated by the creation and application of energy-efficient methods and technology in a number of industries, including manufacturing, construction, research and development, and energy audits. These positions support stability and economic prosperity. Public health and air quality can be enhanced by energy efficiency initiatives including less air pollution from power plants and cleaner mobility. Lower emissions of nitrogen oxides and fine dust translate into less respiratory and cardiovascular issues. Because they frequently have reduced operational expenses, energy-efficient businesses are more competitive in the market. In addition to fostering innovation and bettering products, energy efficiency gives businesses a competitive edge. Energy-efficient buildings and infrastructure are often more resilient to disruptions such as power outages and extreme weather events. Smart energy management systems provide backup power, reduce downtime, and ensure critical services continue to function.

Compliance with Regulations and Guidelines: Businesses must abide by the energy efficiency laws and regulations that are in place in many different countries. Energy-saving initiatives can assist companies in adhering to regulations and avoiding fines. Energy efficiency is regarded as one of the most economical and successful strategies to fight climate change, lower energy costs, and advance sustainable development because of its many advantages. This is the main component of our attempts to switch to greener, more sustainable energy in the future.

10 Energy Consumption Management

Intelligent energy consumption management is the use of technology, data, and automation to optimize how energy is used, distributed, and controlled. The aim is to increase energy efficiency, reduce waste and costs, and minimize environmental impact. The key components and strategies related to smart energy consumption management are energy management system.

Home Energy Management Systems (HEMS): HEMS provides homeowners with real-time energy consumption data, allowing them to make informed decisions about their energy consumption. These systems can also automate the control of appliances and thermostats, reducing energy waste.

Building Energy Management Systems (BEMS): BEMS are used in commercial and industrial buildings to monitor and control HVAC, lighting, and other systems and optimize energy consumption based on usage, weather conditions, and energy prices.

Advanced Metering Infrastructure (AMI): Smart meters provide real-time data about energy consumption. This data is transmitted wirelessly to utilities and customers, allowing for more accurate billing, load forecasting, and insight into energy consumption patterns [14, 19].

Automated Energy-Saving Measures: Automated systems can implement energy-saving measures such as adjusting lighting, HVAC, and equipment settings based on occupancy and usage patterns.

Energy Storage and Analysis: Energy management software and analytics platforms collect and analyze data from a variety of sources, including smart meters, sensors, and building management systems. This helps companies understand when and where energy is being used and identify opportunities for improvement. Battery storage can store excess energy when demand is low and release it again when demand is high or during a power outage. This helps balance fluctuations in energy demand and supply.

Demand Response: Demand response programs allow utilities and consumers to reduce energy consumption during peak demand periods. Automated systems can adjust heating, cooling, lighting, and other systems in response to signals from power companies and grid operators.

Renewable Energy Integration: Smart systems can coordinate the use of renewable energy sources, such as solar panels and wind turbines, to maximize self-consumption and minimize dependence on the power grid.

Microgrid: A microgrid is a localized energy system that can operate independently or in conjunction with the main grid. These enable better control over energy production, consumption, and distribution, increasing resilience and efficiency.

Grid Integration: Smart energy management systems communicate with the grid to balance supply and demand, participate in demand response programs, and support grid stability.

Machine Learning and AI: Artificial intelligence and machine learning algorithms can analyze historical energy data to predict consumption patterns and optimize energy consumption in real time.

Electric Vehicle (EV) Charging Management: Smart charging systems schedule EV charging during off-peak hours to minimize electricity costs and reduce load on the power grid.

Engagement in Behavior: Engaging consumers and building users through feedback mechanisms, incentives, and awareness campaigns can encourage energy-efficient behavior. Smart energy consumption management is essential to achieving sustainability goals, reducing utility bills, and improving energy resilience. Implementing effective solutions and maximizing the benefits of intelligent energy management require collaboration between utilities, consumers, regulators, and technology providers.

11 Importance of Sensors

In order to monitor, regulate, and optimize energy use in real time, sensors are essential components of smart energy systems. These sensors gather data on a range of environmental and system characteristics, giving vital information that enhances waste reduction, energy efficiency, and the sustainability and dependability of energy systems [20, 21]. Data gathering: data about energy production, distribution, and consumption are continuously gathered via sensors and take measurements of variables like wind speed, sun radiation, temperature, humidity, air quality, and power consumption. This information sheds light on energy usage trends and efficiency.

12 Energy Usage Sensors

Electricity Meters: Real-time electrical energy usage is measured using these meters. More sophisticated smart meters offer comprehensive data on power factor, voltage, current, and consumption trends. Gas meters: gas meters monitor the real-time consumption of natural gas and other gases. This is crucial for gas-powered heating, cooking, and industrial activities in residential, commercial, and industrial settings.

Water Meters: Water meters measure water consumption in real time. This data helps you track water usage in homes, businesses, and industrial facilities.

Heat Meter: Heat meter real-time monitoring of energy consumption and systems is based on various sensors that can measure and collect data on various aspects of energy consumption, production, and distribution. These sensors provide critical information to optimize energy efficiency, manage demand, and ensure reliable operation. Here are some common types of sensors used for real-time energy monitoring.

Grid Sensors: Smart grid sensors are critical components of modern power grids, designed to improve efficiency, reliability, and sustainability. These sensors collect real-time data about various aspects of the network, enabling improved monitoring, control, and decision-making. Below are some common types of smart grid sensors.

Voltage Sensors: Voltage sensors monitor voltage levels at various points on the network. These help detect voltage fluctuations that can affect the quality of power to consumers.

Current Sensor: A current sensor measures the flow of current in a network. These are used to monitor load conditions and detect overloads and imbalances.

Power Quality Sensors: These sensors monitor power quality parameters such as harmonics, voltage sags, voltage spikes, and power outages. These help ensure that the power supply to consumers meets quality standards.

Transformer Sensors: Transformer and substation sensors monitor the condition and performance of critical equipment. This data helps prevent equipment failure and optimize maintenance schedules.

13 Renewable Energy Sensors

Solar Panel Sensors: Solar panel sensors monitor solar radiation, panel temperature, and power output, among other aspects of energy generation. This information aids in solar power system operation optimization. Photovoltaic (PV) systems require solar panel sensors, sometimes referred to as solar panel monitoring systems, as key components for tracking and maximizing system performance. These sensors offer information on the effectiveness, output of energy, and condition of solar panels and associated equipment.

Solar Radiation Sensors: These sensors calculate solar radiation or sunshine intensity. The device can estimate the quantity of energy that will be produced and identify shadows and obstructions that may reduce efficiency by tracking the amount of sunshine that reaches the solar panels.

Wind Turbine Sensors: Sensors that track power generation and variables like wind direction and speed are installed in wind turbines. The control and maintenance of wind turbines are facilitated by this data. An essential part of contemporary wind turbines are wind turbine sensors. They are essential for tracking, regulating, and maximizing wind turbine performance. These sensors give wind farm operators' useful information that helps them increase energy production, maintain safety, and prolong the life of their turbines.

14 Real-Time Monitoring

Operators and users may monitor energy usage, grid health, and equipment performance thanks to sensors' real-time insights into energy systems. Errors, inefficiencies, and anomalies can be quickly found with the aid of real-time data. Many types of sensors that can measure and gather data on many elements of energy production, distribution, and consumption provide the foundation for real-time monitoring of energy systems and use. These sensors offer vital data for managing demand, maximizing energy efficiency, and guaranteeing dependable operation.

Here are some common types of sensors used for real-time energy monitoring:

Environmental Sensors: Environmental sensors measure parameters such as humidity, air quality (CO_2 content, VOCs, particulate matter), and ambient temperature. These sensors are important for indoor air quality management and energy-efficient HVAC control.

Pressure Sensors: Pressure sensors measure pressure in real time in a variety of applications such as industrial processes, pneumatic systems, and gas distribution.

Load Cells: Load cells measure force and weight in real time in industrial applications. They are used for load monitoring and optimization in material handling, manufacturing, and construction.

Battery Sensors: In energy storage systems, battery sensors monitor battery state of charge (SoC), health (SoH), voltage, current, and temperature in real time. This data is essential for managing and optimizing energy storage systems. These sensors collect data that can be sent to central monitoring systems, building automation systems, energy management platforms, or IoT platforms [5, 22]. Analyzing this real-time data enables proactive decision-making, load optimization, demand response, and other strategies to improve energy efficiency and system reliability and monitor real-time thermal energy (heat) consumption. They are commonly used in district heating systems, industrial processes, and some residential heating systems. Energy efficiency: sensors enable precise control of energy-consuming devices and systems such as HVAC, lighting, and industrial equipment. By monitoring environmental conditions and occupancy, sensors can adjust settings to optimize energy efficiency without compromising comfort or productivity. In smart grids and smart buildings, sensors detect changes in energy demand and supply. This information can trigger demand response actions such as B. Adjust your heating or cooling system or temporarily reduce energy consumption during peak loads.

Flowmeter: A flowmeter measures the flow rate of a liquid or gas in real time. They are used in a variety of applications, including monitoring water flow in pipes, fuel consumption in vehicles, and gas flow in industrial processes.

Temperature Sensors: Temperature sensors monitor the real-time temperature of a device, environment, or liquid. They are important for HVAC systems, industrial processes, and energy-efficient control strategies.

Current Transformers (CT) and Voltage Transformers (VT): These sensors are used to measure current and voltage in electrical systems. These are essential for power quality monitoring, load analysis, and fault detection.

Battery Sensors: In energy storage systems, battery sensors monitor the battery state of charge (SoC), state of health (SoH), voltage, current, and temperature in real time. This data is essential for managing and optimizing energy storage systems. These sensors collect data that can be sent to central monitoring systems, building automation systems, energy management platforms, or IoT platforms.

Analyzing this real-time data enables proactive decision-making, load optimization, demand response, and other strategies to improve energy efficiency and system reliability. Sensors such as current transformers (CT) and voltage transformers (VT) are used to measure current and voltage in electrical systems. These are essential for power quality monitoring, load analysis, and fault detection. Battery sensors: In energy storage systems, battery sensors monitor the battery state of charge (SoC), state of health (SoH), voltage, current, and temperature in real time. This data is essential for managing and optimizing energy storage systems. These sensors collect data that can be sent to central monitoring systems, building automation systems, energy management platforms, or IoT platforms [4, 15]. Analyzing this real-time data enables proactive decision-making, load optimization, demand response, and other strategies to improve energy efficiency and system reliability.

Resource Optimization: In industrial processes and agriculture, sensors monitor resource consumption and environmental conditions. This information helps optimize resource allocation, thereby reducing energy consumption and waste. In the event of an energy-related emergency, such as a power outage or equipment failure, sensors provide valuable data to quickly diagnose and respond to the situation, minimizing disruption.

Predictive Maintenance: Sensors predict equipment failure by monitoring the condition of critical components. Early detection of problems enables preventive maintenance, minimizing downtime and energy waste. In intelligent networks, sensors measure voltage, current, and power quality. This data helps grid operators manage energy distribution, maintain grid stability, and respond to fluctuations in demand and supply.

Presence Detection: Presence sensors in buildings detect the presence of people within a room or area. This information is used to control lighting, HVAC, and other systems to ensure energy is used only when needed. In residential and commercial environments, real-time energy data from sensors can be provided to users through an easy-to-use interface or mobile app. This will encourage residents to take energy-saving behaviors and make informed decisions about their energy consumption. Electric vehicle grid integration: EV charging infrastructure and vehicle sensors enable smart charging, off-peak charging scheduling, and managing the grid impact of EV deployments.

Overall, sensors can help make energy systems smarter, more efficient, and more responsive to changing conditions. This enables energy users, grid operators, and facility managers to make data-driven decisions that improve energy savings, sustainability, and cost savings.

15 Power Consumption Forecasting Techniques

Predicting power consumption is important for power companies, grid operators, and energy management in various sectors. Accurate forecasting helps optimize energy production and distribution, reduce costs, and ensure reliable supply. Designing a power consumption prediction system requires several important steps and considerations [23]. Here's an overview of the process:

1. **Define goals and scope**. Clearly define the goals of your power consumption prediction system. Are you interested in short-term or long-term forecasting? What level of granularity (hourly, daily, monthly, etc.) do you need? Identify the scope of your project and the specific goals you want to achieve [19].
2. **Data collection:** Collect historical power consumption data. This data must include timestamps and consumption values. We also collect relevant external data, such as meteorological data (temperature, humidity, wind speed), economic indicators (GDP, population), and other factors that can influence consumption.
3. **Data Preprocessing:** Clean and preprocess the collected data. This includes dealing with missing values, outliers, and data inconsistencies. You may need to aggregate or restructure query your data to match your desired prediction granularity.
4. **Features Engineering:** From the data you collect, create relevant features that improve prediction accuracy. For example, you can derive characteristics such as day of the week, time of day, seasonality indicators, and lagged values (previous consumption values).
5. **Model Selection:** Select the appropriate predictive model based on your objectives and data characteristics. Common models include time series models (such as ARIMA, exponential smoothing), machine learning models (such as random forests, XGBoost), and deep learning models (such as LSTM).
6. **Model Training:** Split your data into training and validation sets to train and evaluate your chosen model. Training a model involves tuning hyper parameters, optimizing the model architecture, and evaluating performance using evaluation metrics (such as mean absolute error and mean squared error).

16 Ensembling and Hybrid Approaches

Consider combining multiple predictive models using composition techniques such as wrapping and stacking. Hybrid approaches that combine the strengths of different models often improve accuracy. Ensemble methods [24] are a collection of ML techniques that combine predictions from multiple models to improve accuracy and robustness, providing a powerful approach to energy consumption prediction. Ensemble techniques can reduce bias and errors in individual models by integrating the output of different algorithms such as decision trees, support vector machines, and neural networks. This enables more reliable and accurate predictions, especially for complex and dynamic energy consumption patterns. Additionally, ensemble methods allow for the inclusion of multiple types of data sources, such as historical consumption data, weather conditions, economic indicators, and social factors, resulting in more comprehensive and comprehensive forecasts.

Exogenous variables: If external factors, such as weather data or economic indicators, affect electricity consumption, include these variables in your predictive model. Make sure your data is in sync with your consumption data in terms of timestamps.

Real-Time Data Recording: Implement mechanisms to collect real-time data such as latest power consumption figures and external data sources. This keeps the predictive model up to date with the latest information.

Forecast Period: Define the forecast period. This indicates how far into the future the prediction is made. The choice of forecast period may vary depending on the specific application.

Visualization and Reporting: Develop visualization tools and dashboards to display predicted values, historical data, and all relevant insights. Visualization helps stakeholders understand forecasts and make informed decisions.

17 Various Techniques and Methods Used to Predict Power Consumption

Time series analysis: ARIMA (AutoRegressive Integrated Moving Average): ARIMA models are popular for time series forecasting. It takes into account trends, seasonality, and autocorrelation in historical consumption data.

Exponential smoothing: Exponential smoothing models, such as Holt-Winters, apply a weighted average to historical data and effective for capturing seasonality and trends. Prophet: Developed by Facebook and is a time series forecasting tool that can handle daily and annual seasonality, holidays, and special events. It is user-friendly and suitable for the general public.

18 Regression Analysis

Linear Regression: Linear regression models relate power consumption to various predictor variables, including weather data (temperature, humidity), economic indicators (GDP, population), and historical consumption.

Nonlinear Regression: If power consumption has a nonlinear relationship with the predictor variable, you can use a nonlinear regression model such as polynomial regression.

19 Machine Learning and Artificial Intelligence (AI)

Neural networks: Deep learning neural networks, such as LSTM (Long Short-Term Memory) and CNN (Convolutional Neural Network), can capture complex patterns in power consumption data. These are especially effective for large datasets and unstructured data [25].

Random Forest: Random forest models are ensemble learning techniques that can handle nonlinear relationships and feature importance. These are suitable for both short-term and long-term forecasting.

Gradient Boosting: Gradient boosting algorithms such asXGBoost and LightGBM are used for accurate and efficient predictions and often outperform other methods in competitions and real-world applications [26].

Time Series Decomposition: Seasonal Time Series Decomposition (STL): STL decomposes time series data into seasonal, trend, and residual components, making it easy to model and predict each component separately.

Hybrid Model: ARIMA-X: ARIMA can be combined with external variables (X) such as weather data or economic indicators to improve forecast accuracy. Hybrid machine learning models: Combining traditional time series techniques and machine learning models can yield better results, especially when dealing with complex, multidimensional data.

Long Short-Term Memory Network (LSTM): LSTM networks are a type of recurrent neural network (RNN) designed for sequence data. It can capture long-term dependencies in power consumption data and is suitable for short-term and long-term forecasting.

Clustering and classification: Clustering techniques such as K-means allow you to group similar consumption patterns together, making it easier to make predictions within each cluster.

Exogenous Variables: Incorporating exogenous variables such as weather forecasts, holidays, and economic indicators into predictive models improves accuracy, especially for short-term forecasts. Deep reinforcement learning: For advanced applications, deep reinforcement learning models can be used to optimize energy consumption in real time by making dynamic decisions based on predictions and system constraints.

Choosing the appropriate forecasting method depends on factors such as the type of data, the forecast period (short or long-term), available computing resources, and the level of accuracy required. In practice, combinations of techniques, such as hybrid models and ensemble methods, are often used to improve the accuracy and reliability of predictions.

The specific design of your power usage prediction system will depend on the organization's and application's specific needs and constraints. It's important to collaborate with subject matter experts, data scientists, and IT professionals to create robust and effective predictive solutions. As ML and AI techniques are increasingly integrated into energy consumption forecasting, the need for transparency and interpretability becomes increasingly important. Explainable AI techniques that provide insight into how a model arrives at a particular prediction increase stakeholder confidence and foster trust in the predictive process. Interpretable models enable energy managers, policymakers, and other stakeholders to understand the underlying factors that influence forecasts and make informed decisions based on the insights gained. I can. This transparency is essential to accelerate the adoption of AI-powered predictive tools across industries and sectors.

Incorporating these advanced methods and strategies into energy consumption forecasting will not only increase the accuracy of forecasts but also expand the application of ML and AI technologies in the energy sector. The integration of advanced technologies offers opportunities to revolutionize energy management practices and promote a more sustainable and efficient energy future by continuously improving the quality of forecasts and enabling adaptive models.

20 Power Forecasting Using Machine Learning Techniques

Machine learning techniques have proven effective in predicting energy consumption. These techniques can capture complex relationships in data, adapt to changing patterns, and provide accurate predictions [27, 28].

21 Management of Missing Data

In the past, strategies for addressing missing data have been extensively researched and developed. An imputation approach based on close estimation is one of two types of procedures that eliminates some of the missing data. The first approach, which removes the portion of the data that is missing, is inapplicable since it results in the deletion of important data. In the absence of the data, an unbiased rating would be assigned. As a result, the imputation method is inexpensive. Introduced a number of imputation techniques, including multiple imputation (MI), maximum likelihood estimation (MLE), mean imputation, and last observation carried forward. In essence, mean imputation uses the data set mean to fill in the missing data. However, this method is not suitable for data that is not completely random because it introduces inequality into the data. Another method presented was last observation carried forward, where imputation is based on historical data collected.

The advanced methods presented were multiple imputation (MI) and maximum likelihood estimation (MLE). The multiple imputation method replaces missing data incrementally replacing missing data in each iteration. This method uses statistical analysis based on the observed data to deal with the uncertainty caused by the missing part. An example of a popular MI method is the use of multiple imputation with chained equations (MICE). Maximum likelihood estimation performs a substitution with the assumption made when the parameters and spikes were initially identified from the distribution of the data. Imputation is performed based on the expected parameters. This imputation method was used in probabilistic principal component analysis (PPCA). Both improved imputation methods were compared by where the imputation method was performed on sample data consisting of 87 numerically transformed categorical variables and 29 continuous variables. The study used an RMSE metric to evaluate the performance of the imputation technique. Based on the study, the PPCA method showed a much more promising result compared to MICE, where 65% of the data variables were successfully imputed by PPCA and only 38% were correctly imputed by MICE. This was also supported by the study that compared the performance of six imputation methods, including PPCA and MICE, on a real data set of different sizes. The result showed that MICE could work well on a small dataset, but the MICE method performed poorly on a large dataset.

Here are some machine learning techniques commonly used for power consumption forecasting:

Linear regression models can be used to predict power consumption based on historical data and relevant features. Multiple linear regression allows you to incorporate multiple predictor variables (e.g., temperature and time of day) to improve accuracy.

22 Decision Trees

Decision tree models, such as random forest and gradient boosting, are versatile and can handle both regression and classification tasks. They are useful for capturing nonlinear relationships and feature importance [29]. A random forest is an ensemble method that combines several decision trees with the bootstrap and aggregation technique, often known as bagging, to solve classification and regression problems. The main idea is to mix numerous decision trees rather than depending simply on one to determine the outcome. Decision trees are heavily utilized by random forest as a fundamental learning model. Rows and characteristics from the dataset are randomly chosen to create sample datasets for each model.

23 Support Vector Machines (SVM)

SVM models can be used for regression tasks by finding a hyperplane that best fits the data [30]. SVMs are particularly effective when dealing with high-dimensional data or complex relationships [31]. SVR uses a kernel function to transform the input data into a higher-dimensional space, which allows for more complex relationships to be captured. The algorithm also includes regularization parameters to prevent overfitting. The SVR process involves selecting the appropriate kernel function and regularization parameters, training the model, and making predictions for new data. Overall, SVR is a powerful algorithm that can be used for a wide range of regression tasks [32].

24 Long Short-Term Memory (LSTM)

An artificial neural network with long short-term memory (LSTM) is used in deep learning and artificial intelligence [33]. Because LSTM contains feedback connections, they differ from traditional feedforward neural networks. In addition to analyzing single data points (like photos), such as audio or video, this kind of RNN can also evaluate whole data sequences [34]. Networked, unsegmented handwriting identification, speech recognition, machine translation, robot control, video gaming, and healthcare are a few examples of LSTM applications [35]. The twentieth century has seen the most use of the LSTM neural network.

25 K-Nearest Neighbors (KNN)

KNN is a simple and intuitive algorithm for power consumption forecasting [36]. It makes predictions based on the similarity of historical data points. The choice of the number of neighbors (K) can be tuned for optimal performance. The model may overfit if K is too small because it may be very sensitive to data noise. However, if K is too high, the model can be oversimplified and fail to recognize the underlying trends in the data.

26 Time Series Models

Time series forecasting models like ARIMA (AutoRegressive Integrated Moving Average) and Exponential Smoothing are designed specifically for temporal data. They can capture trends, seasonality, and autocorrelation in power consumption data.

27 Neural Networks

Deep learning techniques, such as feedforward neural networks, recurrent neural networks (RNNs), and long short-term memory (LSTM) networks, can model complex temporal patterns in power consumption data [37]. RNNs and LSTMs are especially suitable for sequential data.

AutoML and Automated Forecasting Tools: Automated Machine Learning (AutoML) platforms, like Google AutoML and Microsoft Azure AutoML, can automate the process of selecting and tuning machine learning models for power consumption forecasting.

Time Series Decomposition with Machine Learning: Combine traditional time series decomposition techniques (e.g., STL decomposition) with machine learning models to forecast each component (trend, seasonality, remainder) separately.

Ensemble methods, such as bagging (Bootstrap Aggregating) and boosting (e.g., AdaBoost, XGBoost, LightGBM), combine multiple models to improve forecasting accuracy.

28 Some Other Models

Hybrid models combine the strengths of different machine learning techniques. For example, you can combine a neural network with a time series model for improved accuracy.

Engineering features from historical data can enhance the performance of machine learning models. For example, include lagged values, moving averages, and weather-related features as input variables.

Hyperparameter Tuning: Hyperparameter tuning techniques, such as grid search or Bayesian optimization, help optimize the parameters of machine learning models for better forecasting results. Regularization methods like L1 and L2 regularization can prevent overfitting in machine learning models, ensuring that they generalize well to unseen data.

Deep reinforcement learning: In advanced applications, deep reinforcement learning models can be used to optimize energy consumption in real time by making dynamic decisions based on forecasts and system constraints.

Selecting the most suitable machine learning technique for power consumption forecasting depends on factors like the nature of the data, the forecasting horizon, available computational resources, and the desired level of accuracy. In practice, it may be beneficial to experiment with multiple techniques and choose the one that provides the best results for your specific application.

References

1. Barbato, A., Capone, A., Rodolfi, M., & Tagliaferri, D. (2011). Forecasting the usage of household appliances through power meter sensors for demand management in the smart grid. In *2011 IEEE International Conference on Smart Grid Communications (SmartGridComm)* (pp. 404–409). IEEE.
2. Mosavi, A., & Bahmani, A. (2019). Energy consumption prediction using machine learning: A review. *Energies, 11*, 1–63.
3. Ziekow, H., Goebel, C., Strüker, J., & Jacobsen, H. A. (2013). The potential of smart home sensors in forecasting household electricity demand. In *2013 IEEE International Conference on Smart Grid Communications (SmartGridComm)* (pp. 229–234). IEEE.
4. Peña-Guzmán, C., & Rey, J. (2020). Forecasting residential electric power consumption for Bogotá Colombia using regression models. *Energy Reports, 6*(1), 561–566. https://doi.org/10.1016/j.egyr.2019.09.026

5. Moreno, M. V., Úbeda, B., Skarmeta, A. F., & Zamora, M. A. (2014). How can we tackle energy efficiency in IoT based smart buildings? *Sensors, 14*(6), 9582–9614.
6. Dey, N., Fong, S., Song, W., & Cho, K. (2018). Forecasting energy consumption from smart home sensor network by deep learning. In *Smart Trends in Information Technology and Computer Communications: Second International Conference, SmartCom 2017, Pune, India, August 18–19, 2017, Revised Selected Papers* (pp. 255–265). Springer.
7. Chitsaz, H., Shaker, H., Zareipour, H., Wood, D., & Amjady, N. (2015). Short-term electricity load forecasting of buildings in microgrids. *Energy and Buildings, 99*, 50–60.
8. Qolomany, B., Al-Fuqaha, A., Gupta, A., Benhaddou, D., Alwajidi, S., Qadir, J., & Fong, A. C. (2019). Leveraging machine learning and big data for smart buildings: A comprehensive survey IEEE. *Access, 7*, 90316–90356.
9. Naganathan, H., Chong, W. O., & Chen, X. (2016). Building energy modeling (BEM) using clustering algorithms and semi-supervised machine learning approaches. *Automation in Construction, 72*, 187–194.
10. Barak, S., & Sadegh, S. S. (2016). Forecasting energy consumption using ensemble ARIMA-ANFIS hybrid algorithm. *International Journal of Electrical Power and Energy Systems, 82*, 92–104.
11. Benavente-Peces, C. (2019). On the energy efficiency in the next generation of smart buildings—Supporting technologies and techniques. *Energies, 12*(22), 4399.
12. Metallidou, C. K., Psannis, K. E., & Egyptiadou, E. A. (2020). Energy efficiency in smart buildings: IoT approaches. *IEEE Access, 8*, 63679–63699.
13. Mariano-Hernández, D., Hernández-Callejo, L., García, F. S., Duque-Perez, O., & Zorita-Lamadrid, A. L. (2020). A review of energy consumption forecasting in smart buildings: Methods, input variables, forecasting horizon and metrics. *Applied Sciences, 10*(23), 8323.
14. Ahmad, N., Ghadi, Y., Adnan, M., & Ali, M. (2022). Load forecasting techniques for power system: Research challenges and survey. *IEEE Access, 10*, 71054–71090.
15. Kuster, C., Rezgui, Y., & Mourshed, M. (2017). Electrical load forecasting models: A critical systematic review. *Sustainable Cities and Society, 35*, 257–270.
16. Niu, D. X., Ma, T. N., & Liu, B. Y. (2017). Power load forecasting by wavelet least squares support vector machine with improved fruit fly optimization algorithm. *Journal of Combinatorial Optimization, 33*, 1122–1143.
17. Ahmad, M. W., Mourshed, M., Mundow, D., Sisinni, M., & Rezgui, Y. (2016). Building energy metering and environmental monitoring—A state-of-the-art review and directions for future research. *Energy and Buildings, 120*, 85–102.
18. Iqbal, N., & Kim, D. H. (2022). IoT task management mechanism based on predictive optimization for efficient energy consumption in smart residential buildings. *Energy and Buildings, 15*(257), 111762.
19. El Khantach, A., Hamlich, M., & Belbounaguia, N. E. (2019). Short-term load forecasting using machine learning and periodicity decomposition AIMS. *Energy, 7*(3), 382–394.
20. Azar, J., et al. (2019). An energy efficient IoT data compression approach for edge machine learning. *Future Generation Computer Systems, 96*, 168–175.
21. Mazlan, N. L., Ramli, N. A., Awalin, J. L., Ismail, M. B., Kassim, A., & Menon, A. (2020). A smart building energy management using Internet of Things (IoT) and machine learning. *Test Engineering Management, 83*, 8083–8090.
22. Li, W., et al. (2018). Implemented IoT-based self-learning home management system (SHMS) for Singapore. *IEEE Internet of Things Journal, 5*(3), 2212–2219.
23. Khan, A. N., Iqbal, N., Ahmad, R., & Kim, D. H. (2021). Ensemble prediction approach based on learning to statistical model for efficient building energy consumption management. *Symmetry, 13*(3), 405.
24. Divina, F., Gilson, A., Goméz-Vela, F., García Torres, M., & Torres, J. F. (2018). Stacking ensemble learning for short-term electricity consumption forecasting. *Energies, 11*(4), 949.

25. Kizilcec, V., Spataru, C., Lipani, A., & Parikh, P. (2022). Forecasting solar home system customers' electricity usage with a 3D convolutional neural network to improve energy access. *Energies, 15*(3), 857. https://doi.org/10.3390/en15030857
26. Di Persio, L., & Fraccarolo, N. (2023). Energy consumption forecasts by gradient boosting regression trees. *Mathematics, 11*(5), 1068. https://doi.org/10.3390/math11051068
27. Olu-Ajayi, R., Alaka, H., Sulaimon, I., Sunmola, F., & Ajayi, S. (2022). Building energy consumption prediction for residential buildings using deep learning and other machine learning techniques. *Journal of Building Engineering, 45*, 103406. https://doi.org/10.1016/j.jobe.2021.103406
28. Shah, S. F., Iqbal, M., Aziz, Z., Rana, T. A., Khalid, A., Cheah, Y. N., & Arif, M. (2022). The role of machine learning and the internet of things in smart buildings for energy efficiency. *Applied Sciences, 12*(15), 7882.
29. Ramos, D., Faria, P., Morais, A., & Vale, Z. (2022). Using decision tree to select forecasting algorithms in distinct electricity consumption context of an office building. *Energy Reports, 8*(3), 417–422.
30. Zhong, H., Wang, J., & Jia, H. (2019). Vector field-based support vector regression for building energy consumption prediction. *Applied Energy, 242*, 403–414. https://doi.org/10.1016/j.apenergy.2019.03.078
31. Duan, J., Tian, X., Ma, W., Qiu, X., Wang, P., & An, L. (2019). Electricity consumption forecasting using support vector regression with the mixture maximum correntropy criterion. *Entropy, 21*(7), 707. https://doi.org/10.3390/e21070707
32. Li, Y. (2019). Prediction of energy consumption: Variable regression or time series? A case in China. *Energy Science & Engineering, 7*, 2510–2518. https://doi.org/10.1002/ese3.439
33. Alsidairi, Z. H. (2018). The prediction of energy consumption using multivariate regression and artificial neural network models: Transport in the GCC. *International Journal of Engineering and Technology (UAE), 7*(4), 98–106.
34. Oprea, S. V., Pîrjan, A., Căruțașu, G., Petroșanu, D. M., Bâra, A., Stănică, J. L., & Coculescu, C. (2018). Developing a mixed neural network approach to forecast the residential electricity consumption based on sensor recorded data. *Sensors, 18*(5), 1443.
35. Shees, A., Hussain, M. T., Tariq, M., Sarwar, A., & Sarwat, A. I. (2023). Integration of artificial intelligence in future smart grids: An LSTM-RNN based approach for optimizing energy efficiency in smart grids. In *2023 IEEE International Conference on Energy Technologies for Future Grids (ETFG)* (pp. 1–6). IEEE.
36. Hong, G., Choi, G.-S., Eum, J.-Y., Lee, H. S., & Kim, D. D. (2022). The hourly energy consumption: Prediction by KNN for buildings in community buildings. *Buildings, 12*(10), 1636. https://doi.org/10.3390/buildings12101636
37. Somu, N., Gauthama Raman, M. R., & Ramamritham, K. (2021). A deep learning framework for building energy consumption forecast. *Renewable and Sustainable Energy Reviews, 137*, 110591. https://doi.org/10.1016/j.rser.2020.110591

Grid Automation: Challenges Issues and Opportunities

S. Vijayalakshmi, Nilima Gaikwad, and S. P. Gayathri

1 Introduction

The traditional electric power grid operates in a one-way fashion, directing electricity from power plants to consumers, a system that has functioned effectively for decades [1]. However, recent years have seen challenges stemming from government deregulation, alongside technical, economic, and environmental concerns. Today's society requires a grid that is more dependable, adaptable, and manageable, all while remaining cost-effective, secure, and compatible. Introducing the 'smart grid': the succeeding phase in the evolution of the electric power industry [2], this innovative system intends to transform electricity generation, transmission, and distribution by allowing power and information to flow in both directions [3].

The implementation of Smart Grid (SG) technologies promises to revolutionize modern businesses by providing efficient solutions to enhance the effectiveness, stability, and resilience of existing power systems [4]. SG facilitates a sustainable power supply by linking together electrical grids and communication infrastructures for long-term benefits. However, SGs face numerous complexities and challenges, particularly concerning capacity, reliability, and security [5]. Consumers are particularly concerned about the possibilities of data breaches and criminal hacking on their personal gadgets. Furthermore, the increasing integration of distributed generation (DG) and electric vehicles (EVs) complicates matters, resulting in synchronization issues, voltage regulation issues, malfunctions, and the presence of

S. Vijayalakshmi (✉)
Department of Data Science, School of Sciences, CHRIST University, Pune, Maharashtra, India

N. Gaikwad
Deapartment of Computer Science, CHRIST University, Pune, Maharashtra, India

S. P. Gayathri
Government Arts College for Women, Dindigul, Tamil Nadu, India

S. Vijayalakshmi et al. (eds.), *Smart Power Systems*, Power Systems,
https://doi.org/10.1007/978-3-031-89459-6_5

harmonic components in current and voltage waveforms. SG has a multitude of challenges regarding communication, security, energy integration, and power quality.

2 Electrical Grid and Smart Grid

2.1 Canonical Electrical Grid System

The electrical grid, often known as the power grid, is a linked network made up of generation, transmission, and distribution units. It permits the transmission of electrical electricity from generation sources to distribution stations [6]. Typically, a significant volume of power is transmitted from the generation station to centralized load centers at voltages of 220 kV or higher. The network formed by these high-voltage lines is termed the super grid. This super grid then supplies power to the sub-transmission network, which operates at voltages of 132 kV or lower [6].

There are two main types of electrical grids: regional grids and national grids [7].

Regional Grids: These grids are created by connecting various transmission systems within a specific area through transmission lines. Power stations in regional grids are typically located near fuel sources, reducing transportation costs. However, they are often situated far from densely populated areas. Electricity generated at high voltages is stepped down using transformers at substations before being supplied to consumers [7]. National Grids: National grids are formed by interconnecting different regional grids. They serve a broader geographical area and facilitate the transmission of electricity across regions [7].

Electric grid systems consist of four major components: energy production facilities, transmission substations, distribution substations, and end consumers. The current operation of electrical grid networks works in the following way: Initially, electricity producing plants generate power, which can come from a variety of sources such as wind energy or nuclear power, for distribution. As electricity travels to users' homes, it is stepped down to a voltage appropriate for domestic use. Finally, residential appliances take electricity from their electric meters.

2.2 What Is Smart Grid?

Different officialdoms and scholars may have different interpretations of the smart grid concept, but there is widespread agreement that it requires integration with an information system for communication to accomplish its "smart" qualities [1].

Various stakeholders offer different perspectives on the smart grid. For example, the Ontario Smart Grid Forum describes it as a modern electric system that employs communication, sensors, automation, and computers to enhance the flexibility, security, reliability, efficiency, and safety of the electricity grid [4]. It aims to empower consumers with greater control over their electricity usage and responsiveness to price fluctuations, integrating diverse energy resources, supporting electric vehicle charging, and promoting connection and coordinated operation for optimized production, distribution, and consumption of electricity, benefiting consumers and the environment [4].

On the other hand, the smart grid is defined by the US Department of Energy as an automated, globally dispersed energy delivery network that features a bidirectional flow of data and electricity [8]. By utilizing distributed computers and networking, it makes it possible to monitor power generation, consumer preferences, and the consumption of specific appliances. This allows for the provision of real-time information and the facilitation of instantaneous supply-demand balancing at the device level [8].

The Canadian Electricity Association emphasizes common themes such as communication, integration, and automation for sustainability, cost-effectiveness, and security, defining the smart grid as a suite of information-based applications enabled by increased grid automation and communication capabilities. These technologies integrate the behavior of connected energy supplies and loads to deliver sustainable, cost-effective, and secure power supplies, combining elements of a traditional distribution network with a bidirectional communication network for sensing, monitoring, and disseminating information on energy consumption [9].

The smart grid represents a technological framework centered on communication and information systems within the energy production, distribution, and consumption processes. Leveraging bidirectional data flow, it establishes an automated and widely dispersed network, offering novel functionalities such as real-time monitoring, operational efficiency, grid resilience, and enhanced integration of renewable energy sources, thereby reducing carbon emissions. Nevertheless, vulnerabilities persist within the smart grid [8]. Disruptions in power generation can jeopardize its stability, potentially leading to significant socio-economic repercussions. Moreover, the exchange of valuable data across smart grid systems poses risks of privacy breaches through theft or manipulation. Consequently, the smart grid has emerged as a prime target for malicious actors, prompting increased attention from governmental, industrial, and academic sectors [10].

2.3 *What Makes a Grid "Smart"?*

A "smart" grid is essentially one that uses digital technology to enable transmission line monitoring and two-way communication between the utility and its customers. Similar to the Internet, the smart grid uses computers, automation, controls, and new technologies to modify the electrical grid digitally in response to rapidly fluctuating power demand [1].

2.4 What Does a Smart Grid Do?

The smart grid offers an unexampled chance to propel the industry like energy forward into a new world marked by enhanced dependability, accessibility, and efficiency, thereby bolstering both our economic prosperity and environmental well-being. Throughout the transition phase, it is imperative to engage in testing, refine technologies, educate consumers, establish standards and regulations, and facilitate information exchange among projects to ensure the realization of the anticipated advantages of it [11].

These benefits include [12]:

1. Enhanced efficiency in electricity transmission.
2. Expedited restoration of electricity.
3. Lower management and operating expenses for utilities, which translates into cheaper electricity bills for customers.
4. Better integration of power generation systems that customers own, like renewable energy sources.
5. Heightened consolidation of big renewable energy.
6. Enhanced incorporation of customer-owned power generation systems, such as renewable energy sources.
7. Augmented security measures.

Currently, electricity disruptions like blackouts can trigger a cascade of failures impacting various sectors such as banking, communications, traffic, and security, particularly posing risks during winter when heating may be compromised. Our electrical power system is more resilient when it has a smarter grid, protecting it from calamities such as strong storms, earthquakes, solar flares, and terrorist attacks. The smart grid enables automatic rerouting during equipment failures or outages, minimizing their impact [4]. It detects and isolates outages swiftly, preventing widespread blackouts. Additionally, smart grid technologies facilitate swift and strategic electricity recovery post-emergency, prioritizing critical services like emergency response. Moreover, the smart grid optimally utilizes customer-owned power generators during utility unavailability, ensuring vital services remain operational during crises. It also addresses the imperative to upgrade or replace aging energy infrastructure, promotes energy efficiency, and fosters consumer awareness about the link between electricity consumption and environmental impact. Furthermore, it enhances national security by leveraging domestically sourced electricity that is more resilient against natural disasters and attacks [11].

2.5 *What Are the Key Technologies Used?*

2.5.1 Integrated Communication

The current grid will be completely transformed by state-of-the-art, fully integrated, two-way communication technologies, which will make it a dynamic, interactive infrastructure for instantaneous power transfer and information interchange. Plug-and-play conditions will be promoted via an open architecture, enabling safe communication, listening, and interaction among grid components [5, 13].

2.5.2 Sensing and Measurement

Smart grid technologies will improve measurements of the power system and make it easier to convert data into insights that can be put to use. They will assess the integrity of the grid and the health of the equipment, enable sophisticated defensive relaying, do away with meter estimations, and discourage energy theft. They will also reduce congestion and allow demand response [5, 13].

2.5.3 Advanced Components

Advanced components are a major factor in determining how the grid behaves. The next generation of these power system devices will make use of the most recent advances in microelectronics, materials science, energy storage, superconductivity, and power electronics. Higher power densities, better power quality and reliability, increased electrical efficiency with major environmental benefits, and improved real-time diagnostics are all expected outcomes of this progress [5, 13].

2.5.4 Advanced Control Methods

Critical components will be monitored using cutting-edge techniques, enabling quick diagnosis and prompt, appropriate reactions to any incident. They will also improve asset management and operational efficiency, as well as market pricing processes [5, 13].

2.5.5 Enhanced Interfaces and Decision Support

Operators frequently just a few seconds to make important choices. To enable grid operators and managers to make well-informed decisions quickly, the contemporary grid will necessitate the widespread, seamless, real-time use of applications and technologies. At every level of the grid, better decision-support interfaces will improve human decision-making abilities [5, 13].

2.6 Smart Grid Architecture

To construct a viable smart grid communication infrastructure, we encounter several hurdles. These include ensuring interoperability and scalability across diverse utility companies and user facilities, as well as integrating emerging technologies like smart meter infrastructure. It's crucial to comprehensively identify the obstacles in designing communication and networking architectures for smart grids and consolidate them into a standardized model [14].

Sood et al. highlighted deficiencies in current IEEE standards governing Distributed Generation (DG) interconnection, contending that they restrict the implementation of potentially beneficial grid applications. They proposed revising IEEE standards to mandate monitoring and control facilities for all DG units, irrespective of their size, to enable effective data communication and management [15].

Chen et al. highlighted the significance of a meticulously crafted information communication framework to facilitate effective electricity utilization. They advocated for smart grids equipped with the capability to collect various data pertaining to electricity generation, consumption, storage, and distribution through a robust communication infrastructure. Utilizing technologies such as grid computing or cloud computing could streamline electricity usage. Consequently, ensuring the resilience and security of smart grids emerges as a central challenge [16].

Before delving into the communication structure of the smart grid, it's crucial to gain an overview of the entire architecture of the smart grid.

Smart grid architecture proposals originate from three primary sources:

1. Government and Organizations: They establish the requirements and frameworks for the smart grid.
2. Industry: They propose strategies for implementing communication infrastructure.
3. Academia: They primarily focus on defining the communication architecture needs and providing solutions.

Nonetheless, a number of national organizations and businesses, including the DOE (Department of Energy), the State of West Virginia, NIST, and others, have offered conceptual frameworks for the smart grid [6, 17].

Market operators, reliability coordinators, generation/load wholesalers, transmission providers, balancing authorities, energy service retailers, distribution providers, and end users (commercial, residential, and industrial) should all be included in the architecture of a smart grid, according to the DOE's Smart Grid System Report [17].

Four primary components should make up a smart grid architecture, according to a West Virginia white paper: advanced control methods, enhanced interfaces and decision support, advanced components, and sensing and measurement [18].

NIST suggested that components of a smart grid architecture include customers, markets, service providers, operations, bulk generation, transmission, and distribution (NIST Framework and Roadmap for Smart Grid Interoperability Standards). One of the most thorough architectural ideas in the most recent smart grid literature is this one [6].

The customer sector can be further segmented into Home Area Networks (HANs), Building Area Networks (BANs), and Industrial Area Networks (IANs), as illustrated in Fig. 2. These networks, which are either wired or wireless and are situated on the premises of the customers (houses, buildings, and industrial locations), facilitate communication between appliances, smart meters, electronics, energy management devices, applications, and customers. Building Automation and Control Networks (BACnet), Home Energy Management Systems (HEMS), and other energy management systems may have an impact on how these networks operate [19].

2.7 *Internet-Based Architecture*

Apart from the basic smart grid architectures proposed by national organizations, other researchers have also showcased smart grid architectures that have particular features that they hope to integrate into the system.

The fact that the existing grid only allows one-way communication between the generation system and downstream distribution nodes was brought to light by Aggarwal et al. On the other hand, a point of consumption for electricity can also function as a point of generation in a smart grid scenario. In order to start the proper activities, communication technologies should make it easier for sensor data to be transmitted to the grid's control center and for control messages to be distributed to various grid sites. Furthermore, the smart grid depends on improved end-user engagement, such as real-time energy meter monitoring [20]. However, this presents various challenges, including:

- Stringent Latency: If the control center misses sensor inputs, it could send erroneous control messages across the grid. Latency typically falls within the range of a few milliseconds, such as 10 ms [20].
- Large Message Volume: As more elements are added to the network, it must handle multiple messages simultaneously without significant latency [20].

Aggarwal et al. [20] suggest an IP-based network built on optical fibers as a way to facilitate communication within the smart grid. They contend that because optical fibers can reach high speeds, an IP-based network serving as the foundation of the smart grid can serve long-term needs and take use of new technologies at lower costs.

A data communication network embedded into the electrical grid, the smart grid, according to Cisco [13], gathers and analyzes data on power transmission, distribution, and consumption almost instantly. Cisco highlights that in the current smart grid, where disparate information "islands" exist, communication is challenging. They promote the Internet design, pointing out advantages including supporting many kinds of systems, connecting a vast number of devices, transferring data across multiple media, and maintaining dependability.

According to Cisco, the Internet Protocol (IP) is the most effective standard suite of protocols for achieving security and interoperability in the smart grid. However, protecting IP-based smart grid connectivity presents substantial issues due to inheriting threats and vulnerabilities from the Internet. It is still up for debate and needs more research to determine whether IP should be used as a communication standard in the smart grid [13].

In conclusion, the Internet-based smart grid communication/networking architecture is highly applicable due to its scalability, security, and interoperability characteristics.

2.8 Power Line Communication Architecture

The transmission of a modulated carrier signal through the wiring system is addressed in [20]. However, as the wires were initially intended for Alternating Current (AC), the power wire circuits have restricted capacity to carry higher frequencies. One prominent application of Power Line Communication (PLC) technology is Broadband over Power Lines (BPL), also referred to as power line Internet. This technology allows broadband Internet access through standard power lines. To access high-speed Internet via BPL, a computer would require connecting a BPL modem to an outlet in a building equipped with the necessary infrastructure [21]. Figure 1 shows the block diagram of power line communication.

In the context of smart power consumption, Liu et al. [22] presented a pilot simulation of a two-way, real-time power line communication (PLC) system. They collected data from AC power lines using PLC terminals. They reasoned that electricity cables were usually owned by utility companies, thus they decided to spread PLC technology. They felt that expanding consumers' residential power communication networks using PLC technology may solve networking issues in an efficient manner at a much lower cost, which would ultimately improve power consumption efficiency [22]. The restriction of channel characteristics, however, was a significant obstacle since PLC was still unable to handle high-speed data transfer. They used an optical composite power cable, which uses power lines as a fiber optic carrier, to get around this [22]. This method uses an alternate transmission media to address the low data rate issue. In the meantime, a protocol stack was presented by Bauer et al. [23] with the goal of incorporating the physical layer of low-speed PLC into a robust communication paradigm.

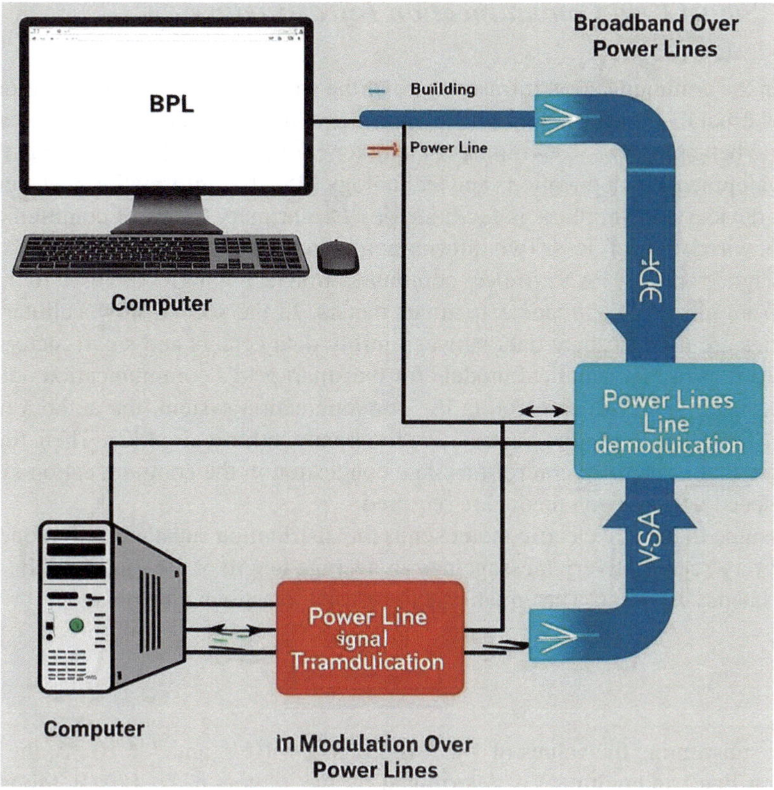

Fig. 1 Understanding power line communication

The NETL [24] highlighted the challenges associated with transmitting data in the smart grid. They emphasized the need for broadband, low-latency, and secure connectivity between transmission substations and control centers. The NETL Smart Grid Implementation Strategy (SGIS) team looked into the viability of broadband over power lines (BPL) as a possible substitute. BPL hasn't been used in transmission subsystems of the smart grid that use high voltage (HV) cables; instead, it has only been implemented in distribution subsystems that use medium voltage wires. Testing was done on a BPL network by NETL in association with American Electric Power (AEP) along a 5-mile, 69 KV line that links three AEP substations. The outcomes demonstrated dependable connection with a usual delay of about 5 ms at speeds greater than 10 MB/S [24].

In summary, power line communication (PLC) itself faces limitations in terms of data transmission speed. To ensure reliable and robust data transmission in the smart grid, improvements must be made either to the transmission media or through the adoption of technologies to enhance the reliability and robustness of low-speed PLC communication.

2.9 *Smart Grid Communication Infrastructure*

Within the communication infrastructure of the smart grid, the communication system is crucial. Improvements to the smart grid infrastructure are being made in order to better analyze vast volumes of data for control and real-time pricing, thanks to developments in applications and technology [25]. The information flow between smart devices and suppliers is facilitated by two primary forms of communication media: wired and wireless. Two information routes are used in the smart grid. In the first, Zigbee and low PAN wireless communication technologies are used to transfer data from electrical appliances to smart meters. In the second flow, cellular technologies are used to share data between utility data centers and smart meters. The introduction of mathematical models for the smart grid's communication architecture is made [2]. Prior to creating the communication system, the authors offer a model for determining the communication bandwidth required [2]. Their findings indicate that in order to control message congestion in the communication system, high-speed transmission media are required.

Assume that every electric meter sends the distribution substation (DS) one message every second. Every message has an average length of 100 bits, and the maximum latency is 10 ms. The equation can be used to estimate the delay:

$$T = \frac{1}{\mu c - \lambda}$$

By substituting the values of $1/\mu = 100$ bits, $\lambda = 10^6$/s, and $T = 10$ ms, the transmission line bandwidth (c) is determined by the authors to be 100.01 Mbps [20]. They counter that if the delay fluctuates greatly, this value is meaningless. For example, if 99% of the messages had a latency of no more than 10 ms. The authors also assume that service time and inter-arrival time are exponentially distributed and that packets have a Poisson distribution at each node. They use an equation to determine the bandwidth:

$$\begin{aligned} p &= P\{W_1 \le t\} \\ &= -pe^{-\mu c(1-p)t} \end{aligned}$$

(a) ZigBee

Zigbee is a popular wireless communication technology known for its affordability, simplicity, moderate data transfer rates, and low power consumption. Zigbee is used in smart grid applications including energy monitoring, home automation, and automated meter reading. Zigbee Smart Energy Profile has been determined by the National Institute of Standards and Technology (NIST) to be the best communication infrastructure for smart grid networks. Zigbee has 16 channels with a maximum output power of 0 dBM, each using a 5 MHz bandwidth. It runs in the 2.4 GHz range. Data speeds of up to 250 Kb/s are supported, with a typical transmission range of 1–100 m. Zigbee operates according to a controlled protocol that is based on IEEE 802.15.4 [14, 26].

(b) GSM

GSM, or Global System for Mobile communication, serves as a platform for transmitting both data and voice services. It functions as a cellular technology connecting mobile phones to the cellular network [14].

(c) Wireless Mesh

A wireless mesh network consists of interconnected nodes that operate collectively as autonomous routers. The self-healing capability of these nodes aids in establishing communication paths by dynamically routing signals through active nodes. The architecture of mesh networks is decentralized, as each node transmits data to its adjacent node. Wireless mesh technology is utilized in various applications, including small business operations and remote areas, due to its cost-effectiveness in providing connectivity [14].

(d) Cellular Network Communication

Cellular networks provide an effective solution for facilitating communication between distant nodes, particularly for utility purposes. They establish dedicated communication paths, allowing for the deployment of smart meters across expansive areas. Various cellular network technologies, such as 2G, 2.5G, 3G, WiMAX, and LTE, are employed to facilitate data exchange between smart meters and utility data centers [14].

2.10 Sensor Networks

Gadze presented a wireless embedded sensor platform with a hierarchical architecture that was tailored for the smart grid. This architecture, which is decentralized and operates at several levels, was created especially to meet the difficulties presented by challenging power situations at the substation and generation levels. Due to elements that are common in substation and generator settings, including as route loss, shadow fading, and ambient noise, wireless sensor communication in these situations is unpredictable [27].

Using in-home Wireless Sensor Networks (WSNs), Erol-Kantarci and Mouftah [28] introduced the Appliance Coordination (ACORD) concept, which lowers home energy consumption expenses. In order to accomplish its goals, this scheme depends on Time-of-Use (TOU) tariffs and Energy Management Units (EMUs). With rates that change during peak, moderate peak, and off-peak times, TOU rates provide invoicing flexibility [28]. EMUs receive and coordinate customer requests [28]. However, several challenges need to be addressed: firstly, changing customer behavior can be challenging, requiring them to be comfortable with accepting changes; secondly, dynamic pricing might lead to load oscillation due to load shifting; thirdly, wireless channels could experience interference, particularly when appliances attempt to coordinate with each other [28].

Due to the expanding problems with power quality, as noted by Bisceglie et al. [29], there is an increasing need for large area voltage monitoring systems. They suggested using self-organizing sensor networks to create a completely decentralized architecture for voltage quality monitoring. Within these networks, every node has the ability to calculate both local and global performance measures through the utilization of local data and data shared with nearby nodes [29].

3 Challenges and Issues

3.1 Storage

One major obstacle confronting electrical engineers today is energy storage. Key concerns revolve around both technological advancements and economic viability. The technological hurdles encompass enhancing the capacity and effectiveness of existing technologies, as well as pioneering new ones. Presently, energy storage facilitates the assimilation of renewable sources, but over time, it might struggle to accommodate all decentralized generation [30].

The second hurdle pertains to economics. Economic challenges differ depending on the intended function of energy storage, be it in generation, distribution, or consumer applications. There may be devised compensation plans to incentivize greater energy storage adoption, yet the repercussions on the field remain uncertain. Moreover, the ownership of storage devices poses a contentious issue. The debate revolves around whether utilities or transmission system operators should be permitted ownership rights [30].

3.2 Communication

Another issue related to the smart grid involves communication, which is a crucial aspect of its design. For the smart grid to operate effectively, seamless communication among its various components and subsystems is essential. As mentioned earlier, sensors must send the data to the main system for the grid to build necessary adjustments [31]. While there are numerous communication methods available for the smart grid, we'll focus on some general challenges rather than delve into specific types. The communication layer of a smart grid faces four main challenges, with the first being its fragmented architecture. Despite the various communication setups possible for a smart grid, a clear frontrunner has yet to emerge. The second obstacle pertains to insufficient bandwidth, hindering the achievement of necessary two-way communication within the smart grid. While the solution seems straightforward—increasing bandwidth—it's a complex endeavor [31] (Fig. 2).

The third challenge pertains to the absence of interoperability among system components. For the smart grid to operate smoothly, all its components must

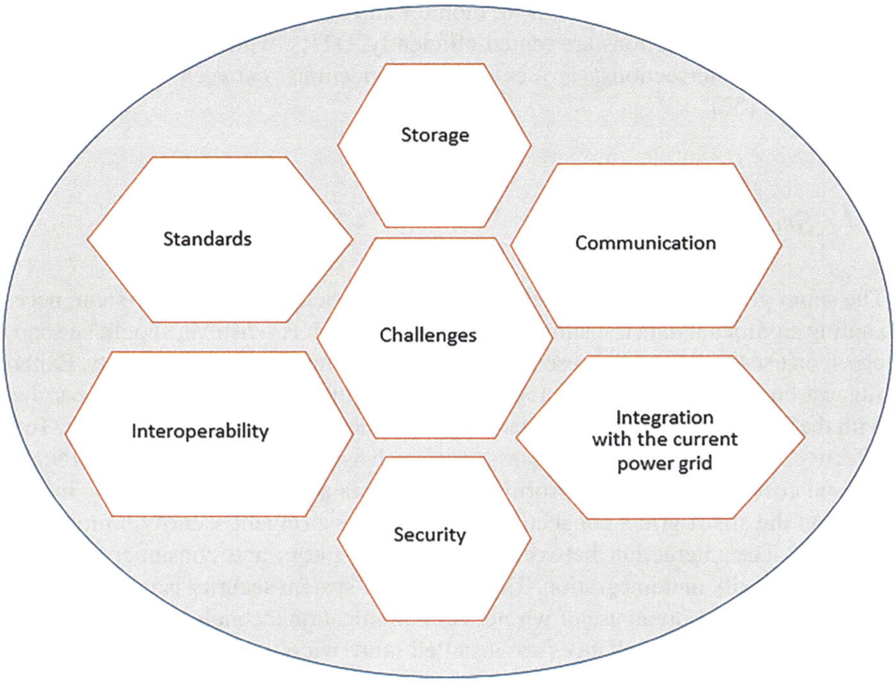

Fig. 2 Challenges

communicate efficiently, a process that requires time and effort. Each element of the system must both transmit information/commands and receive them. The last challenge involves handling the increasing amount of data generated by smart devices. In the contemporary landscape, a multitude of smart devices, ranging from phones to generators, handle vast volumes of data to support various applications [31].

3.3 Integration with the Existing Power Grid

Distribution transformers or DTRs are essential components of the electrical distribution networks. DTRs must develop into intelligent components in and of themselves to guarantee the efficient and dependable operation of the smart grid. To reduce DTR failures, this entails providing them with communication and monitoring tools [32]. For DTRs, overloading is a major concern. In order to reduce these risks, DTRs should maintain voltage and frequency levels from linked generating sources, enable two-way power flow, monitor and communicate, and make sure that electrical interconnections are routed efficiently. DTRs, which are sometimes seen as a grid's intersections, are essential connections between generation and distribution [33].

3.4 Security

The smart grid represents a more streamlined electricity distribution system, necessitating significant data transmission. This data, which is sensitive, should be shared based on user preferences, presenting a major challenge in terms of security. Experts suggest employing web technologies to address this challenge. Drawing parallels with the internet, the smart grid shares a comparable architectural framework. To be effective, the smart grid must possess traits such as flexibility, accessibility, reliability, and cost-effectiveness, mirroring the open data access principles of the internet [34]. As the smart grid's connectivity grows, more stringent security protocols are required. The interaction between different generators and consumers increases with complexity and integration. The problem of system security is made more difficult by the widespread use of wireless communication technologies. Furthermore, as smart meters and appliances are installed more widely, the network boundary is pushed farther inside users' homes, requiring more stringent user security protocols. Through the use of elements including an early warning system, quick and accurate detection, proactive reaction, and automatic recovery mechanisms, the smart grid protection system seeks to resemble the larger smart grid philosophy. However, it is imperative that security be given top priority in all areas, including distribution, consumption, substations, generating, transmission, and scheduling operating systems [35].

The increased interconnectedness and integration within the grid also bring along cyber vulnerabilities. If these problems are not resolved, the current power system's modernization initiatives may be hampered. Unauthorized access to smart metering data, the possibility of attackers remotely disabling equipment, the repudiation of smart metering data, covert power theft, and attacks on smart grid infrastructure that result in power outages are all examples of security problems. Concerns about privacy also exist in light of the metering data leaks of private and sensitive information.

Security and privacy are of utmost importance. For more detailed information, readers are encouraged to consult our survey paper [36].

3.5 Interoperability

The challenges posed by interoperability are significant. The primary issue lies in the uncertainty surrounding the outcomes until the technologies are put into practice. Therefore, the most effective approach to addressing interoperability is through a process of trial and error [37].

While the smart grid promises effortless integration of renewable resources, it also presents certain challenges. Effectively integrating these sources necessitates a comprehensive evaluation of their controls and hardware. This evaluation usually takes place through two methods: formula analysis or simulation studies. However, conducting a field test is both time-consuming and expensive. Additionally, the current lack of distributed energy resources entering the grid means that there aren't sufficient conditions for a field test. Consequently, studying downscaled energy resources in a laboratory setting emerges as a crucial method for gathering information [38].

3.6 Standards

The National Institute of Standards and Technology (NIST), the STATE GRID Corporation of China (SGCC), and the International Electrotechnical Commission (IEC) have worked together to develop roadmaps for the smart grid in order to create a standardized framework and guidelines for the development of smart grid standards. A working group on smart grid strategy was formed by the IEC in April 2009, and it was this group that ultimately produced and released the "IEC Smart Grid Standardisation Roadmap Edition 1.0."

It is essential to compare the standards offered by the SGCC [39], NIST [6], and IEC [40]. The three roadmaps are meant to serve as a useful guide for organizing, creating, building, and managing a smart grid. Furthermore, they all provide a roadmap for smart grid technology standards, with the main objective being the realization of smart grid interoperability.

IEC employs a more technical strategy and provides a top-to-bottom viewpoint. Five main components make up its roadmap: overview, demand, current standards, gaps, and recommendations. NIST, on the other hand, takes a top-down strategy that takes into account stakeholder viewpoints and market situations in the United States. With an emphasis on eight key domains—general, planning, generation, transmission, substation, distribution, utilization, and dispatching—the SGCC roadmap is specifically designed for the Chinese power market. In its roadmap, the SGCC gives smart grid technology top priority.

Overall, while IEC provides a technical viewpoint, NIST offers a market-driven perspective, and SGCC's focus lies within the Chinese power industry, particularly on smart grid technology [39, 40].

3.7 Performance

Due to the existence of heterogeneous systems, extensive deployments, interdisciplinary fields (such as control, communication, power, etc.), and dynamic, non-deterministic systems, smart grid communication is extremely difficult. To provide better, faster, more secure, and reliable controls and communication, efficiency is essential.

3.8 Testbed

Testbeds are vital for smart grids because they help validate test results and facilitate research.

4 Opportunities

4.1 Incorporation of Renewable Energy Sources

The expanding utilization of electricity generated from dispersed energy origins necessitates their effective integration into the grid. Renewable energy origins can be efficiently controlled, mainly delivering power using power electronic converter devices. The energy landscape is shifting toward a more decentralized manner of power generation as renewable energy sources become more commonplace worldwide [38, 41]. The primary goal of global smart grids is to replace centralized electricity production with several decentralized renewable energy sources. Prediction and scheduling are crucial factors to take into account while optimizing renewable energy supplies, as the energy output fluctuates due to weather variations. Thus, having insight into pertinent long-term weather trends greatly enhances optimization efforts [42].

4.2 Introducing Customers Choice

The forthcoming opportunity represents a fresh era of consumer choice. Within the existing grid infrastructure, power generation companies employ various methods to charge and invoice customers. The prevalent method is the flat rate, where the company bills based on the total energy consumption along with a demand charge, reflecting peak energy usage. The significance of demand lies in ensuring sufficient energy supply to meet peak demands. A novel approach to pricing is real-time rates, where prices escalate with increasing energy consumption, not only at the individual level but also within the community [41]. Prices surge as overall energy usage rises, particularly during peak times, typically occurring midday, resulting in higher energy costs compared to nighttime rates.

This introduces customers to not just an alternative pricing model but also a different approach to energy consumption [43]. The new rate system renders off-peak periods more appealing due to their lower pricing, likely prompting customers to shift their energy usage to late night or early morning hours. Those with significant energy requirements will seek savings by adjusting their operations to these more cost-effective times. Such shifts in consumption patterns can significantly alleviate the load on energy producers during peak periods. Reduced demand and a more consistent energy requirement enable energy producers to operate more efficiently [42].

4.3 Leveraging Green Building Standards to Alleviate Burdens

Another potential lies in the construction or renovation of green buildings. As previously noted, increased efficiency correlates with reduced energy demands at specific times. One strategy for decreasing energy requirements involves lessening the load. By optimizing operations, approximately 20–30% of a building's energy consumption can be conserved without altering its structure. Energy audits represent the most prevalent method for optimizing building operations [44]. These audits entail analyzing energy flows within a building and identifying ways to minimize energy input without compromising functionality. Smart grids can facilitate such efforts through the integration of smart sensors. Enhanced sensor technology enables customers to monitor their energy consumption more accurately, thereby simplifying the adherence to green building standards [42].

4.4 Facilitating the Widespread Adoption of Plug-in Hybrid Electric Vehicles Across the Country

A major advancement has been made with the national adoption of electric vehicles (EVs) made possible. Using EVs in conjunction with renewable energy sources has the potential to significantly lower greenhouse gas emissions. In January 2014, transportation systems produced an additional 23% of greenhouse gas emissions worldwide, with electric power generation accounting for 41%. When combined, they make up 64% of greenhouse gases on the planet. Future emissions are expected to drop significantly, even though it might not be possible to replace all power generators and transportation systems. Additionally, EVs themselves can function as batteries [44]. It becomes essential to provide bidirectional power transfer in a smart grid system. Therefore, EVs can be charged during periods of low electricity costs and subsequently utilized as a power source for buildings during high-cost periods. Such applications of EVs can prove highly efficient and effective [42].

4.5 Turning Large-Scale Energy Storage into a Practical Reality

One of the most important parts of the smart grid, which dates back to the early days of the electricity business, is energy storage. Due to its adaptability, it may be used in a wide range of applications. Wind energy integration projects have found great success with the use of Battery Energy Storage Systems (BESS). These systems offer vital voltage and frequency support during fluctuations in generation. Comprising batteries, power electronics, and control systems, BESS exemplifies a robust solution for grid stability [45].

Numerous battery chemistries exist, with sodium-sulfur, lead-acid, and lithium-ion being among the most prevalent. Charging and discharging batteries typically involve direct current (DC). Power electronics play a pivotal role in converting incoming alternating current (AC) to DC for charging the battery and converting outgoing DC back to AC. This bidirectional power electronic interface facilitates the process [42].

References

1. Gao, J., Xiao, Y., Liu, J., Liang, W., & Chen, C. L. P. (2012). A survey of communication/networking in smart grids. *Future Generation Computer Systems, 28*(2), 391–404.
2. Wang, W., Xu, Y., & Khanna, M. (2011). A survey on the communication architectures in smart grid. *Computer Networks, 55*(15), 3604–3629.
3. U.S. Department of Energy. (2010). *Communications requirements of smart grid technologies* (Technical report). US Department of Energy. http://energy.gov/gc/downloads/communications-requirements-smart-grid-technologies
4. Murphy, P., et al. (2010). *Enabling tomorrow's electricity system: report of the Ontario Smart Grid forum* (Technical report). Ontario Smart Grid Forum. https://www.site.uottawa.ca/~rhabash/Smart_Grid_Forum-Report.pdf
5. NETL. (2007). *A systems view of the modern grid.* http://www.netl.doe.gov/smartgrid/referenceshelf/whitepapers/ASystemsViewoftheModernGrid_Final_v2_0.pdf
6. NIST. (2010). *NIST framework and roadmap for Smart Grid interoperability standards, release 1.0.* http://www.nist.gov/public_affairs/releases/upload/smartgrid_interoperability_final.pdf
7. Circuit Globe. *Electrical grid.* https://circuitglobe.com/electrical-grid.html
8. U.S. Department of Energy. (2010). *Smart grid: An introduction* (Technical report). US Department of Energy. https://www.energy.gov/sites/prod/files/2010%20Smart%20Grid%20System%20Report.pdf
9. CEA. (2010). *The smart grid: A pragmatic approach* (Technical report). Canadian Electricity Association. http://www.electricity.ca/media/SmartGrid/SmartGridpaperEN.pdf
10. U.S. Department of Energy. (June). *Smart grid type.* https://www.energy.gov/smart-grid

11. U.S. NIST. (2010, August). *Guidelines for smart grid cyber security* (Vols. 1–3). NIST IR-7628. http://csrc.nist.gov/publications/PubsNISTIRs.html#NIST-IR-7628
12. Clastres, C. (2011). Smart grids: Another step towards competition, energy security and climate change objectives. *Energy Policy, 39*(9), 5399–5408.
13. Cisco, I. (2010). *Internet protocol architecture for the smart grid.* http://www.cisco.com/web/strategy/docs/energy/CISCO_IP_INTEROP_STDS_PPR_TO_NIST_WP.pdf
14. Gao, J., Xiao, Y., Liu, J., Wei Liang, C. L., & Chen, P. (2012). A survey of communication/networking in smart grids. *Future Generation Computer Systems, 28*(2), 391–404. https://doi.org/10.1016/j.future.2011.04.014
15. Sood, V. K., Fischer, D., Eklund, J. M., & Brown, T. (2009). Developing a communication infrastructure for the smart grid. In *Electrical Power & Energy Conference, EPEC, 2009* (pp. 1–7). IEEE.
16. Chen, K., Yeh, P., Hsieh, H., & Chang, S. (2010). 4th International Symposium on Communication Infrastructure of Smart Grid. In *Communications, Control and Signal Processing, ISCCSP 2010* (pp. 1–5). IEEE.
17. U.S. Department of Energy. (2009). *Smart grid system report.* http://www.oe.energy.gov/DocumentsandMedia/SGSRMain_090707_lowres.pdf
18. NETL. (2009, September). *West Virginia smart grid implementation plan.* http://www.netl.doe.gov/energy-analyses/pubs/WV_SGIP_Final_Report_rev1_complete.pdf
19. IEEE. (2010). *IEEE International Conference on Smart Grid Communications.* http://www.ieee-smartgridcomm.org/index.html
20. Aggarwal, A., Kunta, S., & Verma, P. K. (2010). A proposed communications infrastructure for the smart grid. In *2010 Innovative Smart Grid Technologies ISGT* (pp. 1–5). IEEE.
21. Broadridge, R. (1984). Power line modems and networks. In *4th International Conference on Metering Applications and Tariffs for Electricity Supply IEEE Conference, London* (pp. 294–296). IEEE.
22. Liu, J., Zhao, B., Wang, J., Zhu, Y., & Hu, J. (2010). Application of power line communication in smart power consumption. In *IEEE International Symposium on Power Line Communications and Its Applications, ISPLC 2010* (pp. 303–307). IEEE.
23. Bauer, M., Plappert, W., & Wang, C. (2009). Packet-oriented communication protocols for smart grid services over low-speed PLC. In *IEEE International Symposium on Power Line Communications and Its Applications, 2009. ISPLC 2009* (pp. 89–94).
24. NETL. (2010). *Broadband over power lines could accelerate the transmission smart grid.* https://netl.doe.gov/sites/default/files/Smartgrid/06-02-2010_Broadband-Over-Power-Lines.pdf
25. Lewis, R. P., Igic, P., & Zhou, Z. (2009). Assessment of communication methods for smart electricity metering in the UK. In *Proceedings of the 1st IEEE-PES/IAS Conference on Sustainable Alternative Energy (SAE '09)* (pp. 1–4). IEEE.
26. Ullo, S., Vaccaro, A., & Velotto, G. (2010). The role of pervasive and cooperative sensor networks in smart grids communication. In *MELECON 2010–2010 15th IEEE Mediterranean Electrotechnical Conference* (pp. 443–447). IEEE.
27. Gadze, J. (2009). Control-aware wireless sensor network platform for the smart electric grid. *IJCSNS, 9*, 16.
28. Erol-Kantarci, M., & Mouftah, H. T. (2010). Wireless sensor networks for domestic energy management in smart grids. In *25th Biennial Symposium on Communications, QBSC 2010* (pp. 63–66). IEEE.
29. Bisceglie, M., Galdi, C., Vaccaro, A., & Villacci, D. (2009). Cooperative sensor networks for voltage quality monitoring in smart grids. In *PowerTech, 2009 IEEE Bucharest* (pp. 1–6). IEEE.
30. EC. (2013). *DG ENER working paper: The future role and challenges of energy storage.* http://ec.europa.eu/energy/infrastructure/doc/energy-storage/2013/energy_storage.pdf
31. Gungor, V. C., Sahin, D., Kocak, T., Ergut, S., Buccella, C., Cecati, C., & Hancke, G. P. (2013). A survey on smart grid potential applications and communication requirements. *IEEE Transactions on Industrial Informatics, 9*(1), 28–42.

32. Kempton, W., & J. Tomic´. (2005). Vehicle-to-grid power implementation: From stabilizing the grid to supporting large-scale renewable energy. *Journal of Power Sources, 144*(1), 280–294.
33. Rao, N. M., Narayanan, R., Vasudevamurthy, B. R., & Das, S. K. (2013). Performance requirements of present-day distribution transformers for smart grid. In *2013 IEEE Innovative Smart Grid Technologies—Asia (ISGT Asia)* (pp. 1–6). IEEE.
34. Speiser, S., Wagner, A., Raabe, O., & Harth, A. (2013). Web technologies and privacy policies for the smart grid. In *IECON 2013—39th Annual Conference of the IEEE Industrial Electronics Society* (pp. 4809–4814). IEEE.
35. Zhang, T., Lin, W., Wang, Y., Deng, S., Shi, C., & Chen, L. (2010). The design of information security protection framework to support smart grid. In *2010 International Conference on Power System Technology (POWERCON)* (pp. 1–5). IEEE.
36. Liu, J., Xiao, Y., Li, S., Liang, W., & Chen, C. L. P. (2010). Cyber security and privacy issues in smart grids. In *IEEE Communications Surveys & Tutorials, October 27, 2010*, under major revision (submitted for publication)
37. Koh, L. H., Tan, Y. K., Wang, P., & Tseng, K. J. (2012). Renewable energy integration into smart gird: Problems and solutions—Singapore experience. In *2012 IEEE Power and Energy Society General Meeting* (pp. 1–7). IEEE.
38. Fuchs, F. W., Gebhardt, F., Hoffman, N., Knop, A., Lohde, R., Reese, J., & Wessels, C. (2012). Research laboratory for grid integration of distributed renewable energy resources—Design and realization. In *2012 IEEE Energy Conversion Congress and Exposition (ECCE)* (pp. 1974–1981). IEEE.
39. Miao, X., Chen, X., Ma, X., Liu, G., Feng, H., & Song, X. (2012). Comparing smart grid technology standards roadmap of the IEC, NIST and SGCC. In *2012 China International Conference on Electricity Distribution (CICED)* (pp. 1–4). IEEE.
40. TEC SG3. (2010). *TEC smart grid standardization roadmap edition 1.0.*
41. Roozbehani, M., Dahleh, M., & Mitter, S. (2010). *Dynamic pricing and stabilization of supply and demand in modern electric power grids.*
42. Guizani, M., & Anan, M. (2014). Smart grid opportunities and challenges of integrating renewable sources: A survey. In *2014 International Wireless Communications and Mobile Computing Conference (IWCMC)* (pp. 1098–1105). IEEE.
43. Potter, C. W., Archambault, A., & Westrick, K. (2009). Building a Smarter Grid through better renewable energy information, 2009. In *PSCE '09. IEEE/PES power systems conference and exposition* (pp. 1–5). IEEE.
44. Fang, X., Misra, S., Xue, G., & Yang, D. (2012). Smart grid—The new and improved power grid: A survey. *Communications Surveys & Tutorials, 14*(4), 944–980.
45. Tushar, M. H. K., Assi, C., Maier, M., & Uddin, M. F. (2014). Smart microgrids: Optimal joint scheduling for electric vehicles and home appliances. *IEEE Transactions on Smart Grid, 5*(1), 239–250.
46. Nigim, K. A., & Lee, W.-J. (2007). Micro grid integration opportunities and challenges. In *Proceedings of the IEEE Power Engineering Society General Meeting (PES '07)* (pp. 1–6). IEEE.

Empowering Solar Power Generation: The Z-Source Inverter Approach

C. Maheswari, V. Shobana, and M. Savithri

1 Introduction

"Empowering Solar Power Generation: The Z-Source Inverter Approach" presents a new way of thinking about how to use solar energy. Through an exploration of Z-source inverter technology, this all-inclusive manual provides a revolutionary route to optimizing solar power system capabilities. The Z-source inverter offers a flexible way to get over conventional restrictions related to solar energy conversion and integration because of its distinctive topology and working principles. This book explains how Z-source inverters provide improved efficiency, robustness, and adaptability in solar power generation installations through in-depth examination and useful insights. Additionally, it investigates how well they can handle issues like voltage swings, dynamic load demands, and grid instability, opening the door to a more durable and sustainable energy infrastructure. This publication is a vital tool for academics, engineers, politicians, and industry stakeholders who are working to hasten the world's shift to a solar-powered future. Its interdisciplinary approach crosses the fields of physics, engineering, and renewable energy.

Renewable energy, such as solar power derived from the sun, undergoes conversion into electrical energy via solar panels. This process yields direct current (DC),

C. Maheswari (✉)
Department of Electrical and Electronics Engineering, PSG College of Technology, Coimbatore, Tamil Nadu, India
e-mail: cmi.eee@psgtech.ac.in

V. Shobana
Department of Computer Science with Cyber Security, Dr. N.G.P. Arts and Science College, Coimbatore, Tamil Nadu, India
e-mail: shobana@drngpasc.ac.in

M. Savithri
Department of Data Science, Christ University, Bengaluru, Karnataka, India

S. Vijayalakshmi et al. (eds.), *Smart Power Systems*, Power Systems,
https://doi.org/10.1007/978-3-031-89459-6_6

which can then be stored in a battery. A Z-source inverter is active to transition the DC current into alternating current (AC), facilitating single-phase AC supply provision. This project's main goal is to create a Z-source inverter that is especially suited for solar power conversion [1]. To achieve this, components such as PV panels, a boost converter, the Z-source inverter, and a microcontroller are utilized. The project necessitates both 5 V and +15 V DC supplies. To obtain the +15 V AC supply from a 230 V AC source, a step-down transformer is utilized. Subsequently, a bridge rectifier converts the AC supply into DC, with a capacitor employed for AC signal filtering. Further regulation of the DC supply is achieved using 7805 and 7812 regulator ICs, yielding +5 V and +15 V DC supplies, which are common to all circuits involved.

PV panels are utilized to harness DC supply from sunlight, albeit with a small and variable output. To amplify this power source, a boost converter is employed to convert low voltage to high voltage. This circuit incorporates MOSFETs for switching purposes, enabling the energization of the inductor to yield high voltage through frequency modulation. The Z-source inverter is crucial for converting DC supply into AC supply using MOSFETs. By strategically switching the six MOSFETs in timed intervals, AC supply is generated. The inverter circuit comprises three MOSFETs for positive supply switching and another three for negative supply switching. An MOSFET driver circuit is employed to trigger the MOSFET gates, utilizing Optocoupler ICs to isolate high voltage DC from low voltage signals originating from the microcontroller. Additionally, MOSFET driver ICs are utilized to gate trigger the MOSFETs by delivering +15V pulses. The microcontroller, DSPIC30f2010, is responsible for frequency generation to boost the PV panel's supply and for orchestrating MOSFET switching sequences to invert DC supply to AC supply. The hardware model is also constructed and tested, and the results were also compared.

2 Problem Statement and Solution

A more advanced approach for harnessing solar energy is through the utilization of photovoltaic solar cells. Integrating multiple solar panels in series along with a conventional voltage source inverter offers a straightforward method to enhance the output voltage of solar energy systems. However, conventional inverters draw power from the grid and tend to diminish in efficiency over time. Despite its effectiveness, the traditional voltage source inverter design presents several drawbacks, notably in terms of reliability and efficiency. To address these technological challenges, various power electronic circuit topologies have been proposed as potential solutions.

As emerging energy sources, both photovoltaic (PV) arrays and fuel cells demonstrate significant potential across various applications, including residential settings. In the realm of solar power generation systems, there's considerable promise in replacing traditional Voltage Source Inverters (VSI) with Z-source inverters to enhance overall performance.

The newly proposed Z-source inverter introduces a unique operational mode known as shoot-through, a feature not typically found in conventional voltage source inverters. This mode enables an increase in output voltage, offering distinct advantages for both fuel cell and solar panel applications. Notably, this feature presents a more cost-effective and straightforward single-stage method, thereby enhancing the dependability of the converter.

3 Existing Methodology

A power electronic circuit called an inverter is made to transform DC electricity into AC power of the appropriate frequency and amplitude. These devices are widely used in uninterruptible power supply (UPS) and contemporary AC motors. In particular, DC to AC inverters adapt direct current into alternating current in order to create inversion.

Inverters are categorized as DC-AC inverters or AC-AC cyclo-converters based on whether the circuit's output is DC or AC. In specifically, DC to AC inverters produce AC output with varying or set frequency and magnitude. These inverters have the ability to generate three-phase or single-phase AC electricity. The usual range of AC voltage magnitude is 110–380 V AC, and the frequency range is 50 Hz, or 60 Hz, or 400 Hz.

Figure 1 displays the circuit diagram along with the DC-AC converter and filters to reduce harmonic distortion and provide an output AC signal with no distortion. Converters from DC to AC are positioned at the front of the circuit. To produce the intended pure sinusoidal AC voltage, extra filter circuits are used in situations when the yield of AC voltage is more distorted than the reference AC voltage.

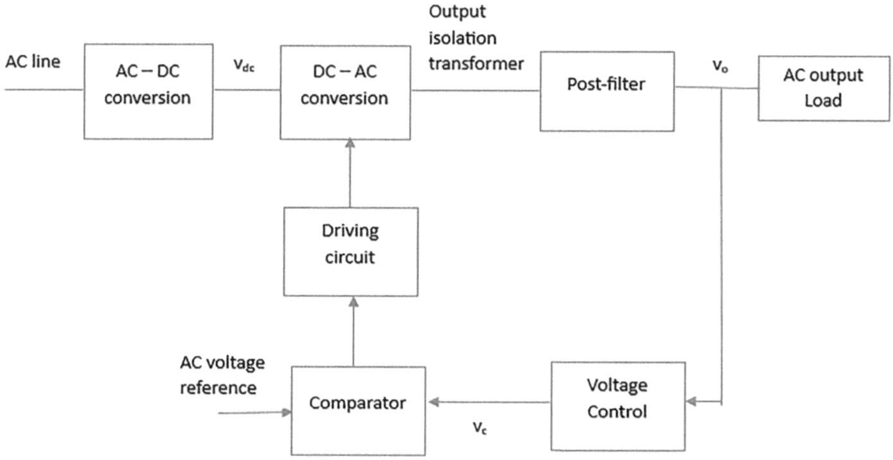

Fig. 1 Power electronic circuit diagram with DC to AC inverter

4 Proposed Methodology

The Z-source inverter-based conversion system for solar power has a unique impedance network in addition to a number of additional benefits that increase its efficiency and dependability when converting solar energy. Its improved capacity to adapt to fluctuations in solar energy generation is one obvious benefit. Conventional voltage source inverters frequently experience difficulties with input voltage variations, which can result in inefficiency and system damage. But because Z-source inverters can handle a larger variety of input voltages by nature, they perform better in this regard. This adaptability adds to the system's overall stability and lifespan in addition to guaranteeing consistent performance in variable sun conditions.

The Z-source inverter architecture also makes it easier to integrate energy storage devices like batteries smoothly, which increases the adaptability and durability of the system. By adding energy storage, the system may maximize energy use and grid interaction by storing excess solar energy during high-generation times and delivering it at low-generation or peak-demand times. In off-grid or hybrid solar systems, where dependable energy storage is necessary for a continuous power source, this capacity is very beneficial. Moreover, the Z-source inverter's adaptability to various solar panel layouts and climatic circumstances is increased by its greater operating voltage range. The resilient performance of the Z-source inverter maximizes the total efficacy and output of the solar power system by ensuring consistent power conversion and delivery even in the face of variable sunlight intensity, shading effects, or panel mismatches. All things considered, the Z-source inverter-based solar power adaptation scheme, with its unmatched efficiency, dependability, and flexibility, marks a substantial development in solar energy technology. Its cutting-edge features and design open the door to a more robust and sustainable energy future, propelling solar power's universal acceptance as a workable approach to balance environmental effect and fulfil the world's expanding energy needs.

The DC foundation or load could be moreover a power source or a current source. The DC source can be any of the following: a battery, diode rectifier, thyristor converter, fuel cell, inductor, capacitor, or a mixture of these due to its versatility.

The general block architecture of the planned Z-source inverter is depicted in Fig. 2, and the functions of each component are explained below.

4.1 Components Description

4.1.1 PV Panel

A preassembled device consisting of six by ten photovoltaic solar cells is called a photovoltaic (PV) module. These modules are the fundamental parts of a solar energy system's photovoltaic array. Harvesting sunlight and turning it into power is

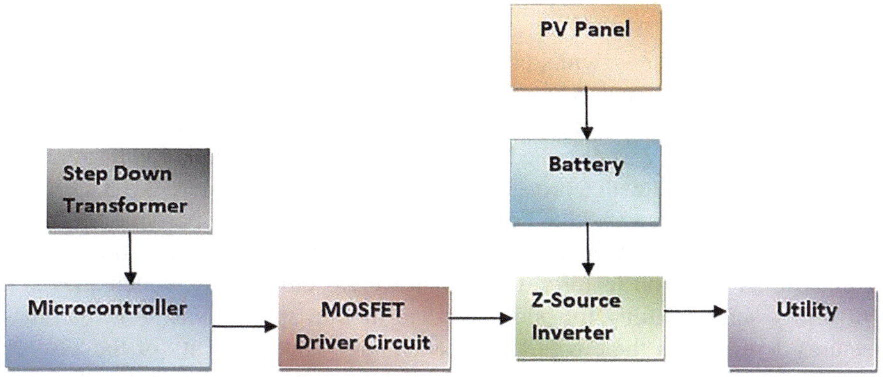

Fig. 2 Overall block diagram of the proposed system

their main job. Then, this electricity can be used to offer solar-generated power in a variety of situations, including residential and commercial ones.

Under standard test circumstances (STC), each module is rated according to its DC affair power, which typically falls between 100 and 365 W (W). Given an identically rated affair, a module's effectiveness dictates its area; a module with an effectiveness of 8 will have double the area of a module with an effectiveness of 16. The specifications of the PV panel in question outline important features that are necessary for its efficient use. When operating under ideal conditions, the panel can produce a substantial amount of electrical energy, as seen by its maximum power output of 5.00 W [2]. When the panel is not connected to any load, its open circuit voltage of 21.00 V indicates the voltage across its terminals, giving information about its potential to generate voltage. The greatest current that the panel can provide under a zero-resistance load is indicated by the short circuit current of 0.34 A, demonstrating its capacity to produce current.

4.1.2 Battery

Lead-acid batteries are known for their simplicity and cost-effectiveness in manufacturing. These secondary batteries have proven their reliability, having been utilized for more than 140 cycles. They are widely manufactured and well-understood technology. When used properly, they offer durability and dependability. Additionally, their self-discharge frequency is minimal compared to other rechargeable battery systems. Furthermore, lead-acid batteries are environmentally friendly, with a remarkably high reclamation rate of 98%. They require low maintenance, as they have no memory effect and no need to refill electrolytes in sealed versions.

Specifications

The Z-source inverter system's physical characteristics and important parameters are described in detail in its specifications, which are necessary to comprehend how it works and integrates into different applications. The system, which has a nominal voltage rating of 12 V, is designed to function within conventional voltage ranges, making it compatible with a broad range of loads and power sources. Its small size of 15.1 × 6.5 × 9.4 cm (5.95 × 2.56 × 3.7 in.) highlights how portable and space-efficient it is, making installation and deployment in a variety of settings simple. Even with its powerful features, the system weighs only 2.7 kg, which adds to its portability and versatility for both fixed and mobile applications. Together, these features demonstrate the efficiency, dependability, and versatility of the Z-source inverter, establishing it as a top choice for effective power conversion and the use of renewable energy in a variety of contexts and industries.

Advantages and Disadvantages

Lead-acid batteries are normally used because of its

(a) Easy and cheap manufacturing: It is easy and inexpensive to manufacture lead-acid batteries. They have been produced all around the world for more than 140 years and are known as dependable, sophisticated secondary batteries with a well-understood technology.
(b) Durability and dependability: Lead-acid batteries are dependable and durable when used properly. Among rechargeable battery systems, their self-discharge rate is among the lowest.
(c) Rapid discharge rates: Because lead-acid batteries have a rapid discharge rate, they can provide the energy bursts required to start engines.
(d) High energy density and long-life cycle: They have one of the longest life cycles among products in their price range and offer a substantial quantity of energy per pound.
(e) Environmental sustainability: Due to its high recycling rate, lead-acid batteries help maintain the environment.

4.1.3 Z-Source Inverter

Power conversion technology has advanced significantly with the Z-source inverter, especially in the area of solar energy utilization [3]. In contrast to traditional inverters, which usually need independent DC-DC converters to control voltage levels, the Z-source inverter is unique in that it can accomplish buck-boost inversion without a DC-DC stage. In addition to streamlining the architecture of the entire system, this simplified design maximizes efficiency by reducing energy losses linked to extra conversion stages. The Z-source inverter's distinctive impedance (Z-) source network is its core component and is essential to enabling effective power conversion in a range of applications. Through the use of this specialized circuitry, the Z-source inverter is able to reliably and flexibly manage a variety of power conversion tasks, such as DC-DC, DC-AC, AC-DC, and AC-AC conversions [4]. Since its adaptability, it can be used in an extensive series of energy systems, which include solar power generation and electric vehicle propulsion, where reliable and efficient power conversion is essential.

The Z-source network has several benefits, one of which is its capacity to enhance voltage when needed. This capability is especially helpful in situations where the power source's input voltage, such as solar panels, fluctuates or drops below what's needed for an efficient power conversion. Even in harsh climatic circumstances, the Z-source inverter can maximize energy harvest and system performance by cleverly adjusting the impedance network to effectively raise the input voltage and assure optimal operation [5]. Further enhancing its attractiveness in real-world applications is the Z-source inverter's intrinsic capacity to manage voltage levels without the need for complicated control algorithms or additional components. The Z-source inverter's resilient and adaptable performance is evident in its ability to optimize power transfer in electric vehicles, stabilize voltage fluctuations in renewable energy systems, and balance load demands in microgrids. This in turn propels the development of clean energy technologies and cultivates a more sustainable energy landscape. All things considered, the Z-source inverter is a revolutionary power adaptation technology that surpasses the constraints of conventional power conversion techniques and provides unmatched efficiency, versatility, and dependability for a wide range of uses. In addition to paving the way for more effective solar power generation, its cutting-edge circuit design and built-in capabilities have the potential to completely transform the field of power electronics and open up innovative avenues for the integration of clean, renewable energy sources and global grid modernization initiatives.

4.1.4 The Pic Microcontroller

Utilizing Complementary Metal-Oxide-Semiconductor (CMOS) technology, the PIC microcontroller is a RISC (Reduced Instruction Set Computer) based microcontroller. It uses distinct buses for data and instruction, allowing programme and data memory to be accessed simultaneously. There are several memory alternatives available for microcontrollers such as the PIC, FLASH memory, EPROM (Erasable Programmable Read-Only Memory), and EEPROM (Electrically Erasable Programmable Read-Only Memory). The DSPIC 30F2010, for instance, is built using FLASH technology, ensuring data retention even when power is turned off. Programming and erasing operations are simplified with the DSPIC 30F2010 compared to other memory types. The PIC microcontroller is notable for being the first RISC-based microcontroller fabricated in CMOS technology, a feature that allows for efficient memory access with separate buses for instructions and data. This architecture enhances the microcontroller's performance and versatility.

Table 1 DSPIC 30F2010 PIN configuration

Pin No.	Pin name	Function
1	VDD	Positive supply voltage
2	VSS	Ground
3	OSC1/CLKI	Oscillator input or external clock input
4	OSC2/CLKO	Oscillator output or clock output
5	MCLR/VPP	Master clear (active low) or programming voltage input
6	AN0/RA0	Analog input 0 or general purpose I/O (GPIO)
7	AN1/RA1	Analog input 1 or GPIO
8	AN2/RA2	Analog input 2 or GPIO
9	AN3/RA3	Analog input 3 or GPIO
10	VSS	Ground
11	AVDD	Analog positive supply voltage
12	AVSS	Analog ground
13	AN4/RA4	Analog input 4 or GPIO
14	AN5/RA5	Analog input 5 or GPIO
15	VDD	Positive supply voltage
16	PGD/RB0	Programming data line or GPIO
17	PGC/RB1	Programming clock line or GPIO
18	RB2/INT0	GPIO or external interrupt 0
19	RB3/INT1	GPIO or external interrupt 1
20	RB4/INT2	GPIO or external interrupt 2
21	PWM1H/RB5	PWM output (high side) or GPIO
22	PWM1L/RB6	PWM output (low side) or GPIO
23	PWM2H/RB7	PWM output (high side) or GPIO
24	PWM2L/RB8	PWM output (low side) or GPIO
25	PWM3H/RB9	PWM output (high side) or GPIO
26	PWM3L/RB10	PWM output (low side) or GPIO
27	PGEC/RB11	Programming and debugging emulation clock line or GPIO
28	PGED/RB12	Programming and debugging emulation data line or GPIO

DSPIC 30F2010 PIN Configuration

See Table 1.

Motor Control PWM Module in DSPIC

For a variety of claim, such as three-phase AC induction motors, three-phase brushless DC motors, and switching reluctance motors, the DSPIC motor control PWM module is specially designed. The motor control PWM module has three or four PWM generators and six or eight output pins, depending on the device [6].

These output pins provide versatility in motor control settings by being able to be set up as separate outputs or as complimentary output pairs. Important PWM

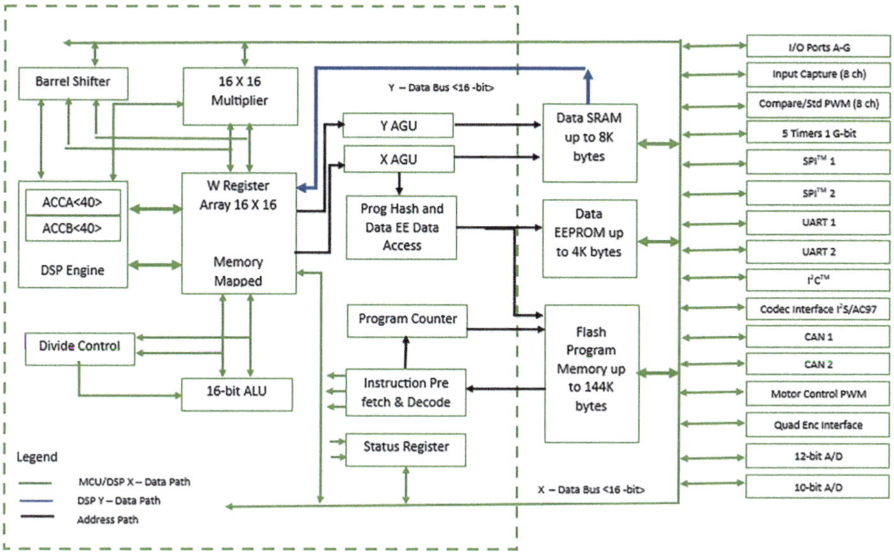

Fig. 3 Dspic30f2010 architecture functional block diagram

operational settings, including output polarity, are configurable and kept in non-volatile memory to guarantee safety and dependability. This function improves the overall reliability of the motor control system and guards against unintentional modifications to PWM settings (Fig. 3). Figure 3 represents the data paths and components like 16 X 16 multiplier, W register array, 16-bit ALU and program counter. It also includes the memory components data SRAM, data EEPROM and flash program memory. The diagram also includes input/output ports, timers and interfaces like UART and CAN. Arrows in the diagram indicates the flow of data, with a legend explain the paths.

Features of DSPIC MICRO Controller

High-Performance Modified RISC CPU

This high-speed, customized RISC CPU is designed with cutting-edge features that optimize performance and adaptability in computing applications. Using a Harvard architecture, it combines variable addressing modes with an instruction set architecture that has been painstakingly tuned by the C compiler to provide efficient execution of complicated instructions. A 16-bit wide data channel and 24-bit wide instructions further enhance its data handling capabilities, enabling quick and accurate data processing. The CPU has sufficient storage for programme instructions thanks to its 48 Kbyte (or 16 K instruction words) onboard Flash programme space, which improves its ability to perform a variety of computational tasks. Furthermore, the addition of 2 Kbytes of RAM offers necessary on-chip data storage, facilitating fast access to non-volatile data necessary for flawless functioning. In addition to these functions, the 1 Kbyte EEPROM guarantees the safe preservation of important information, enhancing the CPU's endurance and strong performance in demanding computing settings. All things considered, this high-performance modified RISC CPU is a state-of-the-art answer created to satisfy the intricate computational requirements of contemporary applications while offering improved dependability and efficiency.

Features of Motor Control PWM Module

- Six channels for PWM output
- Center-aligned modes and independent or complementary output modes
- Three generators for duty cycles
- Time-based dedication
- Polarity of the programmable output
- Deadline management in complementary mode
- Control of output manually
- A/D conversion trigger

Peripheral Features

The system's add-on features expand its capabilities and adaptability to meet a variety of application needs. Rich communication and interoperability with external devices and components are provided by the system's I/O pins, which can operate at high current sink/source levels of 25 mA/25 mA. Precise timing control, made possible by the availability of a pre-scalable programmable timer module, permits synchronized operation with a variety of peripherals and system components [7]. Moreover, precise timekeeping and event monitoring in real-time applications are made easier by the five 16-bit counters and timers that are included, along with the choice to couple 16-bit timers with 32-bit timer modules. The system can more effectively record and analyze external signals or events when 16-bit input functions are available. PWM (pulse-width modulation) may be precisely controlled for motor speed, LED brightness, and other analogue applications thanks to the system's 16-bit compare/PWM output functionalities. Reliable serial communication is ensured by the presence of two FIFO-buffering UART modules, which provide smooth system contact with external devices such sensors, displays, and communication modules. Strong and effective communication in distributed control systems, automotive applications, and industrial automation is made possible by the inclusion of a CAN module with 2.0B compatibility, which improves the system's networking capabilities. All things considered, these auxiliary features improve the system's operation, increase its usefulness, and allow for easy integration into a variety of applications in a variety of industries.

Input Capture Module

The input capture module, described in this section, offers various operational modes tailored for applications necessitating frequency (period) and pulse measurement. This module provides features that are valuable for accurately measuring the frequency and duration of input signals.

USART

The USART, an acronym for Universal Synchronous/Asynchronous Receiver/Transmitter, serves as a vital communication device widely utilized in diverse electronic systems. It is commonly employed in microcontrollers, certain cell phones, barcode readers, and personal computers. Despite its seemingly complex name, the USART plays a fundamental role in facilitating communication between devices, enabling data transmission and reception in both synchronous and asynchronous modes.

IC Voltage Regulators

One type of integrated circuit that is frequently utilized is the voltage regulator (IC). These integrated circuits (ICs) for regulators combine multiple circuit elements into a single IC package, such as control devices, comparator amplifiers, reference sources, and overload protection. They provide regulation for flexible voltage settings, fixed positive voltage, or fixed negative voltage. These regulators may operate with ratings of power ranging from milliwatts to tens of watts, and load currents spanning from several hundred milliamperes to tens of amperes.

Fixed positive controlled voltages within 5 and 24 V and fixed negative regulated voltages within the same voltage range are provided by the series 78 regulators.

4.1.5 MOSFET Driver Circuit

MOSFET

Source (S), gate (G), drain (D), and body (B) are the four terminals of a MOSFET, also referred to as a metal-oxide-semiconductor field-effect transistor. A field-effect transistor-like three-terminal device is created when the MOSFET's body is linked to the source terminal. MOSFETs are widely employed in many different applications in both analog and digital circuitry.

The MOSFET driver section consists of

- Optocoupler section
- Current boosting section

Optocoupler Section

Opto-isolators, sometimes referred to as optical isolators, optocouplers, photocouplers, or photo MOSs, are electronic devices that allow a signal to be sent between various circuit parts while preserving electrical isolation. This is accomplished by sending the signal between a transmitter and a receiver along a brief optical transmission line. By first converting an electrical signal to a signal that is optical and then back into an electrical signal, the opto-isolator ensures electrical isolation between the connected components by effectively breaking electrical contact along the transmission channel.

Current Boosting Section

In electronics, a current source is a component that either emits or receives electric current. It serves as a voltage source's complement. The term "constant-current sink" can also refer to sources that receive power from a negative voltage source. Regardless of changes in the load resistance or other environmental circumstances, these sources can sustain a constant output current.

5 Results and Discussions

The proposed Z-source inverter is designed and simulation was done using MATLAB SIMULINK, and the simulation results obtained using MATLAB, a powerful language for technical computing, showcase the performance of the system under study. Programming, calculation, and visualization are all seamlessly integrated in MATLAB's user-friendly environment, enabling the articulation of issues and their resolutions in well-known mathematical notation. Through MATLAB, engineers and researchers can effectively analyze and interpret data, enabling the optimization and validation of various systems and processes.

5.1 Photovoltaic Circuit

In the context of the Z-source inverter architecture, this simulation model is a crucial tool for assessing and improving the performance of the solar power generating system. Through Simulink integration of the photovoltaic circuit, scientists and engineers are able to assess the efficiency and dynamic behavior of the solar energy conversion process on a broad scale. With the use of this model, it is possible to thoroughly examine variables including solar irradiance levels, panel properties, and system response under various operating scenarios. Moreover, it makes it easier to investigate control plans and system setups that maximize energy harvesting and grid integration effectiveness.

5.2 MPPT Subsystem

In order to maximize the interaction among a solar array, such as solar panels, and a bank of batteries or utility grid, MPPT functions as an essential electrical DC to DC converter. MPPT systems work by effectively converting the increased voltage DC output produced by solar panels, and occasionally wind generators, into the smaller voltage needed for efficient battery charging. MPPT systems optimize energy harvesting and raise overall system efficiency by continuously monitoring and modifying the solar panels' operating point to provide their highest power output. SIMULINK modelling for the MPPT subsystem offers a thorough framework for evaluating and optimizing the MPPT algorithm's performance in relation to the overall solar energy generation system. With the help of this model, engineers and researchers may simulate different MPPT algorithms, assess how well they work in diverse environments, and adjust their settings to extract the most energy possible. Additionally, it makes it easier to incorporate MPPT control schemes with the photovoltaic circuit and Z-source inverter, allowing for a comprehensive approach to system optimization and design.

6 Simulation Results

6.1 Input Voltage and Current Waveforms

MATLAB Simulink is used to simulate the intended Z-source inverter model. The simulation results in contrast with the hardware module, and the output wave shapes are shown below.

The Z-source inverter's steady input voltage as well as current waveforms is displayed in Fig. 4.

6.2 DC Converter Voltage

The simulation output wave form for a DC link with respect to time.

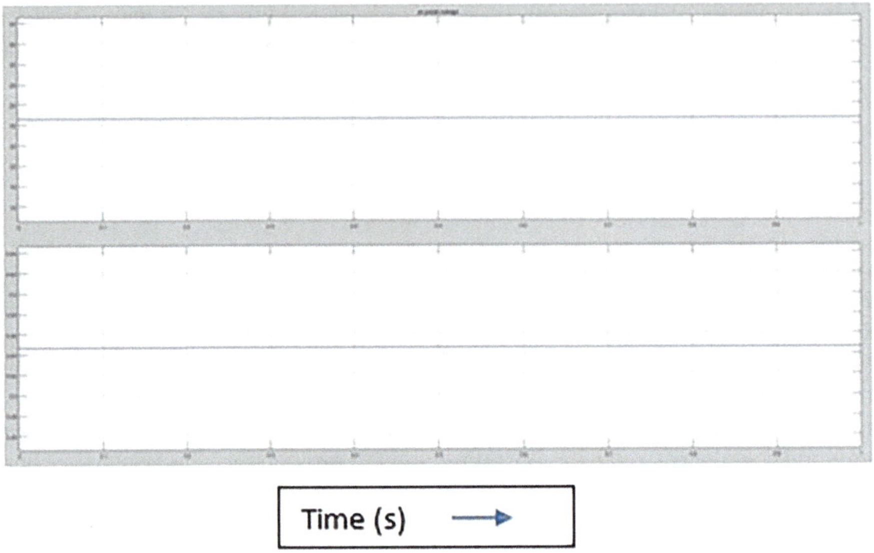

Fig. 4 Performance of Input Voltage and Current

6.3 Inverter Phase Voltage

The simulation results for crucial parameters providing information about the Z-source inverter system's performance. A thorough visualization of the input voltage and current characteristics shows how these crucial parameters change throughout the course of system operation. Understanding how the Z-source inverter interacts with the power source—such as solar panels or batteries and evaluating the system's total energy harvesting capacity depend on this data.

7 Hardware Set Up

The suggested Z-source inverter hardware module's hardware block diagram, shown in Fig. 5, gives a thorough rundown of all of its essential parts and functional design. First, electrical energy is extracted from the source, which is usually solar panels or another renewable energy source. After being collected, this DC voltage is fed into a DC-DC converter, where it is subjected to the appropriate voltage control and conditioning to guarantee compatibility with components further down the line. The DC voltage is simultaneously directed to a battery backup system, where it is kept for later use. Energy storage is integrated into the system to improve its reliability and allow for seamless operation in times of high demand or low energy generation. Additionally, the DC voltage saved can be used directly for certain applications or as a backup power source as necessary. The Z-source inverter then

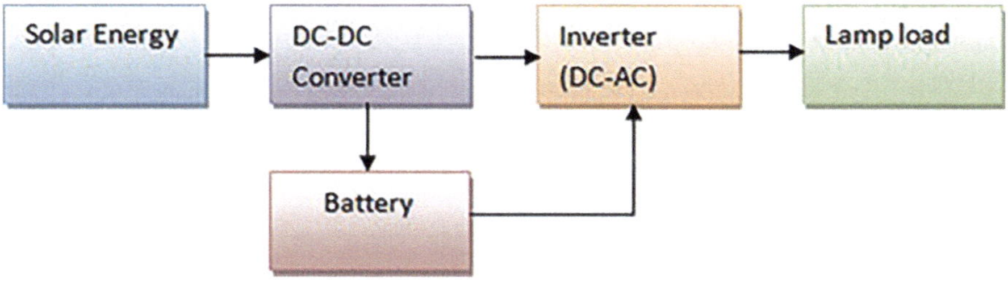

Fig. 5 Block diagram of hardware model

Table 2 Specification of components used in hardware module

Components	Specifications
Solar panel	12 V, 5 W
Battery	12 V, 7A/H
Transformer	(0–15 V) (0–45 V)
MOSFET driver IC	TLP 350
MOSFET	IRF840
Microcontroller	PIC30F2010

receives the DC voltage and converts it into AC voltage that can power a variety of loads. An important component of this process is the inverter, which uses cutting-edge Z-source topology to enable dependable and effective DC to AC conversion. Through the use of this special design, an inverter can manage input voltage variations and provide a consistent AC output even in a variety of operating environments. In the end, the load is powered by the AC output that the inverter generates, giving electrical power for a variety of uses. The Z-source inverter hardware module provides a stable and adaptable way to capture renewable energy and supply clean, dependable power to satisfy various energy requirements, whether it is used for homes, businesses, or factories.

7.1 System Specifications

The specifications of the components used in the hardware module for the proposed Z-source inverter are given in the following table (Table 2).

7.2 Solar Panel

The specifications and rating of the solar panel used for the proposed inverter are given in Table 3.

Table 3 Ratings of the solar panel used in the Z-source inverter

Nominal	12 V
Number of cells (max)	36
Open circuit voltage (Voc)	22 V
Max power volts (Vmp)	18 V

Table 4 Ratings of the battery used in the Z-source inverter

Voltage	12 V
Current	7AH
Height	3.7
Weight	4.8(lb)

Table 5 Ratings of the transformer used in the Z-source inverter

Input voltage	230 V AC
Output voltage	15 V
Output current	2 A
Mounting	Vertical mount type
Winding	Copper

7.3 Lead-Acid Battery

Ratings of the lead-acid battery used for this inverter is given in Table 4. These specs point to a small and light battery system that can be used in cars or for a variety of other uses, including small-scale renewable energy storage, portable devices, and more. It is compatible with a broad variety of devices and systems because to its 12 V voltage, which is in line with typical battery voltage specifications. A moderate capacity, appropriate for short-term backup power or device powering, is indicated by the current rating of 7AH. The lightweight design (weighing 4.8 lb) guarantees portability and simplicity of handling, while the tiny proportions (3.7 in. in height) improve versatility and ease of installation (Table 4).

7.4 Transformers

Table 5 provides the details for the rating of the transformer used in Z-source inverter.

8 Step-by-Step Process for Proposed Hardware Model

The step-by-step procedure for the implementation of the Z-source inverter is given below.

- STEP 1: System is turned ON for suitable application.
- STEP 2: Solar energy received from the sun is collected in PV cells.

- STEP 3: The collected voltage is stored in DC batteries.
- STEP 4: Z-source inverter is used to convert DC supply to effective AC voltage.
- STEP 5: This enhanced voltage source is used to glow the lamp.
- STEP 6: End the process.

9 Hardware Output

According to the project is to provide the single-phase AC supply from solar panel using Z-source inverter. In that we are using two transformers for step down the voltage to giving constant supply to PIC microcontroller and MOSFET drive circuit.

PIC microcontroller is used to control the Z-IMPEDENCE and MOSFET driver circuit. The MOSFET driver circuit getting 15 V constant supply from step-down transformer controls the switching process of the MOSFET.

Initially the supply from solar panel is stored in 12 V battery then giving supply to Z-IMPEDENCE for boost up the voltage. The boosted voltage from Z-IMPEDENCE is going to MOSFET for switching the supply DC to AC using MOSFET driver circuit. Thus, the lamp glows through single-phase AC supply from the inverter

10 Conclusion and Future Scope

The ZSI-based solar power conversion system has been modelled and controlled for single-phase AC supply. Specifically designed for a range of applications, that uses three-phase AC induction motors, the DSPIC motor control PWM module, etc. A single-phase AC supply has been modelled and managed for the ZSI-based solar power conversion system. ZSI uses a simple boost control that is field orientated based. In contrast to the traditional two-stage power conversion method, the suggested system utilizes a single-stage ZSI to optimize PV array power extraction and power a single-phase AC supply for diverse applications. The MATLAB Simulink is used to obtain the simulation results, which are then compared to the hardware results. Under a wide range of solar radiation, the created system is thought to be energy efficient and offers great control with rapid response times, minimal switching losses, and good stability capabilities.

11 Future Scope

The future scope for Z-source inverters in solar power systems is promising, driven by their potential to enhance efficiency, reliability, and flexibility in power conversion and integration. As research and development efforts continue to advance, we can expect further innovations and broader adoption of Z-source inverters across various solar energy applications.

Z-source inverters could be employed in solar-powered EV charging stations to efficiently convert solar energy into electricity for charging EV batteries, thereby promoting clean transportation.

In remote areas or regions with unreliable grid infrastructure, microgrids powered by solar energy can provide a viable alternative. Z-source inverters can be instrumental in such microgrid setups by offering voltage regulation, power conditioning, and seamless transition between grid-connected and islanded modes of operation.

References

1. Meshram, S., Agnihotri, G., & Gupta, S. (2012). The steady state analysis of Z-source inverter based solar power generation system. In *2012 IEEE 5th India International Conference on Power Electronics (IICPE)*. IEEE.
2. Sharma, V., Hossain, M. J., Ali, S. M. N., & Kashif, M. (2020). A photovoltaic-fed Z-source inverter motor drive with fault-tolerant capability for rural irrigation. *Energies, 13*(18), 4630.
3. Dubey, M., Sharma, S., & Saxena, R. (2018). Solar power based water pump employing Z-source inverter for PMSM drive. In *2018 8th IEEE India International Conference on Power Electronics (IICPE)*. IEEE.
4. Peng, F. Z., Joseph, A., Wang, J., Shen, M., Chen, L., Pan, Z., OrtizRivera, E., & Huang, Y. (2005). Z-source inverter for motor drives. *IEEE Transactions on Power Electronics, 20*(4), 857–863.
5. Antonello, R., Carraro, M., Costababer, A., Tinazzi, F., & Zigliotto, M. (2017). Energy–efficient autonomous solar water–pumping system for permanent-magnet synchronous motors. *IEEE Transactions on Industrial Electronics, 64*(1), 1.
6. Liang, X., & Ilochonwu, O. (2011). Induction motor starting in practical industrial applications. *IEEE Transactions on Industry Applications, 47*(1), 271–280.
7. Adamowicz, M. (2011). LCCT-Z-source inverters. In *10th International Conference on Environment and Electrical Engineering (EEEIC)*. IEEE.

Smart Wide Area Monitoring of System Operations

Sivakumar Vengusamy, Mafas Raheem, and Justin Gilbert Alexius Silvester

1 Introduction to Smart Wide Area Monitoring

The term "Smart Wide Area Monitoring (SWAM)" describes the sophisticated, technologically based method of keeping an eye on and controlling large-scale systems, such as transportation, water distribution, and electricity grids. It entails gathering, analysing, and interpreting data in real time from a large geographic region. Subsequently, this data is utilized to guarantee the integrity and dependability of vital infrastructure, optimize system operations, and make well-informed decisions.

SWAM is important because it can give real-time information on the health and performance of these intricate systems, which may lead to proactive decision-making and improved overall dependability. It is impossible to overestimate the significance of smart wide area monitoring. For modern communities to run smoothly, sophisticated systems like transportation and electrical networks are essential. Any interference with or malfunction in these systems might have serious repercussions, including monetary losses and risks to public safety. In order to prevent problems in complex systems, recognize them, and take prompt, efficient action in response, smart wide area monitoring is essential.

In a time when essential infrastructure is quickly changing due to urbanization and technological breakthroughs, traditional approaches of controlling and monitoring these systems are no longer sufficient. These infrastructures' size, complexity, and interconnectedness necessitate a more advanced and flexible strategy. This requirement is met by smart wide area monitoring, which uses cutting-edge technology to construct intelligent monitoring systems, including the Internet of Things (IoT), sophisticated sensors, big data analytics, and artificial intelligence.

S. Vengusamy (✉) · M. Raheem · J. G. A. Silvester
School of Computing, Asia Pacific University of Technology and Innovation (APU), Kuala Lumpur, Malaysia
e-mail: dr.sivakumar@apu.edu.my; raheem@apu.edu.my; justin.gilbert@apu.edu.my

S. Vijayalakshmi et al. (eds.), *Smart Power Systems*, Power Systems,
https://doi.org/10.1007/978-3-031-89459-6_7

Early identification of abnormalities and possible problems in essential infrastructure is made possible by SWAM. This reduces downtime and boosts overall system dependability by enabling proactive action before these problems worsen. Massive volumes of data are produced by SWAM, and when correctly analysed, these data can offer insightful information to decision-makers. System activities are performed more effectively and efficiently thanks to this data-driven methodology. By recognizing weaknesses and giving critical infrastructure the capacity to respond to and recover from disruptions more efficiently, SWAM improves the overall resilience of the system. To safeguard against cyber-attacks and guarantee the accuracy of the data being watched, smart monitoring systems can integrate cutting-edge cybersecurity safeguards.

2 Overview of System Operations Monitoring

Fundamentally, system operations monitoring is the ongoing observation and management of a certain system. To make ensuring the system runs within safe and effective bounds, this involves monitoring a variety of parameters and variables, including voltage, current, temperature, pressure, and flow, in the context of large-scale systems. System operations monitoring has historically depended on basic alarm systems, recurring inspections, and human data collecting. These approaches, however, have shown to be inadequate as systems have increased in complexity and size. The problems caused by complex and networked infrastructures need a more sophisticated strategy. System operations monitoring is the process of continuously observing and managing important system parameters to guarantee a reliable, safe, and effective system. This procedure is essential to many industries, including manufacturing, water management, transportation, and energy. The main objectives of system operations monitoring are as follows:

(a) **Efficiency optimization:** Monitoring ensures that systems run as efficiently as possible by helping to optimize the utilization of resources. This is especially important for energy-intensive sectors or those requiring complex logistical processes.
(b) **Predictive maintenance:** Ongoing observation makes it possible to see possible problems before they become serious enough to cause system breakdowns. This helps with predictive maintenance, which lowers downtime and increases equipment longevity.
(c) **Safety and compliance:** Monitoring makes ensuring that systems run safely and in accordance with laws and regulations. This is particularly important for companies where worker safety is a top priority.

In the past, manual data gathering and sporadic inspections were part of traditional monitoring. This method's speed, precision, and capacity to manage big datasets were all limited; in contrast, advanced monitoring has developed because of the introduction of smart technologies and system operations monitoring. Advanced systems provide a more comprehensive and proactive approach to monitoring by combining real-time data collecting, automation, and advanced analytics.

Several parameters are examined across several industries. In power grids, parameters including voltage, current, frequency, and power factor are constantly monitored. Phasor measurement units (PMUs) are essential tools for observing and interpreting the electrical grid's dynamic behaviour. For efficient traffic management, vehicle speed, occupancy rates, and traffic flow are tracked. Air traffic control systems and aircraft performance metrics are included in the monitoring process in the aviation sector. To guarantee effective water distribution and treatment, flow rates, pressure, water quality, and reservoir levels are tracked. To ensure ideal manufacturing processes, monitoring includes variables such as temperature, pressure, machine performance, and product quality.

In circumstances where prompt response is needed, real-time monitoring is crucial. Real-time monitoring, for example, enables operators in the energy industry to react quickly to abrupt changes in supply or demand, averting possible blackouts or system breakdowns. A crucial element in modern system operations monitoring is automation. Without human input, automated systems can react to pre-set circumstances or triggers and making snap judgements and modifications. This lowers the possibility of human mistakes while also speeding up reaction times. System operations monitoring is becoming more and more dependent on data analytics. Large datasets may be analysed by advanced analytics techniques like machine learning and artificial intelligence to find trends, anomalies, and patterns. Proactive decision-making based on historical and current data is made possible by predictive analytics. The system operations monitoring with the latest developments gives rise to the trends involving:

(a) **IoT integration:** As more devices are connected for more thorough data collecting, the Internet of Things (IoT) is becoming more and more important in system operations monitoring.
(b) **Edge computing:** Increasingly popular, edge computing processes data closer to the data source to cut down on latency and speed up reaction times.
(c) **Digital twins:** More precise monitoring, analysis, and simulation are made possible by the idea of building digital twins, or digital copies, of physical systems.

Thus, one of the most important parts of managing complex infrastructures is system operations monitoring. The convergence of automation, data analytics, and smart technologies will further improve the dependability and efficiency of system operations in a variety of sectors as technology develops.

3 Smart Technology in Wide Area Monitoring

Smart technology has a revolutionary impact on monitoring. It improves the way we monitor, control, and manage vital systems by using the potential of digitalization, automation, and data analysis. The following are some significant ways that smart technology is essential to the monitoring of system operations:

Real-time data: Real-time data collection from a variety of sensors and devices dispersed across a large region is made possible by smart technology. A central monitoring system receives this data, which gives it a constant and up-to-date picture of the system's functioning.

Enhanced Visualization: Operators and decision-makers can more easily comprehend complicated data patterns thanks to the advanced data visualization capabilities that smart systems offer. Making decisions more quickly and intelligently is aided by this.

Optimization: Whether it's water, power, or transportation resources, smart technology can assist make the most use of them. Costs are reduced, and efficiency is raised as a result.

Advanced analytics: Intelligent monitoring systems employ sophisticated analytics, such as artificial intelligence and machine learning, to forecast system behaviour and identify irregularities before they become serious problems. By being proactive, downtime is reduced, and system resilience is increased.

We will explore the essential elements of smart wide area monitoring in the following sections of this chapter which serve as the foundation for a successful smart monitoring system and are crucial to guaranteeing the dependability and durability of vital infrastructure.

4 Components of Wide Area Monitoring

4.1 Networks of Sensors

The use of cutting-edge sensor networks for extensive data collecting is one of the keystones of smart broad area monitoring. These sensor networks, which can include an industrial complex, power grid, or transportation network, are made up of a large number of sensors and measuring devices that are carefully positioned across the monitored region. These sensors' job is to continually collect information on a number of important factors. Depending on the particular application, a wide range of sensors may be included in sensor networks. Numerous factors, such as voltage, electrical current, atmospheric pressure, humidity, flow, and air quality, may be measured using these sensors. Phasor measuring units (PMUs), for example, are essential for real-time voltage and current monitoring in the setting of a power grid.

4.2 Communication Infrastructure

A robust and dependable communication infrastructure is the foundation of intelligent wide area surveillance. It makes it easier for data to be sent from sensors to a central monitoring and control system, allowing for real-time reaction and analysis. The smooth transfer of data from sensors to a central data repository needs to be supported by the communication network. Both wired and wireless technologies, like satellite connections, cellular networks, and fibre optics, may be involved in this. For efficient handling of data traffic, the communication infrastructure should possess adequate bandwidth and low latency, contingent upon the volume of data and the requirement for real-time monitoring.

4.3 Data Storage and Management

Data generated by smart broad area monitoring is significant. For this data to be helpful in making decisions, effective data management and storage are necessary. Data that has been gathered is kept in databases or data repositories, frequently in the form of time series. In order to handle the increasing amount of data over time, the storage architecture needs to be scalable. Data retention rules, which specify how long data is kept on file as well as when and how it is archived or removed, must be defined by systems. Compliance, analysis, and resource management all benefit from this. Real-time and historical data analysis for anomaly identification and system optimization is made possible by the close integration of data storage with data analytics tools and platforms.

Together, the above components constitute the basis of smart wide area monitoring systems, facilitating the effective gathering, transfer, and administration of data that is vital to the stability and dependability of large-scale systems.

5 Data Analytics for System Operations Monitoring

System operations monitoring involves the surveillance and analysis of critical components within a specific network, software application, or broader IT infrastructure at periodic intervals over time. This extensive procedure tries to identify abnormalities, prevent any issues, and maintain maximum operational efficiency. It includes measuring error rates, system responsiveness, and server performance and resource utilization. System operations monitoring plays a vital role in protecting mission-critical systems from interruptions and guaranteeing their continuous reliability as organizations depend more and more on complex digital infrastructures [1].

One of the main difficulties in system monitoring lies in the enormous volume and diversity of data produced by modern technology. Traditional monitoring tools often struggle to handle and comprehend this overwhelming amount of data, which increases the risk of overlooking important abnormalities or potential issues. Additionally, as systems expand in size and complexity, the difficulty of early detection and root cause analysis intensifies, making it more challenging for organizations to maintain systems' optimal performance. Moreover, the dynamic technological environment of today adds a level of complexity. Changing workloads, usage patterns, and the ongoing evolution of hardware and software components are all factors that affect systems. Static monitoring strategies struggle to adapt to these changes, resulting in a reactive rather than a proactive approach to system management [2].

Implementation of data analytics in system operations monitoring brings a change in perspective on how companies handle system monitoring. Fundamentally, data analytics entails a systematic analysis of vast datasets to determine significant patterns and insights. This corresponds to a more sophisticated and flexible approach to system monitoring that tackles the difficulties presented by the modern technological environment. In addition, data analytics implementation has aided in early anomaly detection, predictive maintenance, dynamic monitoring strategies, and root cause analysis.

Anomaly detection involves identifying pattern and events that differ from a system's anticipated or typical behaviour. Data analytics creates a baseline of typical system behaviour using statistical methods and machine learning algorithms. Deviations from this baseline, for example, sudden spikes, unexpected trends, or irregular patterns will be flagged as anomalies. These algorithms will improve their accuracy in differentiating between typical and abnormal behaviour over time by learning from previous data.

Predictive maintenance refers to predicting when system components are prone to failure or repair, enabling organization to arrange preventive interventions before critical issues arise. Data analytics will aid by utilizing past data on the functionality and failure trends of system components. Machine learning models analyse this data to identify trends and correlations, predicting when specific components are likely to degrade or fail. By recognizing the variables leading to failure, organizations can optimize maintenance schedules and extend lifespan of critical assets.

Dynamic monitoring involves adjusting monitoring strategies in real time to adapt to changes in system environment. With data analytics, continuous analyse on system behaviour and performance metrics can aid organizations to set up automated processes that adjust monitoring thresholds, parameters, and alerting mechanisms in response to changing conditions. Machine learning models can help detect the patterns to optimize monitoring strategies according to needs of a specific moment [3].

Root cause analysis is the process of identifying and resolving the root causes of issues in system. Data analytics plays a crucial role in correlating data across various system variables. Analytics tools assist in identifying root cause of an issue by tracing the relationship between various variables and events. Organizations can identify the series of events that led to the issue by analysing past data, which makes the resolution process more effective.

5.1 *Data Analytics for Wide Area Monitoring*

Wide area monitoring involves the systematic observation and analysis of a large geographical area or a distributed network to ensure the optimal functioning and reliability of critical infrastructures. Wide area monitoring is specifically valuable in enhancing the adaptability of complex systems by providing a holistic view of the entire landscape, facilitating efficient decision-making. Data analytics are implemented in wide area monitoring some of its significant challenges [4].

Challenges	Description	Data-driven solution
Data overload	Large volumes of data are produced via wide area monitoring from several sources, like sensors, satellites, and communication networks. Real-time analysis of this stream is very difficult and might result in information overload and the loss of important insights	Data analytics techniques, including machine learning algorithms, assist in processing and interpreting vast datasets. Analytics tools cut through the noise, collecting actionable insights, and making sure that important information is not overlooked by identifying patterns, anomalies, and trends
Early detection of anomalies	Detecting anomalies in wide areas is challenging because it can be difficult to differentiate abnormalities from typical fluctuations in huge datasets	Advanced analytics tools excel in anomaly detection by establishing baseline behaviours and recognizing deviations. Machine learning algorithms are very efficient at identifying subtle irregularities, functioning as an early warning system to facilitate proactive action and stop these issues from getting worse
Dynamic environments	Wide area monitoring deals with environments that are naturally dynamic, depending on fluctuating conditions, varying weather patterns, and changing usage demands. Traditional monitoring approaches find it difficult to adjust to these changing environments	Data analytics introduces adaptive modelling that continuously learns from incoming data. Machine learning models can dynamically adjust to changing conditions, providing a more sophisticated understanding of the environment. This adaptability improves wide area monitoring's capacity to respond quickly to dynamic conditions
Network congestion	Network congestion in wide area monitoring can make it more difficult for data from several sensors and monitoring equipment to be transmitted quickly, especially in cities or busy areas	Data analytics can optimize data transmission by prioritizing critical information and compressing data. Intelligent algorithms can help manage network traffic effectively, ensuring that essential monitoring data reaches the central system without delays

Challenges	Description	Data-driven solution
Data quality and integration	Diverse data sources with different formats, quality levels, and update frequencies are used in wide area monitoring. It might be quite challenging to integrate and ensure the quality of this varied data	Data pre-processing and cleansing methods can be included in data analytics systems to manage missing data, standardize formats, and enhance overall data quality. Data from various sources can be combined using integration algorithms to create a unified dataset that is ready for analysis

Strategic use of data analytics for wide area monitoring transforms challenges into opportunities for increased adaptability and efficiency. Organizations may achieve proactive control and foresight over large and complex landscapes by utilizing machine learning, predictive modelling, and real-time analysis capabilities [5].

6 IoT in Wide Area Monitoring

Internet of Things has transformed wide area monitoring by seamlessly integrating connected devices and sensors to gather real-time data across large geographical areas. IoT facilitates the collection of data from various sources for wide area monitoring, which enables an in-depth knowledge of the environment. Integrating the Internet of Things (IoT) into wide area monitoring not only offers insightful data on the functioning of large-scale systems, but it also helps create more intelligent and environmentally friendly solutions for industrial and urban settings.

6.1 Role of IoT in System Operations

The Internet of Things (IoT) plays a crucial role in modernizing and optimizing system operations across different industries. IoT is a game-changer in system operations, bringing together machines, sensors, and devices to establish an intelligent, data-driven ecosystem. The roles of IoT in system operations include real-time monitoring, remote control and management, and increasing production efficiency.

S. No.	Roles	Description	Examples
1.	Real-time monitoring	Real-time monitoring facilitated by IoT provides the ability to continuously collect and analyse data to gain instantaneous insights into system performance	In manufacturing sectors, sensors embedded in machines can transmit data on operational parameters such as temperature, pressure, and vibration
			This provides a comprehensive view of system health and enables data-driven insights for optimization. Manufacturers can quickly identify anomalies or signs of equipment failure thanks to this real-time data
			As a result, timely maintenance can be initiated, reducing downtime, and preventing costly production interruptions
2.	Remote control and management	System operations significantly improved through remote control and management with the help of IoT. It reduces the need for on-site interventions, minimizes downtime, and enhances the overall efficiency of the manufacturing process	For example, in a manufacturing facility, conveyor systems are commonly used to transport materials or products between different stages of the production line
			The speed of the conveyor belt can be remotely adjusted based on production requirements or to manage the flow of materials
			Through the utilization of real-time data and remote control functionalities, industrial machinery organizations may enhance machine performance, prolong equipment lifespan, and optimize their operations
3.	Increasing production efficiency	IoT plays a major role in increasing production efficiency especially in the field of logistics and supply chain management	RFID (radio frequency identification) technology and IoT-enabled sensors allow organizations to track the movement and condition of their goods throughout the supply chain
			For example, in retail sector, RFID tags on products allow inventory levels to be tracked in real time. This guarantees that shelves are sufficiently stocked, lowering the possibility of overstock or stockout scenarios
			As a result, organization can optimize supply chain logistics and inventory management

6.2 *IoT Devices and Sensors for Wide Area Monitoring*

Wide area monitoring with smart devices and sensors has resulted in a new era of data-driven insights, efficiency, and responsiveness across multiple domains. These devices are essential for collecting data in real time, allowing thorough analysis, and supporting well-informed decision-making in wide area monitoring.

6.2.1 Environmental Sensors

Environmental sensors are designed to measure temperature, humidity, air quality, atmospheric pressure, light, and other environmental factors. For example, wireless sensors for temperature and humidity can be positioned strategically over a large geographic area to provide real-time environmental data. This is particularly beneficial in sectors like agriculture, where monitoring temperature and humidity levels in crop fields helps optimize irrigation and pest control, leading to increased agricultural productivity [6].

6.2.2 Smart Meters

Smart meters can facilitate wide area monitoring of electricity monitoring. These devices provide real-time data on energy usage patterns, helping utility companies optimize power distribution, manage peak loads, and enhance the overall efficiency of the electrical grid. Better resource planning and sustainable energy management are made possible by smart meters, which provide a more comprehensive overview of energy consumption over a large network.

6.2.3 Drones Equipped with Sensors

The combination of drone technology and advanced sensors enables a range of applications across various sectors, offering detailed insights and actionable information [7].

Drones equipped with high-resolution cameras, multispectral cameras, thermal cameras, or even LiDAR (Light Detection and Ranging) sensors can be used for aerial imaging, where this allows detailed mapping and surveying of large area. This would be valuable in agriculture sector for capturing data on crop health, soil conditions, and water distribution.

Besides, they can also be utilized for inspecting critical infrastructure including pipelines, bridges, and power lines. High-resolution imagery and thermal data help identify potential issues, corrosion, or structural anomalies, facilitating preventive maintenance and minimizing downtime.

6.3 Data Integration

Data integration techniques play a critical role in combining, processing, and deriving insights from diverse data sources within the context of integrating IoT for wide area monitoring. Some of the data integration techniques that can be used are data aggregation, data normalization and standardization, APIs (Application Programming Interfaces), data fusion, machine learning and artificial intelligence, edge computing, and cloud-based integration platforms [8].

S. No.	Data integration technique	Description
1.	Data aggregation	Involves combining multiple data points into a single, summarized representation. This can reduce the volume of data transmitted and enhance efficiency In wide area monitoring, aggregating data from multiple sensors can provide a comprehensive overview of environmental conditions without overwhelming the data processing infrastructure
2.	Data normalization and standardization	Ensures that data from different IoT devices are in standardized format and scale, allowing for seamless integration and analysis This is applicable for data that have units and specific formats such as timestamp
3.	Application programming interfaces	APIs enable various software programs and network to interact and exchange data with one another This facilitates integration between IoT devices and data processing platforms, enabling the seamless transfer of data and ensuring interoperability
4.	Data fusion	Data fusion combines data from multiple sources to provide a more detailed and accurate depiction of the monitored environment, enhancing the overall situational awareness
5.	Machine learning and AI	Machine learning algorithms can analyse and model patterns within integrated IoT data, providing predictive insights and identifying correlations This can enhance wide area monitoring by forecasting trends, detecting anomalies, and optimizing responses based on past data
6.	Edge computing	Edge computing move information processing directly to the source, eliminating the need to send vast amount of unprocessed information to a centralized system This technique can be employed to pre-process data on the IoT devices themselves, providing immediate insights and reducing latency
7.	Cloud-based integration platforms	Cloud-based integration platforms, such as AWS IoT or Azure IoT, provide centralized services for managing, processing, and analysing IoT data These platforms offer scalable solutions for handling large volumes of IoT data, providing analytics, and enabling seamless integration with other cloud-based services

7 Smart Grid Applications

Smart grid applications refer to the diverse set of technologies and systems that leverage radical information and connectivity technology to improve productivity dependability as well as sustainability of electric power grids. These applications play a vital role in modernizing traditional power infrastructure to meet the evolving demands of the twenty-first century. Here, we'll explore some key smart grid applications and their impact on the energy landscape.

7.1 Advanced Metering Infrastructure (AMI)

One fundamental component of smart grids is advanced metering infrastructure, which involves the deployment of smart meters. These devices provide real-time data on energy consumption via two-way communication between utilities and consumers. This facilitates demand response programs, allowing consumers to adjust their electricity usage during peak times to reduce strain on the grid [9].

7.2 Grid Automation

Smart grid technologies enable automation in power distribution and management. This includes the use of sensors, intelligent switches, and communication networks to monitor and control the flow of electricity. Automation enhances grid reliability by minimizing response time to faults, reducing outage durations, and improving overall system resilience [10].

7.3 Renewable Energy Integration

The integration of renewable energy sources, such as solar and wind, is a critical aspect of smart grid applications. Smart grids facilitate the efficient integration of these intermittent energy sources by providing real-time monitoring and control, enabling better grid management to accommodate the variability of renewable power generation [11].

7.4 Energy Storage Management

Smart grids enable effective integration and management of energy storage systems. Energy storage plays a key role in balancing supply and demand, storing excess energy during periods of low demand and releasing it during peak demand times. This helps in optimizing the utilization of renewable energy and improving overall grid stability [12].

7.5 Electric Vehicles (EVs) Integration

With the growing adoption of electric vehicles, smart grids provide solutions for managing the increased demand on the grid. They enable intelligent charging infrastructure, allowing for load balancing, demand response, and integration with renewable energy sources. This ensures efficient charging without overloading the grid [13].

7.6 Demand Response Programs

Smart grids facilitate demand response programs that encourage consumers to adjust their electricity consumption in response to price signals or grid conditions. This helps in reducing peak demand, optimizing grid operation, and promoting energy efficiency [14].

7.7 Cybersecurity Measures

As smart grids rely heavily on information and communication technologies, ensuring cybersecurity is paramount. Various cybersecurity measures, including encryption, secure communication protocols, and intrusion detection systems, are implemented to protect smart grid infrastructure from cyber threats [15].

These applications collectively contribute to the development of a more flexible, reliable, and sustainable electricity infrastructure. As smart grid technologies continue to evolve, ongoing research and development efforts focus on addressing challenges and improving the overall performance of these applications.

8 Case Study: Smart Wide Area Monitoring for Power Grids [16, 17]

Smart wide area monitoring involves the integration of advanced technologies to enhance the monitoring, control, and management of large geographical areas. Case studies and real-world implementations in this field showcase the practical applications and benefits of deploying such systems.

8.1 Background

Power grids are intricate networks that interconnect generating stations, electrical substations, high-voltage distribution lines, and loads. The city of Metropolis faces increasing challenges in managing its power grid due to the growing population and evolving energy landscape. To address these issues, the city's utility company decides to implement a smart wide area monitoring system for its power distribution network.

The power sector serves as the cornerstone of a nation's economy, playing a crucial role in everyday operations across various sectors. Reliable, high-quality, and affordable power supply is essential for sustaining economic activities. However, with the proliferation of distributed generation and expansion of power networks through the installation of new transmission lines, modern power systems are becoming increasingly complex. This complexity poses challenges for the operation and control of the power network, with inadequate monitoring and control potentially leading to catastrophic failures, commonly referred to as blackouts, caused by a series of outages. To address these challenges, the power network is evolving towards advanced solutions like Wide Area Monitoring Protection and Control (WAMPAC) systems. WAMPAC enables the implementation of smart grids, which are bidirectional networks capable of self-healing in the event of a failure. These advancements are crucial for ensuring the stability and efficiency of power systems in meeting the growing demand.

8.2 Objectives

- Improve grid reliability and reduce downtime.
- Enhance situational awareness for real-time decision-making.
- Implement predictive maintenance to prevent equipment failures.

8.3 WAMPAC System

The introduction of WAMPAC systems has bolstered the security of the power network and enhanced its reliability. To mitigate catastrophic failures, it is imperative to detect and monitor system security indices promptly, analyse critical real-time information triggering dynamic security needs, and implement wide area information, control schemes, and actions. This section presents a literature survey on the optimal placement of phasor measurement units, wide area monitoring, wide area control, and wide area protection.

8.4 Indian Power Grid

The Power Grid Corporation of India boasts an extensive network spanning approximately 134,018 circuit kilometres of transmission lines operating at various voltage levels, including 800/765 kV, 400 kV, 220 kV, and 132 kV EHVAC, as well as 500 kV HVDC, complemented by 214 substations. As of December 2016, the total transformation capacity stands at around 278,862 MVA. India holds membership in the very large power grid operators, a non-profit organization established in 2004 to address common interests among its members. The planning and development of the transmission system in India are overseen by the power system wing of the Central Electricity Authority (CEA).

The industry faces challenges like increased power demand, emissions reduction, supply reliability, and energy efficiency. Transitioning from traditional to modernized grids is necessary, with decentralized renewable sources and enhanced demand-side management using FACTS controllers like SVC, TCSC, STATCOM, etc. Wide area monitoring and PMUs play crucial roles in monitoring the power network, facilitating applications such as phase angle, voltage stability, line thermal, power oscillation, and damping monitoring. Power system stability ensures the system can operate under normal conditions and recover equilibrium after disturbances.

In wide area monitoring systems, neural networks, particularly recurrent neural networks (RNNs), are utilized for dynamic system identification and state prediction. Effective tuning of RNN weights is crucial for accurate predictions. Particle Swarm Optimization (PSO) is commonly employed for this purpose, but it can occasionally be limited by local optima, hindering its ability to reach the global optimum.

The power system faces issues in monitoring voltage and frequency deviation, communication delays, and natural disasters, resulting in major blackouts. Enhanced monitoring is essential. The existing energy management system experiences time delays and synchronization issues. Implementing WAMPAC can tackle these challenges, significantly impacting power network reliability. Securing all networks is vital to avert catastrophic failures.

8.5 Potential Implementation

(a) **Sensor Deployment** [18]

The utility company strategically installs a network of sensors across the power grid. These sensors include smart meters, phasor measurement units (PMUs), and intelligent electronic devices (IEDs) at substations. These devices collect real-time data on voltage, current, power factor, and other critical parameters.

(b) **Data Integration and Analytics** [19]

The collected data is transmitted to a central control centre, where it is integrated and analysed using advanced analytics and machine learning algorithms. This provides a holistic view of the entire power grid, identifying patterns, anomalies, and potential issues.

(c) **Predictive Maintenance** [20]

Machine learning models are employed to predict equipment failures based on historical data and real-time monitoring. This enables the utility company to schedule proactive maintenance, minimizing downtime and reducing the risk of catastrophic failures.

(d) **Situational Awareness** [21]

Visualization tools and dashboards are implemented to provide operators with real-time situational awareness. This allows them to respond promptly to incidents, reroute power, and take preventive measures to avoid cascading failures.

9 Future Trends in SWAM

Smart Wide Area Monitoring (SWAM) is a crucial aspect of modern infrastructure management, especially in the context of energy systems, transportation networks, and smart cities. As technology continues to advance, several future trends are likely to shape the landscape of smart wide area monitoring, enhancing its capabilities and impact.

(a) **Integration of Artificial Intelligence (AI) and Machine Learning (ML)**

Future smart wide area monitoring systems will increasingly leverage AI and ML algorithms to analyse vast amounts of data in real time. These technologies can help identify patterns, predict potential issues, and optimize the performance of wide area systems. AI and ML will play a pivotal role in enhancing the accuracy and efficiency of anomaly detection and fault prediction [22].

(b) **Advanced Sensor Technologies**

The development of more sophisticated and diverse sensor technologies will be a significant trend. These sensors will not only provide more accurate and granular data but will also be capable of monitoring a wider range of parameters. This could include environmental factors, security metrics, and even social data, contributing to a more comprehensive understanding of the monitored areas [23].

(c) **Edge Computing for Real-time Processing**

The shift towards edge computing will become more prominent in smart wide area monitoring. By processing data closer to the source, edge computing reduces latency and allows for faster decision-making. This is particularly critical for applications that require real-time response, such as power grid management and traffic control systems [24].

(d) **Interconnected and Interoperable Systems**

The future of smart wide area monitoring lies in interconnected and interoperable systems. This means that different monitoring systems will seamlessly communicate and share data, facilitating a holistic view of the monitored area. Interconnected systems enable better collaboration among various sectors, such as energy, transportation, and public safety [25].

(e) **Quantum Computing for Complex Analyses**

As quantum computing technology matures, it holds the potential to revolutionize the analysis of large-scale data sets in smart wide area monitoring. Quantum computing could significantly accelerate computations, enabling more complex simulations, optimizations, and analyses that were previously impractical with classical computing [26].

(f) **Resilience and Redundancy**

Future smart wide area monitoring systems will emphasize resilience and redundancy to ensure continuous operation even in the face of disruptions or failures. This may involve the implementation of decentralized architectures and backup systems to guarantee the reliability of monitoring services.

In essence, the future of smart wide area monitoring is marked by the integration of advanced technologies, increased connectivity, and a focus on enhancing efficiency, security, and reliability. As these trends unfold, the ability to monitor and manage large-scale systems will continue to improve, contributing to the development of more sustainable, resilient, and intelligent urban environments.

References

1. Piette, M. A., Kinney, S. K., & Haves, P. (2001). Analysis of an information monitoring and diagnostic system to improve building operations. *Energy and Buildings, 33*(8), 783–791. https://doi.org/10.1016/s0378-7788(01)00068-8
2. Teti, R., Jemielniak, K., O'Donnell, G., & Dornfeld, D. (2010). Advanced monitoring of machining operations. *CIRP Annals, 59*(2), 717–739. https://doi.org/10.1016/j.cirp.2010.05.010
3. Shin, S.-J., Woo, J., Rachuri, S., & Seo, W. (2019). An energy-efficient process planning system using machine-monitoring data: A data analytics approach. *Computer-Aided Design, 110*, 92–109. https://doi.org/10.1016/j.cad.2018.12.009
4. Shang, C., & You, F. (2019). Data analytics and machine learning for smart process manufacturing: Recent advances and perspectives in the Big Data Era. *Engineering, 5*(6), 1010–1016. https://doi.org/10.1016/j.eng.2019.01.019

5. Zima, M., Larsson, M., Korba, P., Rehtanz, C., & Andersson, G. (2005). Design aspects for wide-area monitoring and control systems. *Proceedings of the IEEE, 93*(5), 980–996. https://doi.org/10.1109/JPROC.2005.846336
6. Nurellari, E., & Srivastava, S. (2018). A practical implementation of an agriculture field monitoring using wireless sensor networks and IoT enabled. In *2018 IEEE International Symposium on Smart Electronic Systems (iSES) (Formerly iNiS), Hyderabad, India* (pp. 134–139). IEEE. https://doi.org/10.1109/iSES.2018.00037
7. Sehrawat, D., & Gill, N. S. (2019). Smart sensors: Analysis of different types of IoT sensors. In *2019 3rd International Conference on Trends in Electronics and Informatics (ICOEI), Tirunelveli, India* (pp. 523–528). IEEE. https://doi.org/10.1109/ICOEI.2019.8862778
8. Levy, A. Y. (2000). Logic-based techniques in data integration. In J. Minker (Ed.), *Logic-based artificial intelligence. The Springer International Series in Engineering and Computer Science* (Vol. 597, pp. 575–595). Springer. https://doi.org/10.1007/978-1-4615-1567-8_24
9. Robert, S., Pawel, S., Wojciech, D., & Lukasz, K. (2022). Advanced metering infrastructure and energy storage for location and mitigation of power quality disturbances in the utility grid with high penetration of renewables. *Renewable and Sustainable Energy Reviews, 157*, 111988.
10. Blonsky, M., Nagarajan, A., Ghosh, S., McKenna, K., Veda, S., & Kroposki, B. (2019). Potential impacts of transportation and building electrification on the grid: A review of electrification projections and their effects on grid infrastructure, operation, and planning. *Current Sustainable/Renewable Energy Reports, 6*, 169–176.
11. Bagherian, M. A., & Mehranzamir, K. (2020). A comprehensive review on renewable energy integration for combined heat and power production. *Energy Conversion and Management, 224*, 113454.
12. Li, X., & Wang, S. (2019). Energy management and operational control methods for grid battery energy storage systems. *CSEE Journal of Power and Energy Systems, 7*(5), 1026–1040.
13. Das, H. S., Rahman, M. M., Li, S., & Tan, C. W. (2020). Electric vehicles standards, charging infrastructure, and impact on grid integration: A technological review. *Renewable and Sustainable Energy Reviews, 120*(109), 618.
14. Wang, F., Xiang, B., Li, K., Ge, X., Lu, H., Lai, J., & Dehghanian, P. (2020). Smart households' aggregated capacity forecasting for load aggregators under incentive-based demand response programs. *IEEE Transactions on Industry Applications, 56*(2), 1086–1097.
15. Tufail, S., Parvez, I., Batool, S., & Sarwat, A. (2021). A survey on cybersecurity challenges, detection, and mitigation techniques for the smart grid. *Energies, 14*(18), 5894.
16. Moxley, R., & Dolezilek, D. (2011). Case studies: Synchrophasors for wide-area monitoring, protection, and control. In *2011 2nd IEEE PES International Conference and Exhibition on Innovative Smart Grid Technologies* (pp. 1–7). IEEE.
17. Nalini, A., Sheeba, P. E., Sivanesan, S., Rama, S., & Bhuvaneswari, T. (2018). Wide area monitoring system for an electrical grid. In *2nd International Conference on Energy and Power, ICEP 2018* (pp. 381–388). Elsevier.
18. Qiu, M., Gao, W., Chen, M., Niu, J. W., & Zhang, L. (2011). Energy efficient security algorithm for power grid wide area monitoring system. *IEEE Transactions on Smart Grid, 2*(4), 715–723.
19. Chintakindi, R., & Mitra, A. (2020). Execution of real-time wide area monitoring system with big data functions and practices. In *2020 IEEE 9th Power India International Conference (PIICON)* (pp. 1–6). IEEE.
20. Pech, M., Vrchota, J., & Bednář, J. (2021). Predictive maintenance and intelligent sensors in smart factory. *Sensors, 21*(4), 1470.
21. Hoag, J. C. (2012). Wide-area smart grid situational awareness communications and concerns. In *2012 IEEE Energytech* (pp. 1–7). IEEE.
22. Ullah, Z., Al-Turjman, F., Mostarda, L., & Gagliardi, R. (2020). Applications of artificial intelligence and machine learning in smart cities. *Computer Communications, 154*, 313–323.
23. Kalsoom, T., Ramzan, N., Ahmed, S., & Ur-Rehman, M. (2020). Advances in sensor technologies in the era of smart factory and industry 4.0. *Sensors, 20*(23), 6783.

24. Mehmood, M. Y., Oad, A., Abrar, M., Munir, H. M., Hasan, S. F., Muqeet, H. A. U., & Golilarz, N. A. (2021). Edge computing for IoT-enabled smart grid. *Security and Communication Networks, 2021*, 1–16.
25. Gupta, P., Kandari, R., & Kumar, A. (2021). An introduction to the smart grid-I. In *Advances in smart grid power system* (pp. 1–31). Academic Press.
26. Bhatia, M., & Sood, S. K. (2020). Quantum computing-inspired network optimization for IoT applications. *IEEE Internet of Things Journal, 7*(6), 5590–5598.

Efficient Methods for Price Prediction of Energy

K. E. Hema Priya, S. Saraswathi, P. Gowri, and K. M. Nandhini

1 Introduction

Energy costs are quite important in various industries and economies. This prediction can be essential for various stakeholders in the energy industry, including consumers, producers, and policymakers. Accurate energy price prediction is essential for making informed decisions such as resource allocation, budgeting, and risk management. In this article, we will explore different methods for energy price prediction, including traditional statistical approaches, machine learning models, and hybrid models.

The significance of energy price forecasting can be viewed from three key perspectives:

(a) Forecasting energy prices with accuracy allows for dynamic cost control, allowing businesses and organizations to adjust their strategies and operations in response to changing energy prices.
(b) Accurate energy price forecasts help investors and businesses comprehend trends in the market and spot business possibilities.
(c) Effective cost of energy forecasting aids in efficient market surveillance and offers a scientific basis for policy decisions.

Forecasting models in the field of energy are categorized into two groups: the traditional model and AI-driven model. Notably, due to the significant

K. E. Hema Priya (✉) · P. Gowri
Sri Krishna Arts and Science College, Coimbatore, Tamil Nadu, India

S. Saraswathi
Nehru Arts and Science College, Thirumalayampalayam, Tamil Nadu, India

K. M. Nandhini
Bannari Amman Institute of Technology, Sathyamangalam, Tamil Nadu, India

S. Vijayalakshmi et al. (eds.), *Smart Power Systems*, Power Systems,
https://doi.org/10.1007/978-3-031-89459-6_8

advancements in artificial intelligence (AI) in recent years, many researchers have turned to AI algorithms for energy prediction tasks. As a result, various models have emerged, and scholars have conducted reviews and analyses of energy prediction models from diverse angles and viewpoints.

1.1 Importance of Energy Price Prediction and Factors Affecting Energy Price Prediction

Energy price prediction holds significant importance for various stakeholders in multiple aspects of energy industry and broader economy. Here are some key reasons why energy price prediction holds importance:

Cost Management: Accurate energy price forecasts allow businesses to plan and budget effectively for their energy expenses, helping them control costs and improve financial stability. When predicting energy prices, cost management relates to a company's capacity to organizations, and individuals to effectively plan and control their energy-related expenses. This process involves using energy price forecasts and related strategies to optimize spending, minimize financial risks, and enhance overall financial stability.

Investment Decisions: Investment decisions in energy price prediction refer to the process of using forecasts of energy prices to make informed choices about allocating capital to various energy-related assets and projects.

Hedging: Hedging is one of the primary strategies for mitigating risk in energy price prediction. It involves using financial instruments such as futures, options, or swaps to establish contracts that lock in future energy prices. By doing so, businesses can shield themselves from adverse price movements. For example, a company can enter a futures contract to buy energy at a fixed price, ensuring cost predictability even if market prices rise.

Market Strategy: Market strategy in energy price prediction involves developing and implementing plans and tactics to optimize business operations and financial decisions based on forecasts of energy prices.

Energy Market Regulation: Energy price predictions are used by regulatory authorities to set price caps, establish pricing mechanisms, and monitor market behaviour. Accurate forecasts help regulators make informed decisions regarding market interventions and ensure fair and transparent energy markets.

Energy Efficiency: When predicting energy prices, the term "energy efficiency" refers to the process of maximizing energy consumption and minimizing waste in order to reduce energy expenditures. Businesses, organizations, and individuals may make educated choices about energy-saving investments and actions thanks to accurate projections.

Renewable Energy Integration: The strategic integration of sources of renewable energy (such as the sun, wind, water, and thermal) into the energy infrastructure and consumption habits based on predictions of the price of energy is known as renewable energy integration. Accurate predictions are essential for optimizing the integration of renewable energy while considering its variable nature and economic viability.

Infrastructure Planning: Accurate predictions are crucial for optimizing infrastructure investments and ensuring a reliable, cost-effective, and efficient energy supply.

Market Regulation: Regulations play a critical role in shaping how energy prices are determined, traded, and reported, and they often consider forecasts of energy prices.

Environment Impact: Refers to the consequences of energy-related decisions, policies, and market dynamics on the environment, particularly in relation to greenhouse gas emissions, air and water quality, natural resource conservation, and other ecological factors. Accurate energy price predictions play a crucial role in shaping these impacts by influencing decisions regarding the use of various energy sources, the adoption of clean technologies, and environmental policies.

Economic Impact: Energy prices have a significant influence on economies at all levels—local, regional, national, and global. It refers to the various economic consequences and effects that result from the anticipation and actualization of energy price forecasts.

Energy Security: In the context of energy price prediction, energy security pertains to the assurance of a consistent and dependable energy supply at reasonable costs while reducing susceptibility to disruptions and price fluctuations. Precise energy price forecasts play a vital role in evaluating and mitigating energy security risks in a proactive manner.

2 Traditional Statistical Methods for Energy Price Prediction

In order to model and estimate future energy costs, standard statistical approaches for this purpose often depend on historical data and a number of statistical techniques.

2.1 Time Series Analysis

It is a commonly used and effective technique for predicting energy prices. Energy price data often exhibit temporal dependencies and seasonality, making time series analysis a natural choice. Gather historical energy price data, which can include electricity prices, natural gas prices, oil prices, or any other relevant energy commodities. Verify that all the information is available at the right time, such as per hour, per day, or monthly, and is free from missing values or outliers. If necessary, to stabilize variance or eliminate trends, use procedures like logarithmic or differencing transformations. Depending on the characteristics of the data, choose an appropriate time series model.

Calculate the predictive parameters, including the autoregressive and shifting average orders (p, d, q for ARIMA), seasonality factors, and any other model-specific parameters. Fit the model that was selected to the historical data, then use statistical tests and diagnostics to evaluate the model's degree of fit. Common diagnostics include checking for residuals' normality, stationarity, and autocorrelation. Use the model to anticipate future energy costs when it has been verified. These projections might be probabilistic or point forecasts, depending on the model. Evaluate the prediction accuracy using apposite performance metrics such as Mean_Absolute_Error (MAE), Mean_Squared_Error (MSE), Root_Mean_Squared_Error (RMSE), or other relevant metrics.

Clearly and simply explain the outcomes of the time series research and projections to the appropriate clients, decision-makers, and traders. Time series analysis is a useful technique for analysing and predicting price changes in the energy markets. Participants in the energy sector may use it to control risks, make educated decisions, and streamline their operations.

2.2 *Exponential Smoothing State Space Models (ETS)*

Exponential smoothing state space models (ETS) are a periodic prognostication technique group that are particularly useful in energy price prediction and other fields where the data exhibit trends and seasonality. These replicas provide a flexible outline for apprehending different patterns in the data while allowing for forecasting future values.

2.2.1 Components of ETS Model

Error (E): The error term represents the noise or randomness in the data, and it follows a normal distribution with mean zero and its variance that may change over time.

Trend (T): The (T) component follows the underlying direction or movement in the time series. It can be additive (A) or multiplicative (M), depending on whether the trend is linear or nonlinear.

Seasonal (S): The (S) component accounts for periodic method that repeats at fixed intervals, such as day-to-day, monthly, or yearly. It can also be additive or multiplicative.

State Space Representation: The ETS model is formulated in a state space framework, which consists of an observation equation that links the observed data to the components (E, T, S) and a state equation that describes how the components evolve over time.

2.2.2 State Space Equations

- **Observation Equation:**
- For additive error (A) and additive seasonality (A):

$$\text{Observation}\left(Y\_t\right) = \text{Level}\left(l\_t\right) + \text{Trend}\left(b\_t\right) + \text{Seasonal}\left(s\_t\right) + \text{Error}\left(e\_t\right)$$

- For additive error (A) and multiplicative seasonality (M):

$$\text{Observation}\left(Y\_t\right) = \text{Level}\left(l\_t\right) + \text{Trend}\left(b\_t\right)^{*} \text{Seasonal}\left(s\_t\right) + \text{Error}\left(e\_t\right)$$

- Similar equations exist for other combinations (e.g., multiplicative error and additive/multiplicative seasonality).
- **State Equation**:
- Level (l_t), Trend (b_t), and Seasonal (s_t) evolve over time according to equations that depend on their previous values and the current observation.
- **Parameter Estimation:**
- Model parameters (e.g., smoothing parameters, initial values) are estimated using optimization techniques like maximum likelihood estimation. These parameters determine the model's flexibility and ability to adapt to the data.
- **Model Identification**:
- The choice of the specific ETS model (e.g., ETS(A,A,A), ETS(A,M,A), etc.) depends on the data's characteristics, which can be assessed through visual inspection and autocorrelation, and partial autocorrelation graphs are examples of diagnostic tools.
- **Model Fitting and Forecasting**:
- Once the model is identified and parameters are estimated, it can be fitted to the historical data. This involves using the state space equations to update the components (E, T, S) at each time step. The fitted model can then be used to make forecasts for future time points. Forecasts can be produced for various time horizons, depending on the forecasting needs.
- **Model Evaluation**:
- Use relevant measures to assess the model's predicting ability such as Mean Absolute Error (MAE), Mean Squared Error (MSE), or Root Mean Squared Error (RMSE).
- Conduct residual analysis to assess the quality of model predictions.
- **Model Selection and Updating:**
- Periodically assess the model's performance and update it as needed. This may involve re-estimating parameters or selecting a different ETS model if the data's characteristics change over time. Exponential smoothing state space models are valuable in energy price prediction because they can adapt to different data patterns, including linear or nonlinear trends and additive or multiplicative seasonality. This flexibility allows them to provide accurate forecasts for energy prices,

which are often influenced by various factors and exhibit complex temporal patterns.

2.3 Regression Models

A set of statistical and machine erudition methods called regression representations is used to forecast energy prices by establishing correlations between energy prices and one or more predictor variables. These models help predict energy prices based on historical data and the influence of various factors

- **Data Collection and Preprocessing**:
- Gather historical data on energy prices (the dependent variable) and potential predictors (independent variables). Predictors can include weather data, supply and demand factors, economic indicators, geopolitical events, and more.
- Ensure the data is cleaned and preprocessed, addressing issues like missing values, outliers, and data transformations if necessary.
- **Model Selection**:
- Based on the characteristics of the data and the anticipated correlations between the variables, select a suitable regression model. Typical regression models for forecasting energy prices.

2.4 Linear Regression

Linear_regression is a commonly used statistical scheme for energy price prediction. It models the relationship between energy prices (the reliant on inconstant) and one or more predictor variables (independent variables) in a linear fashion. It makes the assumption that the dependent and independent variables have a linear relationship.

- **Multiple Regression**:
- Extends linear regression to incorporate multiple independent variables.
- **Polynomial Regression**:
- Captures nonlinear relationships by using polynomial functions.
- **Ridge Regression and Lasso Regression**:
- Used for handling multicollinearity and feature selection.
- **Time Series Regression**:
- Cartels regression analysis through time series techniques to capture temporal dependencies.
- **Feature Selection and Engineering**:

- Determine and pick the pertinent predictor factors that have the most potential to affect energy costs. To more accurately depict relationships, feature engineering may entail the creation of additional variables or transformations. To comprehend the data and the relationships between the variables, do an exploratory data analysis.
- **Model Building**:
- Match the training set of data to the chosen regression model. The coefficients, or weights, corresponding to each predictor variable will be estimated by the model. To determine the relevance of predictor variables, evaluate the model's goodness of fit using statistical metrics such as *R*-squared, adjusted *R*-squared, and *p*-values.
- **Model Evaluation**:
- Use a separate validation dataset or cross-validation to evaluate the replica's recital. Common evaluation metrics include Mean Absolute Error (MAE), Mean Squared Error (MSE), Root Mean Squared Error (RMSE), and others.
- Examine residual plots and diagnostics to ensure the model assumptions are met.
- **Prediction and Prognostication**:
- The model may be used to forecast and make predictions after it has been validated. Intended for energy price prediction, this typically involves projecting future price values based on the predictor variables' values.
- **Regularization (Optional)**:
- Regularization techniques such as Lasso or Ridge regression can be used to prevent extreme results and put limitations to the model coefficients in situations of multicollinearity or overfitting.
- **Model Interpretation**:
- Interpret the regression model to understand the relationships between predictor variables and energy prices. Coefficients provide insights into the strength and direction of these relationships.
- **Model Updating and Maintenance**:

- Update and improve the model often in response to fresh data and shifting market conditions. This ensures that the model remains accurate and relevant. Regression models in energy price prediction provide valuable insights for traders, energy companies, and policymakers by quantifying the impact of various factors on energy prices. These models help in making informed decisions, managing risks, and optimizing resource allocation in volatile energy markets.

3 Hybrid Models

Hybrid model is an efficient price prediction of energy that refers to a combination of various techniques and methods that is designed to increase the exactness of energy price predictions. These representations leverage the strengths of different approaches to overcome the limitations of individual methods. In the framework of energy price estimate, a hybrid model typically blends traditional statistical techniques, machine learning algorithms, and potentially other data-driven or domain-specific methods to provide extra robust and exact estimates. The aim is to augment the efficiency of price forecast in energy markets by leveraging the complementary aspects of different forecasting approaches.

Understanding Hybrid Models
Hybrid models for energy price forecast represent a fusion of various prediction methods and data sources to enhance predictive capabilities. These models recognize that no single approach can comprehensively capture the multifaceted factors influencing energy prices. Instead, they integrate different methodologies, often including statistical, machine learning, and domain-specific techniques, to create a more robust forecasting framework.

Challenges and Considerations
While hybrid models offer significant advantages, they also come with challenges. These include the need for substantial data, computational resources, and expertise in model development and maintenance. Additionally, interpreting the outputs of hybrid models can be complex due to their diverse nature, making it essential to maintain transparency and interpretability.

3.1 Combining Statistical and Machine Learning Models

The energy sector is marked by its inherent volatility, influenced by factors ranging from supply and demand dynamics to geopolitical events and weather patterns. Predicting energy prices accurately is a complex task that requires harnessing various modelling techniques. One powerful approach is the development of hybrid models that blend statistical and machine erudition devices. In this essay, we determination explore how combining these two types of models enhances the efficiency of energy price estimates.

Statistical Models in Hybrid Energy Price Prediction
Statistical models, such as ARIMA (Autoregressive_Integrated_Moving_Average) and exponential smoothing, have long been used for time series forecasting, including energy prices.

However, statistical models have limitations. They may struggle with complex, nonlinear relationships and are less suitable for capturing abrupt changes in energy markets, such as sudden spikes caused by unexpected events. This is anywhere machine culture models come into play.

Machine Learning Models in Hybrid Energy Price Prediction
Because they can handle complicated, data with high dimensions, machine erudition replicas like random forests, Gradient Boosting (GB), and neural networks have become popular for predicting energy prices.

Advantages of machine learning models:

- Nonlinear relationships
- Feature engineering
- Adaptability

Benefits of combining statistical and machine knowledge models in hybrid energy price prediction are evident:

- **Improved Accuracy:** Hybrid models often outperform standalone models, providing more accurate and reliable price forecasts.
- **Robustness:** By leveraging the strengths of both approaches, hybrid models are more robust and can better grip a wide variety of market circumstances.
- **Interpretability:** Depending on the design, hybrid models can strike a balance between interpretability and predictive power.

3.2 Combining Time Series Data with Explanatory Variables

Combining time series data with explanatory variables in hybrid models is a powerful approach for predicting the price direction of energy efficiently. Here's a high-level overview of how you can go about it:

- **Data Collection:** Gather historical time series data for energy prices. This data typically includes daily or hourly price values.

- Collect relevant explanatory variables, such as weather data (temperature, wind speed, sunlight), demand forecasts, supply data (e.g., renewable energy production), economic indicators (e.g., GDP growth), and geopolitical events (e.g., OPEC decisions)
- **Feature Engineering:** Create lag features for the time sequence data, capturing tendencies and seasonality. Engineer additional features from the explanatory variables, such as moving averages or indicators of extreme events.
- **Model Selection:** Choose appropriate machine learning models for time sequence forecasting, such as ARIMA, LSTM, or Prophet. Select regression replicas like linear_regression, decision_trees, or ensemble methods for the explanatory variables.
- **Hybrid Model Architecture:** Train the time series model separately on the historical price data. Train the regression model on the explanatory variables. Combine the predictions from both models using techniques like stacking or weighted averaging.

4 Incorporating External Information

In the context of energy price production, incorporating external information efficiently refers to the practice of integrating various external data sources and relevant information into predictive models and decision-making processes to improve the correctness and dependability of energy price forecasts or production estimates. This approach recognizes that energy prices can be prejudiced by a multitude of issues beyond historical price trends and by incorporating external data efficiently, one can make more informed and precise predictions.

Key components of this concept:

- **External Information:** This refers to data and information sources that are not straight related to ancient energy prices but container impact them.
- **Efficiency:** Efficiency in this context implies the effective and optimal use of external information.
- **Price Production:** This term relates to the estimation or prediction of energy prices, which is critical for various stakeholders in the energy sector, including energy suppliers, consumers, and investors.
- **Modelling and Analysis:** Incorporating external information efficiently typically involves employing advanced modelling methods, such as appliance learning and statistical models, to analyse the relationships between external variables and energy prices.
- **Real-Time Adaptation:** Since external factors influencing energy prices can change rapidly, an efficient approach should also allow for real-time updates and adjustments to the models based on incoming external data.
- Overall, incorporating external information efficiently for energy price production is about leveraging external data sources in a way that enhances the accuracy, reliability, and timeliness of energy price predictions or production estimates, ultimately leading to additional informed and effective managerial in the energy sector.

4.1 Incorporation of Weather Data

The incorporation of weather data, as part of incorporating external information, refers to the practice of integrating meteorological or atmospheric data into predictive models, analytical systems, or decision-making processes to enhance the accuracy, relevance, or insightfulness of various applications. Weather data is considered an external information source because it originates from outside the specific domain or system being analysed, and yet it can have a significant impact on that domain or system.

Here's how weather data can be incorporated:

Data Gathering: Fold historical weather statistics for the region of attention. This data may include temperature, humidity, wind speed, sunlight duration, and precipitation levels, among other relevant meteorological variables.

Feature Engineering: Create weather-related features that are likely to influence energy consumption or production. For example, you can compute space heating and preservation degree days to estimate energy petition for central heating and conserving purposes based on temperature data.

Correlation Analysis: Conduct an association examination to recognize the relations between weather variables and energy prices. This can help recognize which weather aspects consume the furthermost substantial bearing.

Model Integration: Integrate weather data into your energy price prediction model. Common approaches include weather variables as additional features in regression models, such as linear regression or decision trees. Using weather data to enhance time sequence prediction models, like ARIMA or LSTM. Employing machine education techniques that can apprehend complex interactions among weather variables and energy prices.

Normalization and Scaling: Normalize or scale weather data to ensure it's on the same scale as other variables in your model. This helps prevent certain variables from dominating the prediction process.

4.2 Incorporating Macroeconomic Indicators

Incorporating macroeconomic indicators as part of external information when predicting the price direction of energy can provide valuable insights into the broader economic context that influences energy markets. Here are some key macroeconomic indicators and ways they can be used for energy price prediction:

- **Gross Domestic Product (GDP):** GDP growth is an important pointer of financial healthiness. A rising budget characteristically leads to increased energy consumption. Incorporating GDP data can help predict energy demand trends.
- **Inflation Rate:** Changes in inflation rates can impact energy prices. High inflation may lead to higher energy costs as the purchasing power of currency decreases. Monitoring inflation rates can be important for understanding potential price increases.
- **Interest Rates:** Central bank interest rates can influence the cost of borrowing for energy infrastructure projects and impact investment decisions in the energy sector. Energy price models can consider interest rate changes and their effects on energy investments.
- **Commodity Prices:** Energy prices are often correlated with other commodities like oil, coal, and natural gas. Monitoring commodity price trends can help predict energy price movements.
- **Trade Balance:** Energy is often a significant component of a country's trade balance. Analysing trade data, including energy imports and exports, can help predict energy price movements. To incorporate macroeconomic indicators effectively, you can use statistical techniques, econometric models, time sequence examination, and machine knowledge algorithms. These replicas can analyse

historical data and incorporate external economic data to make predictions about energy prices. Regular updates and adjustments to the model based on changing economic conditions are essential for accurate energy price predictions.

4.3 News and Textual Analysis Using NLP

News mentions to appropriate and applicable material about up-to-date proceedings, progresses, and occurrences, often conveyed by reporters and newsflash organizations. It can encompass an extensive range of issues, as well as politics, economics, communal issues, science, and energy-related matters. In the context of incorporating external information for energy price prediction, news can include articles, reports, press releases, and other textual sources that contain information about factors that may influence energy markets and prices. Textual analysis using NLP is the process of extracting meaningful insights and patterns from text_data, such as update trainings and reports, to enhance the accuracy of energy price prediction models. Here's how NLP can be applied for this purpose:

- **Event Extraction:** NLP models can be trained to extract specific events from news articles that are relevant to energy markets. For example, the discovery of a new oil reserve or a major renewable energy project announcement can be important events to consider.
- **Key phrase Extraction:** Identifying key phrases or terms that are frequently mentioned in news articles related to energy can help in understanding the central themes and concerns of the industry.
- **Temporal Analysis:** NLP can help track how news sentiment and topics variation over period. This temporal analysis container can be used to detect tendencies and shifts in the energy market sentiment.
- **Language Translation:** If analysing news from international sources, language translation using NLP can be employed to ensure that relevant information is not missed due to language barriers.
- **Summarization:** Automatic text summarization techniques can create concise summaries of lengthy news articles, making it easier for analysts to quickly grasp the key insights.

5 Validation and Backtesting

Validation and backtesting are crucial steps in the development and evaluation of efficient methods for price prediction of energy. They help assess the performance and reliability of your predictive models. Models perform on data they haven't seen during training. It helps you determine whether your models simplify well to new, unseen statistics. In the context of energy price prediction, validation is essential to ensure that your models can make accurate predictions in real-world scenarios. Backtesting is a specific form of validation commonly used in financial and energy markets. It evaluates how well your predictive models would have performed in the past when applied to historical data.

5.1 Walk-Forward Authentication

Walk-forward authentication and bagging are distinct methods used for different purposes in machine knowledge, particularly in time sequence prediction and ensemble methods.

- **Purpose:** Walk-forward validation is a technique used to measure the presentation of time sequence prediction models, including those used for energy price prediction.
- **How It Works:** It involves training a model on historical data up to a certain point in time, making predictions for a future period, then moving the training window forward and repeating the process. Essentially, you keep updating the model with new data as it becomes available and evaluate its performance iteratively.
- **Advantages:** Walk-forward validation is valuable for assessing how well your model performs in a real-time scenario. It accounts for evolving trends and seasonality in the data and helps you detect whether your model maintains its predictive accuracy as new data arrives.

5.2 Evaluating Prediction Performance

Evaluating prediction performance in validation and backtesting is crucial to ensure that your predictive models or trading strategies are effective and reliable. Remember that validation and backtesting provide insights into how your models or strategies perform in historical data. However, they don't guarantee future performance. Continuous monitoring and adaptation are essential as market conditions can change.

Additionally, the high-quality of assessment metrics would align with your exact objectives and the nature of the prediction or trading task. Careful consideration of these metrics helps you make informed decisions, improve your models or strategies, and manage risk effectively.

These evaluation metrics and considerations help you make informed decisions, improve your models or strategies, and manage risk effectively. Whether you are evaluating predictive models or trading strategies, thorough evaluation is essential for assessing their effectiveness and reliability.

6 Stream Processing

Stream processing is a management technique which improves the continuous data flow by filtering, analysing, transforming, and enhancing the data in real time. It helps various enterprises to combine the data from various sources which includes transactions, weather reports, and other services. Stream processing is required to make responsive applications which helps in facilitating faster decisions to improve user experience from which adaptive applications can also be developed.

Stream processing architecture typically involves the following components:

1. **Source:** Data is collected from various sources such as IoT devices, sensors, databases, or applications.
2. **Ingestion**: The data is ingested into the stream processing system, where it's processed.
3. **Processing**: This is where data is transformed, analysed, and enriched processing can include various steps such as filtering, aggregating, joining, or applying machine learning models.
4. **Storage**: The data which is processed may be stored temporarily for further analysis or to support the downstream applications.
5. **Output/Action**: The output of processing can trigger actions, alerts, or be sent to other systems, databases, or applications for consumption.
6. **Scalability and Resilience:** Architectures must be designed accordingly to handle various data volumes and ensure system resilience and fault tolerance.

The design flow of a stream processing system involves several key steps to ensure efficient and effective real-time data processing.

1. **Define Objectives and Requirements:** Understand the objectives and the specific requirements required of the stream processing system. Identify what kind of data will be processed, the desired outcomes, and performance expectations.
2. **Data Processing:** The data is processed based on the objectives and requirements such as filtering, aggregation, or other processing operations to the ingested data.
3. **Storage and State Management:** Decides whether to store the processed data temporarily or persist it for future analysis.
4. **Real-time Analysis and Action:** Analyse the processed data in real time to extract insights, identify patterns, or trigger the actions or alerts based on the analysis results.
5. **Scalability and Performance Optimization:** Ensure the scheme is designed for scalability to handle increasing data volumes and processing demands. Optimizes the performance by fine-tuning processing algorithms, data structures, and resource allocation.
6. **Monitoring and Metrics**: Implement monitoring mechanisms to track system health, performance, and processing accuracy and collect the relevant metrics to assess the system's efficiency and effectiveness, making adjustments as needed.

7. **Feedback Loop and Iteration:** It gathers feedback from the system's operation and user requirements. Iterate on the design, processing logic, and system architecture to improve its capabilities and performance.
8. **Deployment and Maintenance**: Deploy the stream processing system in the production environment while ensuring smooth integration with existing systems and workflows. Deliver enduring maintenance, bring up-to-date, and assistance to resolve any problems, and maintain system updates to reflect evolving needs.

6.1 Implementation of a Cloud-Based Solution for Real-Time Prediction and Scalability

- Determine the problem you want to solve and gather requirements.
- Choose a cloud provider that fits your needs.
- Set up data collection and storage in the cloud.
- Clean and preprocess the data for analysis.
- Develop machine learning models using historical data.
- Capture and process real-time data streams using cloud-based tools.
- Apply the models to real-time data for predictions and insights.
- Configure scalable resources to handle varying workloads.
- Monitor and optimize performance and security.
- Integrate with other systems and deploy the solution on the cloud platform.
- Continuously improve the solution through feedback and updates.

6.2 IoT in Energy Prediction

IoT (Internet of Things) in energy prediction involves leveraging IoT devices, data collection, and advanced analytics to forecast energy consumption, optimize energy usage, and improve overall energy efficiency.

Key Components of IoT:

1. **Devices and Sensors**: IoT strategies, armed with devices, form the foundation of the IoT ecosystem. These sensors gather diverse data such as temperature, light, motion, and more.
2. **Connectivity:** Connectivity technologies like Wi-Fi, cellular grids, and Bluetooth empower strategies to transmit data to centralized platforms or other connected devices.
3. **Data Processing and Analysis:** Composed data is handled, analysed, and transformed using cloud computing, edge computing, and advanced analytics.

4. **Applications and Services:** The creation of applications and services that address specific needs, ranging from smart homes and cities to industrial automation and healthcare solutions.

Applications of IoT in Energy Prediction:

- **Smart Grids and Demand Forecasting**: IoT devices can monitor and analyse energy consumption patterns for accurate demand forecasting. This helps the utilities to plan for peak demand periods and optimize energy distribution.
- **Renewable Energy Optimization**: IoT sensors on solar boards, gust turbines, and other renewable liveliness sources can monitor their efficiency and environmental conditions.
- By analysing this data, energy production can be optimized for maximum output and efficiency.
- **Energy consumption monitoring and management:** IoT-enabled smart meters and devices in homes, businesses, and industries can track energy consumption. This data can be used to optimize energy usage, reduce wastage, etc.
- **Energy market analysis:** IoT can gather data on market prices, weather conditions, and other factors affecting energy supply and demand. Advanced analytics can help in the prediction of market trends, helping stakeholders make informed decisions and optimize energy trading.
- **Load balancing and energy distribution:** IoT devices can monitor and analyse loads on the grid, which is useful in dynamic load balancing and efficient energy distribution. This helps in preventing loss and saves energy for further use.
- **Weather and environmental impact prediction:** IoT sensors can collect weather and environmental data, which is crucial for predicting energy production from renewable sources. This information helps in planning and adjusting energy generation accordingly.
- **Optimizing energy storage systems**: IoT can monitor energy storage systems like batteries and predict when to store or release energy based on demand patterns and availability of renewable sources, improving overall.

7 Challenges and Limitations

7.1 Overfitting

Overfitting is a critical idea in machine knowledge and statistical modelling which is essential for understanding the balance between creating a model and one that fails to generalize to new, unseen data.

- **Example**: Imagine teaching a computer program to recognize handwritten digits. If you train it too much on a specific set of examples, it might become so specialized that it can only recognize those exact samples and won't perform well on any new, different handwritten digits. This is precisely what overfitting is all about.

The consequences of overfitting in machine learning:

1. **Reduced Generalization**: The model achieves well on the working out data nevertheless ill on hidden data, as it has essentially memorized the noise and specific patterns of the training set.
2. **Poor Predictions:** When applied to new data, the overfitted model may make inaccurate or unreliable predictions due to its overemphasis on training data.
3. **Inefficient Resource Use**: Overfitted models may use excessive computational resources during training, and their complexity can make them impractical for real-world applications.
4. **Limited Applicability:** An overfit model might be too specialized and failed to adapt to different datasets or real-world scenarios, limiting its usefulness.

7.2 Changing Energy Markets and Black Swan Events

The energy sector is undergoing a profound transformation, driven by an array of factors encompassing technological innovation, policy shifts, climate concerns, and evolving consumer preferences. These dynamics are reshaping traditional energy markets, challenging established business models, and fostering the expansion of renewable and bearable energy foundations. Simultaneously, the concept of black swan events, unexpected and impactful occurrences that defy conventional predictions, has gained prominence. When applied to energy markets, black swan events introduce an element of uncertainty, necessitating a reassessment of risk management and strategic planning.

Roles

- **Evolving Energy Markets:** Evolving energy markets play an essential part in the worldwide changeover toward sustainable energy sources. Renewable energy technologies, energy storage solutions, and decentralized energy production are gaining traction, altering the dynamics of supply, demand, and pricing within the energy market.
- **Black Swan Events:** Black swan events act as disruptive catalysts, prompting a re-evaluation of existing energy strategies and policies. These unforeseen incidents, ranging from extreme weather events to sudden geopolitical shifts, can significantly impact energy production, distribution, and consumption patterns.

Consequences

1. **Energy Markets:** The consequences of evolving energy markets include a reduced dependence on fossil oils, lowered greenhouse gas emanations, enhanced energy security, and potential economic growth through renewable energy

investments. Conversely, the shift may also present challenges in grid integration, intermittency of renewable sources, and infrastructure investments.
2. **Black swan events:** The consequences of black swan events in the energy sector can lead to supply disruptions, price volatility, infrastructure damage, and geopolitical tensions. Additionally, they may accelerate or impede the adoption of certain energy technologies and alter regulatory frameworks.

Limitations

- **Predictability:** Black swan events, by definition, are unpredictable and rare, making it challenging to prepare adequately for their occurrence. Traditional risk assessment models often fail to account for these unforeseen disruptions.
- **Response Preparedness:** Developing effective responses to black swan events is complicated due to their unexpected nature. Rapid and adaptive decision-making is essential, but established frameworks may prove insufficient in addressing the unique challenges these events.

Advantages

1. **Diversification_Energy_Sources**: Embracing changing energy sources allows for diversification_energy_sources. Transition toward renewable energy and emerging technologies helps reduce dependency on a solitary energy source, enhancing flexibility and stability in the energy supply.
2. **Energy Security**: A diversified and decentralized energy market enhances energy security. By relying on a mix of sources and distributed generation, regions can better withstand disruptions, whether caused by black swan events or other factors.
3. **Technological Advancements and Innovation**: Evolving energy markets drive innovation and technological advancements. The pursuit of cleaner and more efficient energy sources fosters breakthroughs that can benefit multiple sectors, from energy production to transportation and beyond.
4. **Economic Growth and Job Creation**: The growth of renewable energy sectors can stimulate economic development and create employment opportunities. Investments in renewable energy projects and related technologies spur economic activity and help in job creation.
5. **Resilience against Black Swan Events**: Understanding and preparing for black swan events enhance overall system resilience. By considering various potential disruptive scenarios, the energy sector can develop contingency plans and adapt swiftly to unexpected challenges.
6. **Global Energy Transition Leadership**: Embracing changing energy markets positions a country or region as a leader in the global energy transition.
7. **Societal Health Benefits**: Shifting to cleaner energy sources results in improved air and water quality, contributing to better public health outcomes. Reduced pollution can alleviate health issues associated with traditional fossil fuel-based energy production.

8 Case Studies

8.1 Energy Price Prediction in Deregulated Markets

Energy price prediction in deregulated markets is a critical aspect of modern energy management and trading. Deregulated markets are characterized by competition and the absence of a central authority determining energy prices.

In these markets, source and request dynamics, climate patterns, geopolitical proceedings, supervisory changes, and technological advancements significantly influence energy prices. Predicting these price fluctuations accurately is essential for market participants to brand knowledgeable decisions, optimize operations, and achieve risks effectively.

Roles in Energy Price Prediction in Deregulated Markets

- **Market Participants:**
  1. **Energy Traders and Suppliers**: They use price predictions to make timely and informed decisions regarding buying and selling energy contracts.
  2. **Investors:** Predictions help investors allocate resources strategically, aiming for profitable positions in the energy market.
  3. **Consumers:** Forecasted prices assist consumers in handling their energy feasting and expenditures effectively.
- **Energy Analysts and Researchers:** They develop and refine prediction models, leveraging historical data, market trends, and emerging factors to enhance the accuracy of price forecasts.

  1. They continually assess and validate prediction models to ensure they align with real-world market behaviour.

- **Regulators:** Utilize price predictions to assess the effectiveness of existing policies and regulations in promoting fair competition and consumer welfare.
- **Risk Managers:** Leverage price predictions to identify potential risks associated with price fluctuations, enabling the formulation of risk management strategies to mitigate financial losses.
- **Energy Planners and Grid Operators**: Incorporate predicted energy prices to optimize resource allocation, demand-response programs, and grid operations, ensuring a stable and reliable energy supply.
- **Technology Developers**: Use price predictions to tailor and improve energy-related technologies, like energy storage solutions or demand-side management systems, to meet market demands and maximize efficiency.
- **Consultants and Advisory Firms:** Provide guidance to market participants based on price predictions, helping clients make informed decisions to optimize their energy portfolios and investment strategies.

Case Study

Title: Forecasting Day-Ahead Power Values in a Relaxed Market.

Objective: To prediction day-ahead power prices in a relaxed energy marketplace using a machine learning approach.

Methodology

1. **Data Collection**: Historical data on electricity prices, weather conditions, and market demand.
2. **Feature Engineering**: Creating relevant features such as past price trends, day of the week, weather forecasts, and public holidays.
3. **Model Selection:** Employing a machine learning model like XGBoost, a popular algorithm for regression tasks.
4. **Training and Validation**: Using a helping of the historical statistics for training and validating the model's presentation.
5. **Prediction:** Generating day-ahead price predictions using the trained model for unseen data.

Results: The XGBoost model demonstrated a high degree of accuracy in predicting day-ahead electricity prices. It effectively captured price fluctuations influenced by weather conditions, demand patterns, and historical market trends.

8.2 Renewable Energy Price Forecasting

Renewable energy price forecasting is a critical procedure that includes predicting the future prices of energy resulting from renewable foundations like planetary, airstream, hydro, geothermal, and biomass. Exact price estimates are essential for investors, developers, representatives, and market participants in making informed decisions regarding renewable energy project development, financing, trading, and policy formulation.

The complexities involved in renewable energy price forecasting stem from various factors, including weather patterns, technological advancements, government policies, market dynamics, demand fluctuations, and advancements in energy storage technologies. Sophisticated models and algorithms are employed to analyse historical data and incorporate multiple variables to derive reliable and precise price predictions.

Roles

1. **Energy Developers and Investors:** Utilize price forecasts to plan investments, assess project viability, and make well-versed choices regarding the development and financing of renewable energy projects.
2. **Government and Regulatory Bodies**: Rely on price forecasts to formulate policies, set incentives, and implement regulatory frameworks that encourage renewable energy adoption and promote sustainability.

3. **Utilities and Energy Traders:** Leverage price forecasts to optimize their energy portfolios, trading strategies, and market positioning, maximizing profits in the renewable energy market.
4. **Energy Planners and Grid Operators**: Incorporate price forecasts to plan the addition of renewable energy into the standing grid, optimize resource allocation, and enhance grid stability and reliability.
5. **Technology Manufacturers and Suppliers**: Use price forecasts to align production capacity with expected demand, optimize pricing strategies, and drive research and development efforts in renewable energy technologies.

Advantages

- **Informed Decision-Making**: Provides stakeholders with valued visions to make well-informed results regarding investments, policy formulation, and market participation, enhancing operational efficiency.
- **Risk Mitigation:** Enables proactive risk management by identifying potential price fluctuations, allowing stakeholders to design risk mitigation strategies for their projects and investments.
- **Cost Optimization:** Helps in optimizing operational costs, resource allocation, and project financing, leading to cost-effective development and deployment of renewable energy projects.
- **Sustainable Growth**: Facilitates a sustainable and cleaner energy transition by aiding in the planning and incorporation of renewable energy into the grid, contributing to a greener and more sustainable.

Case Study

Title: Forecasting Solar Energy Prices for Investment Decisions.

Objective: To forecast solar energy prices for a specific region to aid investment decisions in solar power projects.

Methodology

1. **Data Collection:** Historical solar energy production data, weather patterns, technological advancements, and market trends.
2. **Feature Engineering**: Creating features such as sunlight hours, solar panel efficiency, weather forecasts, and energy demand patterns.
3. **Model Selection:** Utilizing a machine learning model like LSTM (Long_Short_Term Memory) for time sequence forecasting.
4. **Training and Validation**: Using historical data to train the model and validate its accuracy in predicting solar energy prices.
5. **Prediction:** Generating future solar energy price predictions for investment assessment.

Results: The LSTM model demonstrated a high accuracy rate in predicting solar energy prices, providing valuable insights for potential investors in the solar energy market.

8.3 Peak Load Forecasting

Peak weight forecasting involves predicting the maximum demand for electricity that a power system will experience during a specific period, typically daily or hourly. Accurate peak load forecasting is crucial for energy utilities and grid operators to ensure a reliable and stable energy supply. It helps in planning for sufficient capacity, scheduling maintenance, and making informed decisions regarding energy generation, transmission, and distribution.

Predicting peak load involves analysing historical consumption patterns, weather conditions, economic indicators, and societal events. Various prognostication models, counting statistical methods, machine learning algorithms, and artificial intelligence, are used to develop accurate predictions.

Roles

- **Energy Utilities and Grid Operators:** Utilize peak load forecasts to optimize power generation, ensure grid stability, and plan for capacity expansion to meet peak demand, minimizing energy wastage and operational costs.
- **Energy Planners and Infrastructure Developers**: Use peak load forecasts to plan new infrastructure projects, including power plants, transmission lines, and substations, aligning capacity with expected future demand.
- **Regulatory Bodies:** Rely on peak load forecasts to set regulatory standards, assess utilities' performance, and make policy decisions that encourage efficient capacity planning and resource allocation.

Case Study

Title: Forecasting Daily Highest Electricity Load for Town Area.

Objective: To forecast daily highest electricity load for a town area to aid grid planning and capacity allocation.

Methodology

1. **Data Collection**: Historical electricity consumption data, weather patterns, economic indicators, and past peak load records.
2. **Feature Engineering**: Creating features like historical peak loads, weather forecasts, diurnal of the week, and superior events. **Model Selection:** Using a combination of autoregressive combined touching average (ACTA) and neural system models for forecasting.
3. **Training and Validation**: Using historical statistics to train and authenticate the model's accuracy in predicting daily peak load.

Results: The neural network model demonstrated high accuracy in predicting daily peak electricity load, as long as valuable insights for capacity preparation and grid management.

9 Future Trends

9.1 *Role of AI and Quantum Computing in Future Trends*

9.1.1 Role of AI (Artificial Intelligence)

Enhanced automation and decision-making: AI algorithms enable automation of complex tasks and decision-making processes. Machine learning models learn from data, allowing systems to improve and optimize actions without explicit programming.

Personalized user experience: AI-powered schemes analyse user behaviour and favourites to provide personalized experiences, ranging from content recommendations to tailored product suggestions.

Cybersecurity: AI strengthens cybersecurity by identifying patterns in data to detect and prevent cyber threats. AI can rapidly respond to evolving threats, enhancing overall security.

Autonomous systems: AI is fundamental for the growth of autonomous vehicles, drones, and robotics. These technologies rely on AI algorithms for navigation, obstacle detection, and decision-making.

Automation and robotics: AI will power advanced automation and robotics, enabling machines to perform tasks with minimal human intervention. This will impact manufacturing, logistics, and services industries, foremost to augmented efficiency and productivity.

Natural Language Processing (NLP): AI-powered NLP will improve language translation, sentiment analysis, chatbots, and voice assistants, enhancing communication and accessibility across diverse languages and cultures.

9.1.2 Role of Quantum Computing

Exponential processing power: Quantum computing that proposes an exponential, computational control while compared to classical computer and serves as a instrumental for solving multifaceted problems in numerous domains such as optimization, cryptography and material science.

Optimization besides logistics: Quantum algorithms can optimize complex systems, supply chains, and logistics, leading to substantial efficiency improvements in various industries.

Cryptography and security: Quantum computing poses both a challenge and a solution to cryptography. It can break current encryption standards, driving the need for quantum-safe encryption methods.

Financial modelling: Quantum computing can enhance financial modelling and risk assessment by rapidly processing vast amounts of financial data and running complex simulations for investment strategies.

Complex problem solving: Quantum computing will tackle complex problems that classical computers struggle with, including optimization, drug discovery, molecular modelling, climate simulations, and financial modelling. It will significantly accelerate scientific advancements.

Quantum internet: Quantum computing will enable the development of a quantum Internet, facilitating secure quantum communication and distributed quantum computing, paving the way for a new era of interconnectedness and communication.

9.1.3 Future Synergy

Hybrid solutions: Combining AI and quantum computing can lead to powerful hybrid systems. AI algorithms can be enhanced using quantum processing to handle complex and large-scale datasets, pushing the boundaries of machine learning and data analysis.

Quantum AI algorithms: Research is ongoing to develop quantum algorithms specifically designed for AI tasks, potentially unlocking new capabilities and accelerating AI advancements.AI (Artificial Intelligence) is revolutionizing technology by automating complex tasks, personalizing user experiences, improving healthcare diagnostics, aiding sustainability efforts, enhancing cybersecurity, and powering autonomous systems.

In the future, combining AI and quantum computing will lead to powerful hybrid systems, and specialized quantum algorithms for AI tasks will accelerate advancements in both fields, determining the future of skill and culture.

9.2 *Impact of Renewable Energy Transition on Price Predictability*

The change to renewable liveliness is reshaping the dynamics of energy markets globally. As we move away from fossil fuel-dependent systems to embrace clean and sustainable energy foundations like solar, wind, hydro, and geothermal, the landscape of energy price predictability is undergoing a substantial transformation. Renewable energy technologies, coupled with advancements in energy storage and grid integration, are challenging traditional models of energy price forecasting. Understanding the impact of this transition on price predictability is vital for stakeholders in the energy sector, including investors, policymakers, utilities, and consumers.

9.2.1 Impact on Price Predictability

Intermittency and weather dependency: Renewable sources are inherently intermittent and weather-dependent. Predicting energy production from these sources becomes challenging due to variability in weather patterns, cloud cover, wind speed, and sunlight. This unpredictability directly influences the accuracy of price predictions.

Market dynamics and supply variability: The integration of renewable sources affects traditional supply-demand dynamics. Renewable energy's abundant availability during certain periods, such as sunny or windy days, can lead to temporary oversupply, causing fluctuations in energy prices that are harder to predict using conventional methods.

Policy and regulatory influences: Government policies, incentives, and regulations significantly impact the renewable energy sector. Sudden policy shifts or changes in subsidy structures can alter price predictability, making long-term forecasts uncertain.

9.2.2 Future Trends

Advanced predictive analytics: Amalgamation of advanced predictive analytics and machine learning algorithms that consider weather data, grid load, technology advancements, and policy shifts improves the accuracy of renewable energy value predictions.

Innovative forecasting models: Development of innovative forecasting models that account for the recurrent nature of renewable bases and incorporate real-time data to make short-term predictions more precise, ensuring grid stability and efficient energy trading.

Blockchain and decentralized energy markets: Implementation of blockchain technology to create transparent and decentralized energy markets. Smart contracts and blockchain-enabled platforms can streamline transactions and enable more accurate and automated pricing mechanisms.

Hybrid energy systems: The rise of hybrid energy systems integrating multiple renewable sources and energy storage. Forecasting models will need to evolve to accommodate the complexity of these hybrid systems and predict their combined output and impact on prices.

10 Conclusion

The changeover to renewable liveliness is redefining how we perceive and predict energy prices. The irregular nature of renewable sources, coupled with technological advancements and policy influences, presents both challenges and opportunities in accurately forecasting energy prices. Embracing innovative predictive models and adapting to changing market dynamics will be key to navigating the evolving landscape of renewable energy price predictability.

The traditional models of price forecasting are being challenged due to the intermittent and weather-dependent nature of renewable sources, supply variability, technological advancements, and policy dynamics. This paradigm shift necessitates a re-evaluation of how we predict energy prices and plan for a sustainable energy future.

To address these challenges, there is a growing need for advanced predictive analytics and innovative forecasting models. Machine learning algorithms and real-time data integration determination show a vital role in improving the correctness of predictions. Additionally, the adoption of blockchain technology and decentralized energy markets will bring transparency and efficiency to the pricing mechanisms, further enhancing predictability.

Interconnected Intelligence: Navigating Through Power Quality Checking and Control Using Smart Intelligence-Based Methods

Shreyansh Padarha, Sakshi Modi, and J. Lekha

1 Introduction

In the dynamic landscape of contemporary energy systems, the emergence of smart grids stands as a testament to the ongoing evolution of power distribution. As the economic burden of power quality problems continues to rise, there is a pressing need for innovative approaches to mitigate their impact. This chapter embarks on a comprehensive exploration, delving into the multifaceted dimensions that characterise smart grids and their profound implications for the global energy paradigm. Smart grids represent a transformative leap from traditional power distribution models, integrating technically superior methodologies while improving effectiveness, longevity, and better perishability. This chapter explores the structure of smart grids, explaining how sensors, communication networks, and algorithms work together to make them function.

At the heart of this technological revolution lies the strategic utilisation of data. The discourse unfolds to illuminate the critical role of data acquisition and preprocessing techniques, addressing the inherent challenges posed by the voluminous and dynamic nature of data in smart grid environments. Smart intelligence-based methods and applications, rooted in artificial intelligence, machine learning, and data analytics, emerge as potential game-changers in this domain. Power quality, a cornerstone of effective energy distribution, is scrutinised in detail, with a focus on diverse methodologies ranging from Fourier theory to advanced machine learning-based fault detection systems. The exploration extends beyond technical intricacies to encompass a comparative analysis of global smart grid development. Developed nations, emerging markets, and developing economies each contribute a unique narrative to the unfolding story of smart grids.

S. Padarha (✉) · S. Modi · J. Lekha
School of Sciences, CHRIST University, Lavasa, Maharashtra, India

S. Vijayalakshmi et al. (eds.), *Smart Power Systems*, Power Systems,
https://doi.org/10.1007/978-3-031-89459-6_9

This chapter aims to unravel the layers of complexity surrounding smart grids, providing readers with a nuanced understanding of their evolution and the transformative impact on the broader landscape of contemporary power systems.

2 Traditional Power Quality Monitoring and Control Measures

Power quality monitoring and control stand as an indispensable cornerstone within the realm of power systems. Conventional means of power quality oversight encompass the utilisation of age-old methodologies for scrutinising power quality, approaches anchoring voltage and current assessment, as well as well-established paths to mitigate power quality issues. Prior to delving into smart process and knowledge retrieval systems, it is crucial to have a foundational understanding of the longstanding history of power quality processes that span across centuries. The classical techniques employed for power quality scrutiny encompass the venerable Fourier analysis, the versatile wavelet analysis, and the time-frequency analysis [1].

2.1 *Fourier Transform Analysis*

Fourier transform is a mathematical transformation method that aids the process of identifying, segregating, and analysing different kinds of power quality (PQ) errors and fluctuations. The FT (Fourier Transform) can correctly identify the magnitudes and phase of harmonic components present in the signal [2]. Information from the Fourier transform spectrum applied on a car's power quality control can help identify the frequency components of vibrations, aiding in targeted power quality control measures [3]. However, it does not give the time information of these harmonics in the analysed signal. To overcome this limitation, wavelet transform of a disturbance signal can detect these disturbances accurately the magnitude and time information [4].

2.2 *Wavelet Transform Analysis*

In power systems, Wavelet analysis stands as a robust and widely adopted tool employed for identifying and analysing diverse power quality (PQ) events like imbalance, harmonics, etc. It has gained significant popularity in the domain of PQ disturbance detection and analysis within power systems. The wavelet transform, a mathematical technique of utmost importance, equips us with the capability to scrutinise signals in both the time and frequency domains, enabling a multifaceted understanding of signal characteristics [4]. Wavelet transform's distinctive ability lies in its capacity to disassemble a signal into various frequency bands, a feature that facilitates the precise identification of temporal and spectral attributes within the signal. This invaluable attribute empowers the accurate detection and characterisation of a wide range of PQ disturbances, including voltage sag, swell, transients, and harmonics [5]. The wavelet transform, by providing detailed magnitude and time information, excels in its capacity to pinpoint the onset and evolution of these disturbances, thus enabling robust PQ analysis and management [6].

This makes wavelet analysis a pivotal tool in maintaining high-quality power delivery and safeguarding sensitive equipment in power systems [7].

For a continuous signal $x(t)$, the CWT function is given as:

$$\mathrm{CWT}(a,b) = \int_{-\infty}^{+\infty} x(t)\cdot\psi_{a,b}(t)\,\mathrm{d}t = \frac{1}{\sqrt{a}}\int_{-\infty}^{+\infty} x(t)\cdot\psi\left(\frac{t-b}{a}\right)\mathrm{d}t \tag{1}$$

where the continuous variables are a and b, the scale and translation factor, respectively [8].

2.3 *Classic Voltage and Current Measurement*

The computing process with respect to the current and voltage constitutes a cornerstone in the realm of power quality control, serving as an indispensable foundation for ensuring the reliability and stability of electrical systems. Among the noteworthy methodologies employed for the precise measurement of voltage and current, we find the application of Rogowski coils, the precision of Hall effect sensors, and the utility of current transformers [9]. These established techniques, highlighted in Table 1, contribute towards advancing power quality control and management, thereby ensuring the optimal performance of electrical systems [13–16].

2.4 *Mitigating Limitations of Classic Power Quality Systems*

As for the time-tested strategies to rectify power quality concerns, they encompass the implementation of passive filters, the dynamic intervention of active filters, and the innovative combination of hybrid filters [17]. But, in the contemporary power sector, traditional power quality control methods remain relevant, essential, and are ideally complemented by the emerging paradigm of smart grids. Power quality issues continue to impose substantial annual costs globally, encompassing lost productivity, equipment damage, and customer dissatisfaction. In India, an APQI study estimated a loss of 20,000 Crore Rs annually, due to downtime and power quality control issues [18]. Traditional methodologies, deeply embedded in the history of power quality control, provide valuable insights into the health of electrical grids,

Table 1 Prominent current and voltage measurement techniques

Techniques	Description	Advantages and disadvantages	References
Rogowski coils	Low-power AC current transformer	**Advantages**	[4, 10]
		Low magnetic losses, low insertion impedance	
		Disadvantages	
		Accuracy affected by proximity, burden, and positioning. Proposed tests for accuracy evaluation, particularly with distorted signals	
Hall effect sensors	Deliver voltage signal proportional to magnetic field	**Advantages**	[11, 12]
		Suitable for AC or DC fields	
		Disadvantages	
		Noisy, temperature-dependent, affected by core saturation and DC offset, less suitable for AC currents. More challenging to engineer for high accuracy	
Current transformers	Widely used for AC current measurement in power systems	**Advantages**	[12]
		High accuracy, wide current range	
		Disadvantages	
		Core saturation in measuring distorted signals. Proposed linearized flux gate solution (LEM 'UltraStab' CTs) with minimal error over a broad frequency range	

offering a historical perspective and a foundation of knowledge. These classical techniques, including Fourier analysis, wavelet analysis, and time-frequency analysis, allow us to scrutinise power quality and understand its intricacies. They can still be employed for retrospective analysis and as a reference point for power quality control. However, the dynamic nature of power systems demands real-time adaptability, which smart grids with their AI-driven capabilities can provide [19, 20]. For example, the Fourier transform may provide a detailed frequency analysis, but it can be computationally intensive and may require complex signal processing, potentially increasing the resource demands for power quality control. Smart grids, driven by artificial intelligence, machine learning, and data analytics, present a complementary approach that addresses the limitations of traditional methodologies by offering real-time monitoring, assessment, and regulation of power quality. By leveraging intelligent algorithms, these systems optimise power performance, minimise downtime, and ensure the consistent delivery of high-quality electrical energy. The coexistence of traditional and smart approaches ensures a comprehensive, robust, and efficient power quality control system that meets the demands of the contemporary power landscape effectively.

3 Expert Systems: AI Breakthrough in Power Systems

AI (Artificial Intelligence), characterised by its unparalleled proficiency in the rapid analysis of extensive datasets, is instrumental in propelling power systems from a reactive to a proactive operational paradigm. In this subsection, we delve deep into the role of AI as a predictive and preemptive tool, which plays a pivotal role in averting the escalation of power quality deviations into critical anomalies. AI was integrated into power systems, through the advent of expert systems, a system that emulated the decision-making ability of a human expert by reasoning through a large database of knowledge in terms of if-then rules. These systems were used for diagnosis, control, design planning, and scheduling of power systems [21]. There has been a paradigm shift in the power industry from conditional rule-based (if-then-else-then) expert systems to AI and ML enabled power control systems in the past three decades [22, 23].

3.1 Load Handling in Distribution Expansion Planning

3.1.1 ANN-Based Load Forecasting

AI-based load forecasting leverages historical data, real-time information, and predictive models to anticipate electricity demand accurately as shown in Fig. 1.

Application of ML (Machine Learning) algorithms has improved the efficiency and accuracy in load management by identification of contributing factors and patterns [24, 25]. By considering variables like weather conditions, seasonal trends, and economic indicators, AI-driven systems can provide highly precise load forecasts. This allows utility companies to optimise resource allocation, plan for infrastructure upgrades, and ensure grid reliability.

Taking two implicated variables in power control systems, the association between load and temperature exhibits a nonlinear pattern. This pattern is often likened to a hockey stick or the Nike swoosh symbol. On the left side, as the temperature decreases due to heating requirements, the load increases. Conversely, on the right side, the load rises as the temperature increases due to cooling demands.

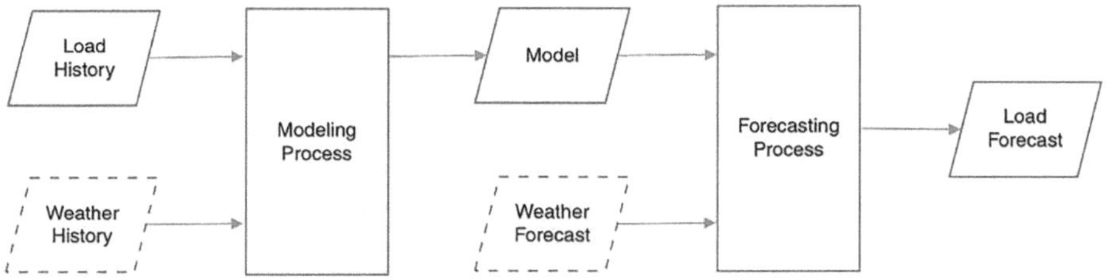

Fig. 1 Typical load forecasting model

3.1.2 Load Allocation

Traditionally, rule-based approaches have been implemented for load distribution and allocation, and these frameworks use past heuristic properties and variables to model and allocate power resources. These systems too have been built on the artificial intelligence development language PROLOG. Such an expert system was employed in Taiwan's power distribution system, in 1989 incorporating heuristic rules for minimum-loss and minimum-cost designs, ensuring high reliability in the selected expansion plan [26].

Smart intelligence is also being incorporated to optimise power allocation for direct consumers (end users). A 2022 study examined load allocation in residential homes and managed it through AI-based IoT technology, with a focus on optimising energy consumption and cost efficiency [27]. The study employed the Genetic Elephant Herd Optimization (GEHO) algorithm to schedule and control home appliances, shifting their operation to off-peak hours and reducing peak demand. By utilising various Operation Time Intervals (OTIs) and AI techniques, the research minimised energy expenses and enhanced the overall efficiency of load allocation, ultimately contributing to a more sustainable and cost-effective power management system for smart homes. The average costs in households were substantially lower when Genetic Elephant Herd Optimization (GEHO) algorithm was used to schedule and control appliances, while the costs were higher when power allocation was happening unplanned [27].

3.2 *Power Generation Scheduling*

Power engineering has a complexity known as the generator scheduling problem which influences the finances and management of power systems. It is the problem of having enough generators for usage to meet the per diem load demand with reserve capacity. This problem has been scrutinised by many researchers but the primary challenges like choosing the correct generator for commitment, developing schedules that conform to network and operational limitations, considering nonlinear features and challenges like premeditating the most cost-efficient generator timetable still persist [28, 29]. Using integer programming and dynamic programing like methods, algorithms have been created for creating appropriate schedules. Unfortunately nonlinear features are tough to incorporate in these algorithms as they often rely on simplified constraints.

Wong and Tsang [30] devised an AI-based algorithm for automating the process of generator scheduling, accounting for operational constraints and nonlinear ramp rates. Their method converted the complexity as a tree-search, using heuristics to find the optimal schedule by exploring various paths and pruning faulty ones, thus balancing load demand with generation levels efficiently [30]. An enhanced greedy generation scheduling approach that uses an automated energy operation model and is based on a classic heuristic algorithm approach is presented [31].

3.3 Real-Time Fault Diagnosis and Alarm Handling

Neural networks and varied deep learning architectures are extensively used for power systems fault diagnosis and alert systems. Real-time fault diagnosis pipelines and frameworks have been in existence since the 1990s, but modern systems are much more complex and efficient, capturing multiple data sources and generating possible response mechanisms, and quasi-real-time fault processing system was implemented in a study by Jinjie Ji et al. [32, 33].

Fuzzy logic and ANNs have been used in tandem, for real-time fault detection and diagnosis in power systems across various studies [34–37]. One study, by Gustavo Pérez Alvarez underscores the significance of these intelligent monitoring and supervision systems in boosting the reliability and effectiveness of electricity generation and distribution [38]. These systems autonomously make decisions in real time, offer maintenance recommendations, and determine the best actions to mitigate or eliminate changes or faults within the power system.

The study also highlights the importance of data management infrastructure and the integration of auxiliary systems to minimise the impact of failures and improve overall system performance. The pipeline presented in the Alvarez Study, where the smart decision relies on fuzzy logic for triggering the solution phase. Fuzzy logic enables the system to work with degrees of truth, which is particularly useful when dealing with uncertain or vague data. It allows the system to make decisions based on partial truths rather than strict binary true/false values. It also allows for handling imprecise information, while ANN evaluates expert system results to select the best solution.

While contemporary alarm systems and fault detection mechanisms in power systems are intricate, they do possess inherent limitations. AI plays a crucial role in addressing these constraints. Langstrand et al. tackled the challenge of 'alarm flooding' a situation characterised by an overwhelming deluge of information during plant disturbances or abnormal operating conditions, in a study by using deep neural networks to effectively filter expected alarms. This research exemplifies how AI holds promise in enhancing alarm management during emergencies and reducing the burden of excessive information for operators in scenarios such as nuclear power plants [38, 39].

Artificial neural networks are also being utilised for future alarm prediction in data-driven process control systems. They learn from historical data, aiding plant operators in preventing failures. Different models, such as LSTM and FCN, are applied to provide alarm predictions with various time windows, ensuring timely corrective actions. In a 2022 study, LSTM performed the best for alarm predictions, giving a 94% accuracy at 10 min intervals [40].

3.4 Power Quality Control

Over the last decade, there has been a substantial surge of interest in the field of Power Quality (PQ) research, with the complexity of diagnosing PQ issues requiring a profound level of engineering expertise that spans across numerous domains [41–44]. The challenge lies in the multidisciplinary nature of the required expert knowledge, encompassing domains ranging from electric drives to harmonics and signal analysis. To address these intricacies, innovative tools rooted in artificial intelligence (AI) have come to the forefront, attracting considerable attention from researchers in the electric power domain, leading to a substantial body of literature on AI applications in this field [45–57]. AI techniques have found utility in various facets of electric power like security, system diagnostics, and load predictive modelling along with power electronics and electrical machine applications [58, 59].

4 Smart Grid: Data Gathering and Preprocessing Methods

The concept of smart grids has evolved over the years and is characterised by various key attributes that contribute to power quality management. Several definitions of smart grids have been proposed, and they can be understood from both a solution perspective and a components perspective [60]. The climate group defines a smart grid as a combination of software and hardware tools facilitating efficient power routing, real-time information exchange with customers, and demand side management, aiming to enhance energy efficiency and data capture [61]. Adam and Wintersteller emphasise the use of digital technology to augment and optimise energy consumption, integrate renewable sources, and engage customers through smart metering [62]. Similarly, Miller outlines the multifaceted role of the smart grid, including enabling consumer participation, accommodating diverse energy sources, fostering new markets and services, ensuring power quality, optimising asset utilisation, and enhancing resilience against disruptions [63]. In essence, a smart grid refers to an electrical network that employs information and communication technology to autonomously gather and analyse data, encompassing information on power provider and consumer interactions. This approach is aimed at improving the economic viability, efficacy, green nature, and dependability of electricity generation and delivery.

Innovative concepts in smart grid technology necessitate the integration of sensitive electrical and electronic components. Maintaining electromagnetic harmony among these parts is essential for an uninterrupted and high-quality energy supply in future deregulated grid environments. The reliability of smart grids heavily relies on power quality, which covers two primary types of disturbances, namely events and variations. Events include sudden voltage fluctuations, dips, swells, and interruptions, while variations entail gradual voltage shifts, harmonics, flickers, and imbalances. The quality of interaction between the grid and connected devices determines the actual power quality levels [64].

4.1 *Attributes of Big Data in Smart Grid*

Within the smart grid domain, the SGAM (Smart Grid Architecture Model) offering a comprehensive framework to comprehend this complex system. SGAM integrates three dimensions, blending domains, zones, and layers, with domains encompassing areas such as generation, transmission, and distribution.

Zones delineate the structure of power system management, including enterprise, market, station, and process. The interoperability layers consist of components like business models, information, and communication. This architectural overview assists in identifying the constraints and commonalities among existing smart grid standards [65]. Similar to the widely accepted 5 Vs of big data, the smart grid big data also have similar attributes [66, 67]:

- **Volume** (enormous amounts of data from devices like smart metres).
- **Velocity** (real-time data exchange requirements).
- **Variety** (diverse data formats).
- **Veracity** (concerns regarding data quality and accuracy).
- **Value** (the challenge of extracting valuable insights from vast data sets).

4.2 *Sources of Data in Smart Grids*

Intelligent grids represent a fusion of energy and information, generating extensive data covering various processes such as generation, transfer, distribution, and utilisation of electricity. The data storage contains electrical details obtained from distribution stations, distribution switch stations, electricity metres, as well as non-electrical information such as marketing data, meteorological observations, and regional economic metrics [68]. This amassed data serves crucial functions such as power plant scheduling, subsystem operation, critical power tools maintenance, and strategic guidance for marketing initiatives. These data sources are typically classified into three main categories: measured data (captured through sensors and smart metres for system monitoring), business data (encompassing marketing strategies and competitor analysis), and external data [68, 69]. The rise of traits associated with large datasets and their abundance of data sources in smart grids have necessitated advanced techniques for data acquisition and preprocessing. Adopting data-driven technologies is vital for extracting meaningful insights and ensuring the effective operation of smart grids.

4.3 Data Preprocessing Techniques for Smart Grids: Enhancing Data Quality and Utility

Data preprocessing plays a crucial role in improving data quality for smart grid analytics. It is a crucial step in data churning and interpretation which includes refining, altering, and readying unprocessed data.

In smart grid settings, preprocessing methods are employed to derive valuable insights from the extensive data produced by smart grid devices. A survey of data preprocessing methods used in the smart grid domain found that the most commonly used techniques include imputing missing values, identifying outliers, reducing data dimensions, scaling data, transforming data, and partitioning data [70–73].

Missing data is a challenge encountered when preprocessing data for grids. Missing data can occur due to a variety of reasons such as sensor failure or communication errors. Missing data can be handled through imputation techniques such as mean imputation, regression imputation, and k-nearest neighbour imputation [74]. Another difficulty is dealing with outliers in the data. Outliers can occur due to measurement errors or other anomalies in the data. Outliers can be mitigated by using box plot, z-score method, and clustering techniques [74, 75].

Data reduction techniques are used to reduce the size of large datasets while preserving their essential characteristics. PCA (Principal Component Analysis) is a statistical technique that can be used to identify patterns in large datasets by reducing their dimensionality while retaining most of their variability [76, 77].

Data scaling is another important technique. It is used in smart grid data preprocessing to normalise the range of values of different attributes in a dataset. Normalisation helps to ensure that all attributes are treated equally during analysis and that no attribute dominates the others due to its scale [76]. Using data transformation, raw data is transformed into a more suitable format for analysis. For example, logarithmic transformation can be used to correct the distribution shape of severely skewed data [76]. After which data partitioning is performed where large datasets are divided into smaller subsets. Cross-validation is a commonly used partitioning technique that involves dividing a dataset into two parts: training set and testing set [78].

4.4 Modelling and Optimising Smart Grids (SGs) Using Data Analytics Techniques

Analysing extensive and complex data sets is essential for deriving meaningful insights. Within smart grids, this analysis is vital for utilising the substantial data produced by smart grid devices to uncover patterns and trends, leading to improved grid performance. Numerous analytical methods are applied to various smart grid applications, including load prediction, fault identification, preventive maintenance, and managing power quality (Table 2).

Table 2 Data-driven techniques in SGs [71, 73]

Algorithm	Applications	Category
Artificial neural networks (ANN)	Load forecasting, fault detection	Supervised
Support vector machines	Stealthy false data injection detection	Supervised
PCA (principal component analysis)	Data reduction, feature extraction	Dimensionality reduction
Random forest	Smart grid stability prediction	Supervised
Decision trees	Fault detection, diagnosis, small-signal stability analysis	Supervised
K-nearest neighbours	Predicting power consumption	Supervised
Hidden Markov models	Islanding prediction, anomaly detection	Unsupervised
Genetic algorithms (GA)	Optimising demand response in smart grids	Supervised
Particle swarm optimization (PSO)	Energy consumption monitoring	Semi-supervised
Naïve Bayes	Demand side management	Supervised

5 Transformation of Smart Grids: From Traditional to Next-Generation Smart Grids

Future power industry practices are poised to be based on smart grid technologies. Comparing them to conventional electrical grids, their goals are to be more dependable, efficient, and robust. Using state-of-the-art technologies like sensors, transmission networks, and data analytics, smart grids improve grid performance and customer satisfaction. Next-generation smart grids (NGSG) outperform previous smart grids (SG) through the use of modern and more automation and control methodology. The NGSG guarantees effective energy production, transmission, and distribution by deeply integrating data-driven methodologies. Agent-oriented demand side management, Fog computing to manage consumption, agent-based energy transformation, internet of things (IoT) equipped inverter, intelligent control, etc. are just a few of the sophisticated technical aspects that come with the NGSG. Predictably, it will ensure its consistent operation and reduce the range of nonlinearity effects [71].

5.1 Compound Market Grid Annual Growth Rate

Smart grids are multimodal networks that provide instantaneous information sharing between providers and consumers. The deployment of technologies like energy storage and smart metres in conjunction with their implementation will play a significant role in ensuring a reliable and efficient supply of increasingly environmentally friendly electricity. The market for smart grids worldwide was valued at 49.7 billion dollars in 2022 and is expected to reach over 130 billion dollars by 2028 [79].

Data-driven technologies are now influencing the evolution of conventional power systems, leading to the emergence of an ordinary smart grid (SG) and a subsequent-generation smart grid (NGSG). Advanced technologies such as smart metres, sources of environmentally friendly energy, and sophisticated transmission systems are integrated by the NGSG, enabling seamless communication and enhancing power production quality. This transition has resulted in significant annual growth rates for countries embracing NGSGs, fostering their journey towards a sustainable future with minimised carbon emissions and lifelong sustainability.

5.2 Difference Between Traditional SG and NGSG Along with Their Architectural Framework

Over the past few years, the landscape of power generation and distribution has undergone a transformative shift towards more sustainable and efficient solutions. As a testament to this transition, a comprehensive comparison between the traditional Smart Grid (SG) and the emerging Next-Generation Smart Grid (NGSG) becomes imperative to understand the magnitude of changes. This comparative analysis encompasses various critical aspects, ranging from power generation methods to security technologies, offering an insightful look into the evolving field of smart grids. The table below presents a side-by-side evaluation of the key characteristics that set apart the traditional SG from the NGSG, highlighting the monumental transition in the power grid sector (Table 3).

The traditional smart grid (SG) and the next-generation smart grid (NGSG) exhibit significant differences in their architectural frameworks as well. The

Table 3 Comparing smart grid frameworks

Traditional SG	NGSG
Centralised power grid	Decentralised power grid
Large power plants	Distributed energy resources
Transmission and distribution lines	Microgrids
Unidirectional power flow	Bidirectional power flow
Limited use of renewable energy sources	Extensive use of renewable energy sources
Limited use of energy storage systems	Prominent utilisation of energy storage devices
Digital electronics, microprocessors	IoT components and devices made possible by edge computing
SGs provide automated protection	Blockchain technology of the next generation will support the security system

traditional SG operates as a centralised power grid, relying on large power plants to generate electricity, which is then transmitted to consumers through an extensive network of transmission and distribution lines. It is separated into three primary segments: the physical surface, which deals with the production, distribution, and transmission of electricity, and the communication layer and application layer. The communication layer encompasses the infrastructure for data exchange among various components of the SG, and the application layer includes software applications responsible for managing and controlling the grid [80, 81]. In contrast, the NGSG represents a shift towards a decentralised power grid that leverages distributed energy resources (DERs). Its architectural framework introduces several key components to enable advanced functionalities:

- **Advanced Metering Infrastructure (AMI):** This component allows utility companies and customers to communicate instantaneously by giving contemporaneous data on energy consumption.
- **Distribution Automation (DA):** DA automates the distribution of electricity by integrating sensors and control systems, allowing for effective monitoring and management of electricity flow.
- **Distributed Energy Resources (DER):** Natural energy sources that can be smoothly connected into the grid, supporting sustainable energy generation, are examples of DER.
- **Demand Response (DR):** Demand Response (DR) incentivises consumers to reduce their energy use during peak hours, giving facilities control over peak energy demand [82].

The NGSG's architectural framework reflects a departure from the centralised model of the traditional SG and embraces decentralisation, renewable energy sources, and advanced communication and control systems, aligning with the goals of sustainability, reliability, and efficiency in modern power grids.

5.3 Smart Grid Initiatives in Different Countries

The International Smart Grid Action Network (ISGAN) has conducted a survey among experts in 18 member countries on smart grid drivers and associated technologies in 2020 [83]. The purpose of the project was to choose the top six drivers and top five technologies, as well as to thoroughly assess the current economic levels of each member country by continent. This research aims to assist in the early establishment of policies for promoting and implementing smart grids, as well as to statistically assess the status in each nation in order to give reference data for detecting temporal changes in drivers and technologies, as well as the primary promotion strategy in each country.

The European Union has also been actively involved in smart grid projects. The European Technology and Innovation Platform for Smart Networks for Energy Transition has identified several smart grid projects that are being put into practice across Europe [84]. These projects aim to develop innovative solutions for incorporating into the grid sustainable energy sources, improving energy efficiency, and ensuring grid stability. They use AI, data-driven technologies, and power quality management to achieve these goals.

The operation of various programs used in the power management system, including demand response, demand management, and energy quality management, is thoroughly and critically analysed in this review.

The table shows the initiative that uses data-driven technologies for smart grid projects across various countries (Table 4).

6 Conclusion

Several definitions of smart grids have been proposed, and they can be understood from both a solution perspective and a components perspective, as proposed by Vivek Agarwal and Lefteri H. Tsoukalas [60]. The evolution of power systems, as illustrated in the varied sections, reflects a transformative journey from traditional rule-based approaches to the integration of advanced AI and ML technologies. The application of AI and ML in load forecasting, power generation scheduling, real-time fault diagnosis, and power quality control has significantly enhanced the efficiency, reliability, and sustainability of smart grids. Different neural network architectures can be integrated to solve power systems problems. The evaluation stages in smart grid projects involve a comprehensive assessment of reliability and performance, employing sophisticated methods to ensure optimal functionality. A comprehensive evaluation model integrates economic, environmental, and social aspects, establishing a multi-layered framework. For example, Faster R-CNN is tailored for industrial objects detection or LSTM is used for sequential time sensitive problem statements. Albeit there are challenges and hindrances in distribution expansion planning, emphasising the shift from rule-based load allocation to AI-driven optimisation strategies [100–101]. Moreover, there exists a critical role of data acquisition and methods for initial processing in smart grids, addressing big data characteristics and highlighting methods to enhance data quality. The use-cases extend to modelling and optimising smart grids using diverse data analytics techniques, showcasing the significance of support vector machines and artificial neural networks. Few models utilise the Analytical Hierarchy Process (AHP) to analyse the intricate relationship between layers, facilitating a holistic understanding of smart grid performance [93]. Another approach employs the Fuzzy Analytical Hierarchy Process (FAHP) method for reliability assessment [94]. This method involves the systematic evaluation of criteria, incorporating fuzzy logic to handle uncertainties and complexities in smart grid [95–96] reliability assessment. The emergence of next-generation smart grids centres around their decentralised, renewable energy-focused architecture and the transformative impact on the power industry [97–99]. The substantial growth forecast in the smart grid market underscores

Table 4 Smart grid initiatives in different countries

Country	Project name	Location	Description	Year started
South Africa [85]	Renewable Energy Independent Power Producer Procurement Programme (REIPPPP)	Various locations across South Africa	Competitive bidding program to procure renewable energy from independent power producers	2011
China [86]	State Grid Corporation of China (SGCC)	Various locations across China	State-owned enterprise responsible for building and operating power grids in China	2002
Japan [87]	Smart Grid Demonstration Project	Various locations across Japan	Government-funded project to promote investments in smart power grids	2010
Germany [88]	E-Energy Project	Various cities across Germany including Berlin, Frankfurt, and Mannheim	Smart grid pilot project aimed at integrating renewable energy sources into the grid	2010
India [89]	National Smart Grid Mission (NSGM) and Smart City Mission (SCM)	Various locations across India including Puducherry, Chandigarh, and Ahmedabad	NSGM aims to modernise the power sector by implementing smart grid technologies while SCM aims to develop sustainable cities with smart grid solutions	2010
United States [90, 91]	California Solar Initiative (CSI), Texas Smart Grid Initiative (TXSGI), and New York State Smart Grid Consortium (NYSSGC)	Various locations across California, Texas, and New York	CSI aims to install 3,000 MW of solar capacity by 2016, TXSGI aims to modernise the grid with advanced technologies, and NYSSGC aims to develop a comprehensive smart grid roadmap in New York State	CSI: 2007, TXSGI: 2008, NYSSGC: 2008
Canada [91]	Ontario Smart Grid Fund (OSGF), Quebec Smart Energy Strategy, and British Columbia Hydro Smart Grid Program (BC Hydro)	Various locations across Ontario, Quebec, and British Columbia	OSGF aims to support the development of a modern electricity system in Ontario	OSGF: 2011, Quebec Smart Energy Strategy: 2006, BC Hydro: 2011
Australia [91]	Smart Grid, Smart City Project (SGSC) and Victorian Smart Grid Trial (VSGT)	Various locations across Australia including Sydney and Victoria	SGSC aimed to demonstrate the benefits of smart grids in urban areas while VSGT aimed to test the feasibility of integrating natural resources into the grid	SGSC: 2010, VSGT: 2009

(continued)

Table 4 (continued)

Country	Project name	Location	Description	Year started
United Kingdom [92]	Low Carbon London Project and Manchester Energy Project (MEP)	Various locations across London and Manchester, respectively	While MEP sought to create a low-carbon energy system for Manchester, the Low Carbon London Project sought to determine whether it was feasible to incorporate sources of clean energy into the grid	Low Carbon London Project: 2010, MEP: 2013
France [92]	Lyon Confluence Project and Grenoble Eco-Cité Project (GECO)	Lyon Confluence district in Lyon and Grenoble respectively	Lyon Confluence Project aimed to develop a sustainable district with smart infrastructure while GECO aimed to develop a low-carbon energy system in Grenoble	Lyon Confluence Project: 2003, GECO: 2012

the pivotal role of data-driven technologies in shaping the future of power systems, contributing to a more sustainable and efficient energy landscape.

References

1. Granados-Lieberman, D., Romero-Troncoso, R. J., Osornio-Rios, R. A., Garcia-Perez, A., & Cabal-Yepez, E. (2011). Techniques and methodologies for power quality analysis and disturbances classification in power systems: a review. *IET Generation, Transmission & Distribution, 5*(4), 519–529.
2. Venkatesh, C., & Siva Sarma, D. V. S. S. (2010). *Techniques used for detection of power quality events—A comparative study*. National Power Systems Conference (NPSC).
3. Zhu, C., Huang, H., Liu, H., Song, C., Ma, F., & Liu, Z. (2014). On-line vibration monitoring and diagnosing of a multi-megawatt wind turbine gearbox. In P. Velex (Ed.), *International Gear Conference 2014: 26th–28th August 2014, Lyon* (pp. 1089–1098). Chandos Publishing.
4. Redfern, M. A. (2007). *Power quality analysis using Fourier theory*. International Conference on Power Electronics and Drive Systems.
5. Pattnaik, S. S., Parida, S. K., & Panda, S. K. (2010). Wavelet transform based detection of power quality disturbances. *International Journal of Computer Applications, 1*(9), 1–5.
6. El-Sharkawy, M. A., & Abido, M. A. (2009). Power quality disturbance classification using wavelet transform energy entropy and support vector machines. *Electric Power Systems Research, 79*(11), 1498–1508.
7. Phadke, A. G., & Thorp, J. S. (2009). *Computer relaying for power system* (2nd ed.). Research Studies Press.
8. Aggarwal, R., & Kim, C. H. (2007). Wavelet transform in power system: Part 1—General introduction to the wavelet transform. Power Engineering Journal, 14(2), 81–87.
9. Kulkarni, S. V., & Khaparde, S. A. (2010). A review on voltage and current sensors for power system applications. *IEEE Transactions on Power Delivery, 25*(9), 2534–2545.
10. Mingotti, A., et al. (2022). Effect of proximity, burden, and position on the power quality accuracy performance of Rogowski coils. *Sensors, 22*(1), 397.
11. Texas Instruments Precision Hub Blog. (2020). *Six ways to sense current and how to decide which to use*.

12. JakeG84. (2017). *Current monitoring—CT vs hall effect vs Rogowski coil?* Electrical Engineering Stack Exchange.
13. Mingotti, A., Peretto, L., & Tinarelli, R. (2020). Smart characterization of Rogowski coils by using a synthesized signal. *Sensors, 20*(12), 3359. https://doi.org/10.3390/s20123359
14. Arar, S. (2021). *Hall effect current sensing: Open-loop and closed-loop configurations*. All About Circuits. https://www.allaboutcircuits.com/technical-articles/hall-effect-current-sensing-open-loop-and-closed-loop-configurations/
15. Pollefliet, J. (2018). Current, angular position, speed transducers. In J. Pollefliet (Ed.), *Power electronics* (pp. 18.1–18.34). Academic Press. https://doi.org/10.1016/B978-0-12-814641-5.50003-2
16. Burkard, J., & Biela, J. (2020). *Hybrid transformers for power quality enhancements in distribution grids: Comparison to alternative concepts* [Conference Paper]. https://api.semanticscholar.org/CorpusID:211094082
17. Singh, B., & Singh, B. (2016). A review on power quality improvement techniques using passive, active, and hybrid filters. *Renewable and Sustainable Energy Reviews, 60*, 1318–1334.
18. Forum of Regulators (FOR), India. (2015). *Power quality regulations in India*.
19. Faruque, M. D. O., et al. (2015). Real-time simulation technologies for power systems design, testing, and analysis. *IEEE Power and Energy Technology Systems Journal, 2*(2), 63–73. https://doi.org/10.1109/JPETS.2015.2427370
20. Luo, A., Xu, Q., Ma, F., & Chen, Y. (2016). Overview of power quality analysis and control technology for the smart grid. *Journal of Modern Power Systems and Clean Energy, 4*(1), 1–9. https://doi.org/10.1007/s40565-016-0185-8
21. McCarthy, J., Minsky, M. L., Rochester, N., & Shannon, C. E. (2006). A proposal for the Dartmouth summer research project on artificial intelligence, August 31, 1955. *AI Magazine, 27*(4), 12.
22. Kazibwe, W. E., & Sendaula, H. M. (1992). Expert system targets power quality issues. *IEEE Computer Applications in Power, 5*(2), 29–33. https://doi.org/10.1109/67.127821
23. Tzafestas, S. G. (2011). Expert control systems: An introduction with case studies. In *Control systems, robotics, and automation* (Vol. XVII, pp. 343–356). Encyclopedia of Life Support Systems (EOLSS) National Technical University of Athens.
24. Sheikh, S. K., & Unde, M. G. (2012). Short-term load forecasting using ANN technique. *International Journal of Engineering Sciences & Emerging Technologies, 1*(2), 97–107. https://doi.org/10.7323/ijeset/v1_i2_12
25. Hong, W.-C. (2009). Electric load forecasting by support vector model. *Applied Mathematical Modelling, 33*(5), 2444–2454. https://doi.org/10.1016/j.apm.2008.07.010x
26. Roopa, Y. M., SatheshKumar, T., Khatoon Mohammed, T., Turukmane, A. V., Krishna, M. S. R., & Krishnaiah, N. (2022). Power allocation model for residential homes using AI-based IoT. *Measurement: Sensors, 24*, 100461. https://doi.org/10.1016/j.measen.2022.100461
27. Chen, J.-L., & Hsu, Y.-Y. (1989). *An expert system for load allocation in distribution expansion planning*. National Taiwan University.
28. Baldwin, C. J., Dale, K. M., & Dittrich, R. F. (1959). A study of the economic shutdown of generating units in daily dispatch. *Transactions of the American Institute of Electrical Engineers Power Apparatus and Systems, Part III, 78*(4), 1272–1282.
29. Bard, J. F. (1988). Short-term scheduling of thermal-electric generators using Lagrangian relaxation. *Operations Research, 36*(5), 765–766.
30. Wong, K. P., & Tsang, C. P. (1989). A heuristic search method for short-term thermal generator scheduling. In J. S. Gero (Ed.), *Artificial intelligence in design* (pp. 505–520). Springer.
31. Abdel-Hakim, A. E., & Abo-Elyousr, F. K. (2022). Heuristic greedy scheduling of electric vehicles in vehicle-to-grid microgrid owned aggregators. *Sensors, 22*(6), 2408. https://doi.org/10.3390/s22062408
32. Swarup, K. S., & Chandrasekharaiah, H. S. (1991). Fault detection and diagnosis of power systems using artificial neural networks. In *Proceedings of the First International Forum on Applications of Neural Networks to Power Systems, Seattle, WA* (pp. 102–106). IEEE. https://doi.org/10.1109/ANN.1991.213505

33. Ji, J., Chen, Q., Jin, L., Zhou, X., & Ding, W. (2021). Fault diagnosis system of power grid based on multi-data sources. *Applied Sciences, 11*, 7649. https://doi.org/10.3390/app11167649
34. Sun, J., Qin, S.-Y., & Song, Y.-H. (2004). Fault diagnosis of electric power systems based on fuzzy Petri nets. *IEEE Transactions on Power Systems, 19*(4), 2053–2059. https://doi.org/10.1109/TPWRS.2004.836256
35. Demidova, G., Rassõlkin, A., Vaimann, T., Kallaste, A., Zakis, J., & Suzdalenko, A. (2021). An overview of fuzzy logic approaches for fault diagnosis in energy conversion devices. In *2021 28th International Workshop on Electric Drives: Improving Reliability of Electric Drives (IWED), Moscow, Russia* (pp. 1–7). IEEE. https://doi.org/10.1109/IWED52055.2021.9376389
36. Youssef, O. A. S. (2004). Combined fuzzy-logic wavelet-based fault classification technique for power system relaying. *IEEE Transactions on Power Delivery, 19*(2), 582–589. https://doi.org/10.1109/TPWRD.2004.826386
37. Petrović, I., Nikolovski, S., Glavaš, H., & Relić, F. (2020). Power system fault detection automation based on fuzzy logic. In *2020 International Conference on Smart Systems and Technologies (SST), Osijek, Croatia* (pp. 129–134). https://doi.org/10.1109/SST49455.2020.9264039
38. Pérez Alvarez, G. (2020). *Real-time fault detection and diagnosis using intelligent monitoring and supervision systems*. IntechOpen. https://doi.org/10.5772/intechopen.90158
39. Langstrand, J. P., Nguyen, H. T., & McDonald, R. (2021). Applying deep learning to solve alarm flooding in digital nuclear power plant control rooms. In T. Ahram (Ed.), *Advances in artificial intelligence, software and systems engineering. Advances in intelligent systems and computing* (Vol. 1213, pp. 821–830). Springer. https://doi.org/10.1007/978-3-030-51328-3_71
40. Dix, M., et al. (2022). An AI-based alarm prediction in industrial process control systems. In *2022 IEEE International Conference on Big Data and Smart Computing (BigComp), Daegu, Republic of Korea* (pp. 242–245). IEEE. https://doi.org/10.1109/BigComp54360.2022.00053
41. Key, T. S. (1979). Diagnosing power quality-related computer problems. *IEEE Transactions on Industry Applications, IA-15*, 381–393.
42. Martzloff, F. D., & Gruzs, T. M. (1988). Power quality site surveys: Facts, fiction, and fallacies. *IEEE Transactions on Industry Applications, 24*, 1005–1018.
43. Burke, J. J., Grifith, D. C., & Ward, J. (1990). Power quality—Two different perspectives. *IEEE Transactions on Power Delivery, 5*, 1501–1513.
44. Domijan, A., Heydt, G. T., Meliopoulos, A. P. S., Venkata, S. S., & West, S. (1993). Directions of research in electric power quality. *IEEE Transactions on Power Delivery, 8*, 429–436.
45. Madan, S., & Bollinger, K. E. (1997). Applications of artificial intelligence in power systems. *Electric Power Systems Research, 41*, 117–131.
46. Wildberger, A. M. (1995). Overview of exploratory research in soft computing applications to power systems at the Electric Power Research Institute. In *Proceedings of the IEEE International Conference on Systems, Man and Cybernetics, Vancouver, BC, Canada* (Vol. 3, pp. 2007–2012). IEEE.
47. Liu, C.-C. (1997). Intelligent systems applications to power systems. *IEEE Computer Applications in Power, 10*, 21–24.
48. Rahman, S. (1993). Artificial intelligence in electric power systems: A survey of the Japanese industry. *IEEE Transactions on Power Systems, 8*, 1211–1218.
49. Liu, C.-C., & Dahlgren, R. (1997). Intelligent systems applications in a competitive power industry environment. In *Proceedings of the IEEE Fourth International Conference on Advances in Power System Control, Operation & Management, Hong Kong* (pp. 24–29). IEEE.
50. El-Hawary, M. E. (1998). *Electric power applications of fuzzy systems*. IEEE Press.
51. Warwick, K., Ekwue, A., & Aggarwal, R. (1997). *Artificial intelligence techniques in power systems*. IEE Press.
52. Momoh, J. A., & El-Hawary, M. E. (2000). *Electric systems, dynamics, and stability with artificial intelligence applications*. Marcel Dekker.

53. Zhang, Z. Z., Hope, G. S., & Malik, O. P. (1989). Expert systems in electric power systems—A bibliographical survey. *IEEE Transactions on Power Systems, 4*, 1355–1363.
54. Taylor, T., & Lubkeman, D. (1989). Applications of knowledge-based programming to power engineering problems. *IEEE Transactions on Power Systems, 4*, 345–352.
55. Anis, W. R., & Morcos, M. M. (1998). Electric power quality and artificial intelligence: Applicability and overview. In *Proceedings of Large Engineering System Conference of Power Engineering, Halifax, NS, Canada* (pp. 107–111). IEEE.
56. . (1999). Electric power quality and artificial intelligence: Applicability and overview. *IEEE Power Engineering Review, 19*, 63–65.
57. Morcos, M. M., & Anis Ibrahim, W. R. (1999). Electric power quality and artificial intelligence: Overview and applicability. *IEEE Power Engineering Review, 19*, 5–10.
58. Vas, P. (1999). *Artificial-intelligence-based electrical machines and drives*. Oxford University Press.
59. Bose, B. K. (1994). Expert system, fuzzy logic, and neural network applications in power electronics and motion control. *Proceedings of the IEEE, 82*, 1303–1323.
60. Agarwal, V., & Tsoukalas, L. H. (2011). Smart grids: Importance of power quality. In *Energy-efficient computing and networking* (pp. 136–143). Springer. https://doi.org/10.1007/978-3-642-19322-4_15
61. Compromiso RSE. (n.d.). *Smart 2020: Transformando España hacia la eficiencia*. https://www.compromisorse.com/upload/estudios/000/36/smart2020.pdf
62. Atkinson, R. D., & Castro, D. (2008). Digital quality of life: Understanding the personal and social benefits of the information technology revolution. *SSRN Electronic Journal*. https://doi.org/10.2139/ssrn.1278185
63. Miller, J. (2008). The smart grid: How do we get there? *Smart Grid News*.
64. Bollen, M. H. J., Zhong, J., Zavoda, F., Meyer, J., McEachern, A., & Lopez, F. C. (2010). Power quality aspects of smart grids. In *The International Conference on Renewable Energies and Power Quality (ICREPQ'10), Granada, Spain, 23–25 March 2010*. http://hdl.handle.net/10722/126131
65. CEN-CENELEC-ETSI Smart Grid Working Group Reference Architecture. (2012). *Reference architecture for the smart grid* (Technical report)
66. Zhu, T., Sheng, X., Zhang, Q., Gu, Y., Yi, P., & Li, Y. (2015). Emergent technologies in big data sensing: A survey. *International Journal of Distributed Sensor Networks, 2015*, 902982.
67. SyC Smart Energy. (2021). *IEC SRD 63200—SGAM basics*. https://syc-se.iec.ch/deliveries/sgam-basics/
68. Liu, K., Sheng, W., Zhang, D., Jia, D., Hu, L., & He, K. (2015). Big data application requirements and scenario analysis in smart distribution network. *Proceedings of the CSEE, 35*(2), 287–293.
69. Zhao, T., Zhang, Y., & Zhang, D. (2014). Application technology of big data in smart distribution grid and its prospect analysis. *Power System Technology, 38*(12), 3305–3312.
70. Alghamdi, T. A., & Javaid, N. (2022). A survey of preprocessing methods used for analysis of big data originated from smart grids. *IEEE Access, 10*, 29149–29171. https://doi.org/10.1109/ACCESS.2022.3157941
71. Ahsan, F., Dana, N. H., Sarker, S. K., et al. (2023). Data-driven next-generation smart grid towards sustainable energy evolution: Techniques and technology review. *Protection and Control of Modern Power Systems, 8*(1), 43. https://doi.org/10.1186/s41601-023-00319-5
72. *Big data analytics in the smart grid*. (n.d.). https://resourcecenter.smartgrid.ieee.org/publications/white-papers/sgwp0003
73. Zhang, Y., Huang, T., & Bompard, E. F. (2018). Big data analytics in smart grids: A review. *Energy Informatics, 1*(1), 8. https://doi.org/10.1186/s42162-018-0007-5
74. Fan, C., Chen, M., Wang, X., Wang, J., & Huang, B. (2021). A review on data preprocessing techniques toward efficient and reliable knowledge discovery from building operational data. *Frontiers in Energy Research, 9*, 652801. https://doi.org/10.3389/fenrg.2021.652801

75. Ansari, M. H. K., Vakili, V. T., & Bahrak, B. (2019). Evaluation of big data frameworks for analysis of smart grids. *Journal of Big Data, 6*(1), 109. https://doi.org/10.1186/s40537-019-0270-8
76. Kaur, D. (2020). *Energy forecasting in smart grid systems: A review of the state-of-the-art techniques*. https://arxiv.org/abs/2011.12598
77. Diaba, S. Y., Shafie-khah, M., & Elmusrati, M. (2022). On the performance metrics for cyber-physical attack detection in smart grid. *Soft Computing, 26*(23), 13109–13118. https://doi.org/10.1007/s00500-022-06761-1
78. *Smart grid big data analytics: Survey of technologies, techniques, and applications*. (2021). https://ieeexplore.ieee.org/document/9272794
79. *Topic: Smart grids*. (2023). https://www.statista.com/topics/9349/smart-grids/
80. Misra, S., & Bera, S. (2018). Introduction to smart grid. In *Smart grid technology* (pp. 3–17). Cambridge University Press. https://doi.org/10.1017/9781108566506.003
81. Kabalcı, E., & Kabalcı, Y. (2018). Introduction to smart grid architecture. In *Energy systems in electrical engineering* (pp. 3–45). Springer. https://doi.org/10.1007/978-981-13-1768-2_1
82. Ramanath, A. (2022). *Understanding the architecture of smart grids*. https://eepower.com/technical-articles/understanding-the-architecture-of-the-iot-and-smart-grids/
83. An, A., & Bae, J. (2021, September). *Smart grid case studies, smart grid drivers, and technologies by country, economies, and continent: Analytical report. ISGAN Annex 2*
84. Tripathi, S., Verma, P. K., & Goswami, G. (2020). A review on SMART GRID power system network. In *2020 9th International Conference System Modeling and Advancement in Research Trends (SMART)* (pp. 55–59). IEEE.
85. Relancio, J., Cuellar, A., Walker, G., & Ettmayr, C. (2016). South African CSP projects under the REIPPP programme: Requirements, challenges and opportunities. *AIP Conference Proceedings, 1734*(1), 110002.
86. *A review on SMART GRID Power System Network*. (2020, December 4). https://ieeexplore.ieee.org/document/9337067/
87. Zhang, Y., Chen, W., & Gao, W. (2017). A survey on the development status and challenges of smart grids in main driver countries. *Renewable and Sustainable Energy Reviews, 79*, 137–147.
88. Giordano, V., Meletiou, A., Covrig, C. F., Mengolini, A., Ardelean, M., Fulli, G., Sánchez Jiménez, M., & Filiou, C. (2013). *Smart Grid projects in Europe: Lessons learned and current developments*. Joint Research Centre, European Commission.
89. Sharma, A., Saxena, B. K., & Rao, K. V. S. (2017). Comparison of smart grid development in five developed countries with focus on smart grid implementations in India. In *2017 International Conference on Circuit, Power and Computing Technologies (ICCPCT)* (pp. 1–6). IEEE. https://doi.org/10.1109/ICCPCT.2017.8074195
90. Khosrojerdi, F., Akhigbe, O., Gagnon, S., Ramírez, A., & Richards, G. (2021). Integrating artificial intelligence and analytics in smart grids: A systematic literature review. *International Journal of Energy Sector Management*. https://doi.org/10.1108/ijesm-06-2020-0011
91. SAP Insights. (n.d.). *The smart grid: How AI is powering today's energy technologies*. https://www.sap.com/insights/smart-grid-ai-in-energy-technologies.html
92. IEA. (n.d.). *Executive summary: Unlocking smart grid opportunities in emerging markets and developing economies—Analysis*. https://www.iea.org/reports/unlocking-smart-grid-opportunities-in-emerging-markets-and-developing-economies/executive-summary
93. Li, J., Li, T., & Han, L. (2018). Research on the evaluation model of a smart grid development level based on differentiation of development demand. *Sustainability, 10*(11), 4047. https://doi.org/10.3390/su10114047
94. Mashal, I. (2022). Smart grid reliability evaluation and assessment. *Kybernetes, 52*(9), 3261–3291. https://doi.org/10.1108/k-12-2020-0910
95. Gupta, R., & Chaturvedi, K. T. (2023). Adaptive energy management of big data analytics in smart grids. *Energies, 16*(16), 6016. https://doi.org/10.3390/en16166016

96. Tiwari, S., Jain, A., Ahmed, N. M. O. S., Charu, Alkwai, L. M., Dafhalla, A. K. Y., & Hamad, S. A. S. (2022). Machine learning-based model for prediction of power consumption in smart grid-smart way towards smart city. *Expert Systems, 39*(5), e12832.
97. Kumar, D., & Bhowmik, P. S. (2019). Hidden Markov model based islanding prediction in smart grids. *IEEE Systems Journal, 13*(4), 4181–4189.
98. Jeyaranjani, J., & Devaraj, D. (2022). Improved genetic algorithm for optimal demand response in smart grid. *Sustainable Computing: Informatics and Systems, 35*, 100710.
99. Xue, X., & Tsai, P. W. (2022). Integrating Energy Smart Grid's ontologies through multi-objective particle swarm optimization algorithm with competitive mechanism. *Sustainable Energy Technologies and Assessments, 53*, 102442.
100. Babar, M., Tariq, M. U., & Jan, M. A. (2020). Secure and resilient demand side management engine using machine learning for IoT-enabled smart grid. *Sustainable Cities and Society, 62*, 102370.
101. *Smart grid annual growth rate*. (n.d.). https://www.mordorintelligence.com/

Smart Intelligence Aided Power and Energy Management

S. Vijayalakshmi, Savita Dahiya, and S. P. Gayathri

1 Introduction

The electricity infrastructure, like oxygen, is crucial to modern society but is often overlooked—until a crisis arises. Today, severe weather events all around the world are wreaking havoc on outdated grid infrastructure, causing power outages that endanger health, safety, and economic activity. At the same time, several other factors are exerting strain on century-old grids. The way power is generated is changing very quickly: more solar and wind power, and less coal and fossil fuel. This transition necessitates the development of new procedures and management approaches. The "who" is also changing, with energy being produced not only by established energy corporations, but also by a slew of new competitors and prosumers.

"Smart intelligence aided power and energy management" refers to the application of advanced intelligent technique, such as AI, ML, and information analytics, to optimize the generation, distribution, and consumption of power and energy resources. This approach involves using smart algorithms and predictive models to analyze large volumes of data, forecast energy demands, identify inefficiencies, and automate decision-making processes to improve overall enhanced effectiveness, dependability, and environmental consciousness of power and energy systems. By

S. Vijayalakshmi
Department of Data Science, CHRIST University, Pune, Maharashtra, India
e-mail: s.vijayalakshmi@christuniversity.in

S. Dahiya (✉)
School of Engineering and Technology, Maharishi University of Information Technology, Noida, Uttar Pradesh, India

S. P. Gayathri
Department of Computer Science, Government Arts College for Women, Dindigul, Tamil Nadu, India

S. Vijayalakshmi et al. (eds.), *Smart Power Systems*, Power Systems,
https://doi.org/10.1007/978-3-031-89459-6_10

leveraging these intelligent technologies, organizations can make more well-informed choices to decrease energy expenses and mitigate environmental effects, and enhance the resilience and responsiveness of energy infrastructure.

2 Review on AI in Energy Sector

In [1], researchers proposed employing machine learning (ML) algorithms to analyze daily and seasonal home energy consumption patterns, aiming to pinpoint peak usage times throughout the day and across different seasons. By leveraging clustering algorithms, they categorized homes into distinct behavioral classes, facilitating targeted energy-saving efforts and yielding more precise estimates of potential energy savings. Moreover, they accounted for seasonal variations, recognizing that homes exhibit varying energy usage patterns across different seasons. Subsequently, homes were grouped based on their energy consumption behavior. Individuals can achieve cost savings by shifting energy usage to off-peak hours when tariffs are lower. This approach enables the identification of a subset of inefficient energy users who could benefit from more frequent targeting with energy efficiency initiatives.

In [2] author introduced the "HEM System," a sophisticated approach utilizing Reinforcement Learning (RL) and Neural Fitted Q-learning approach that is self-learning and adaptable. This technique proposes a HEM decision-making system characterized by speed, adaptability, and energy efficiency. By employing a centralized decision-making approach and connecting domestic appliances through organized networks via Wi-Fi, the system can identify peak load situations and automatically adjust power demand, thereby reducing overall power usage. Evaluation of this approach utilized a typical Cold Regions Hydrological (CRH) model, indicating its efficacy in quickly and realistically reducing demand and conserving energy during peak hours while also contributing to the reduction of domestic carbon footprints. Implementation of this model in city blocks with significant residential buildings could result in substantial reductions in total energy usage by modifying energy demand during peak periods. Additionally, by decreasing peak demand, the model could help local power distribution firms enhance their resources and maintain reasonable tariffs. However, limitations of this research include the absence of a preprocessing layer and limitations in decision-making capabilities.

Energy management poses a formidable challenge in smart cities due to the complexity and vital role of energy systems. Consequently, addressing this issue demands considerable attention and effort. Modeling and simulation serve as common methods for assessing the technological and policy implications of smart solutions, as well as for devising optimal strategies to transition from current urban environments to smarter ones [3].

Intelligent home often involves analyzing IoT data from connected devices to improve efficiency and facilitate quick and precise decision-making. The development of efficient smart cities relies on effective energy management within smart homes [4]. This includes practices such as Digital Soil Mapping (DSM), Peak Load Reduction (PLR), and carbon emission reduction initiatives. In urbanized regions of industrialized countries, significant electrical energy is required to support both residential and industrial loads. The information in the report, residential and industrial establishments account for around 39–40% of total energy demand in Canada. The demand for electricity in these households is extremely varied and swings throughout the day, as proven by several load surveys. As a result, finding effective solutions to controlling smart home electrical demand and minimizing energy use during peak hours will help smart cities become energy-efficient as well.

A number of intelligent computing techniques, especially machine learning [5, 6] and artificial neural networks [7], have become known as effective solutions for addressing issues in smart city development [8] and flexible service coordination [9]. These approaches provide efficient energy-related solutions by automating procedures like electric vehicle scheduling [10]. Furthermore, researchers have investigated adding environmental sustainability issues [11] and adaptive control algorithms for systems with unpredictable dynamics [12]. Approaches such as extracting features fusion and multi-modal fusion [13–15] help to provide a deeper comprehension of energy usage patterns. Utilization of these advanced methodologies by researchers from other disciplines demonstrates their versatility and application in a variety of energy associated research scenarios.

According to [16], there is an increasing demand for energy-efficient construction in developing countries. The United States Environmental Protection Agency (USEPA) aims to reduce energy consumption in commercial buildings by 20% between 2020 and 2030, along with a governmental target to decrease pollution by 33% by 2025. Achieving these objectives with the assistance of the Internet of Things (IoT) could lead to significant resource savings and environmental improvement for industries. Smart meters, essential components of smart grids due to their capacity to easily share information, are novel energy meters that continuously monitor energy usage and provide vital data such as frequency, voltage, and phase angle. Furthermore, they feature wireless communication capabilities for exchanging diagnostic information [17].

Various studies have highlighted the effectiveness of Artificial Intelligence (AI)-based solutions in automating, controlling, and managing energy consumption from individual homes to the grid level. For instance, Elkazaz et al. [18] introduced an automated control method, employing genetic algorithms, to optimize the operational efficiency of distributed generation (DG) units in residential settings. Their findings demonstrated a notable reduction in daily household consumption costs. Similarly, Jo and Yoon [19] proposed three intelligent models as Internet of Things (IoT) platform services for smart homes. These models effectively addressed challenges such as network congestion and energy wastage associated with managing smart home devices across separate IoT platforms. By integrating IoT devices like sensors and smart meters, smart grid systems and smart homes create substantial data flows. This data serves as a valuable foundation for AI applications to predict network loads, user consumption behaviors, and accurately delineate consumption patterns for individual energy users [20].

3 Energy System

When we refer to energy systems, we encompass various entities, including equipment, buildings, plants, and even intelligent energy grids, which require energy for operation, maintenance of specific conditions, or energy transfer. Whether these systems are consuming energy or facilitating its transfer, the presence of a smart management system capable of promptly responding to fluctuating inputs, such as shifts in demand, or forecasting future requirements to optimize grid assets, can yield significant benefits."

Artificial intelligence (AI) has developed as an innovative technology with substantial applications in a variety of fields, including energy management. With rising demand for efficient utilization of energy and increasing complexity of power systems, AI has the ability to transform how we manage and utilize energy resources [21–23].

Short-term forecasting of the required electricity generation and fuel consumption by fossil fuel power plants based on available renewable energy sources such as solar or wind, as well as system demand. Static models cannot provide these short-term forecasts; instead, dynamic models with decision-making skills are required. AI can be a great tool for simulating human decision-making and operating smart energy systems autonomously. Through machine learning, computers can teach themselves to make better decisions and operate more correctly, to the point where, given enough data, they can easily outrun people.

4 Introduction to Smart Power System

Advanced power networks reflect the progression of conventional energy infrastructures into contemporary, integrated, and intelligent systems. These networks capitalize on emerging techniques such as AI, ML, IoT, and large data analysis to optimize efficiency, dependability, and environmental sustainability related to electricity creation, conveyance, circulation, and utilization.

Unlike traditional power grids, smart energy systems have continuous surveillance, control, and communication capabilities, allowing for proactive management of the flow of electricity and system operations. Valuable data on energy use, equipment health, and environmental conditions are constantly collected and evaluated by sensors and smart meters distributed throughout the grid infrastructure. This data-driven strategy enables predictive maintenance, fault detection, and optimal asset use, resulting in increased grid resilience and less downtime.

One of the most important characteristics of smart power systems is their capacity to accept distributed sources of energy (DERs), such as wind turbines, solar power plants, energy storage devices, and electric vehicles. These decentralized power sources present both potential and challenges, necessitating creative solutions for grid interconnection, demand-side control, and grid balancing [24] (Fig. 1).

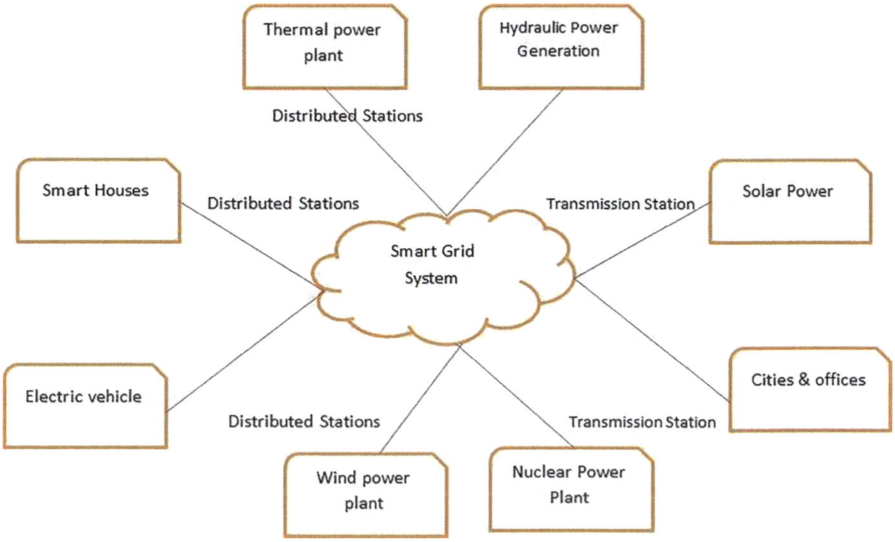

Fig. 1 Smart power system

In addition, smart electric systems include demand response programs and flexible pricing mechanisms, allowing users to actively participate in managing their energy and optimize their use of electricity according to cost and environmental factors. This transition to a more interactive and decentralized energy environment promotes increased energy efficiency, grid flexibility, and renewable power integration.

In short, modern power systems provide a paradigm shift in electricity infrastructure, integrating modern technology, data analytics, and user participation to produce a more effective, trustworthy, and eco-friendly energy future [25, 26].

5 Role of AI in Energy Management System

Artificial Intelligence (AI) revolutionizes energy management by introducing advanced computational techniques to optimize energy production, distribution, and consumption. Introducing algorithms into smart intelligence-aided energy and power management systems has various advantages, including altering how we maximize and manage energy resources. These artificial intelligence models use complex algorithms and data analytics approaches to interpret massive volumes of data from multiple sources within power systems, resulting in more precise predictions, pre-emptive decision-making, as well as effective resource allocation [27–29]. Here's how AI models play an important role in smart intelligence-assisted power and energy management:

- **Predictive Analytics:** AI system evaluates past data to estimate energy consumption patterns, enabling proactive adjustments in energy generation and distribution. This helps utilities anticipate peak demand periods and optimize resource allocation accordingly.
- **Smart Grid Optimization**: AI optimizes the operation of smart grids by dynamically adjusting parameters such as voltage levels, grid topology, and energy flow in real time. This ensures efficient energy transmission and distribution, reducing losses, and enhancing grid reliability.
- **Load Forecasting:** AI algorithms predict future energy consumption based on factors like weather patterns, historical data, and user behavior. These forecasts guide decision-making regarding energy production, distribution, and pricing, optimizing resource utilization and cutting costs.
- **Demand Response:** AI enables automated consumption response system by evaluating current information from smart meters and connected devices. It identifies opportunities to curtail energy usage in time of high demand, rewarding users to reduce consumption or shift usage to off-peak hours.
- **Energy Efficiency Optimization:** AI-driven analytics identify inefficiencies in energy systems and recommend optimization strategies to minimize wastage and improve overall efficiency. This includes equipment maintenance scheduling, equipment replacement decisions, and operational adjustments.
- **Renewable Energy Integration:** By anticipating variable output from energy sources that are renewable such as solar and wind, AI systems make it easier to integrate them into the grid and optimizing their integration with conventional energy sources. AI also helps manage energy storage systems, such as batteries, by optimizing charging and discharging cycles based on demand forecasts and grid conditions.
- **Fault Detection and Diagnostics:** AI-powered systems monitor energy infrastructure for anomalies and faults, enabling rapid detection and diagnosis of issues such as equipment failures or grid disturbances. This proactive approach reduces downtime, improves system reliability, and prevents potential safety hazards.
- **Asset Management and Maintenance:** AI analyzes data from sensors and IoT devices installed in energy infrastructure to assess the health and performance of assets. By predicting equipment failures and recommending maintenance schedules, AI helps minimize downtime, extend asset lifespan, and optimize maintenance costs.
- **Energy Trading and Market Optimization:** AI algorithms analyze market trends, pricing data, and regulatory policies to optimize energy trading strategies. This includes predicting market prices, identifying arbitrage opportunities, and dynamically adjusting energy supply and demand to maximize profitability (Fig. 2).
- **Customer Engagement and Personalization:** AI-powered energy management platforms engage consumers by providing personalized insights into their energy usage patterns, tips for energy conservation, and tailored recommendations for

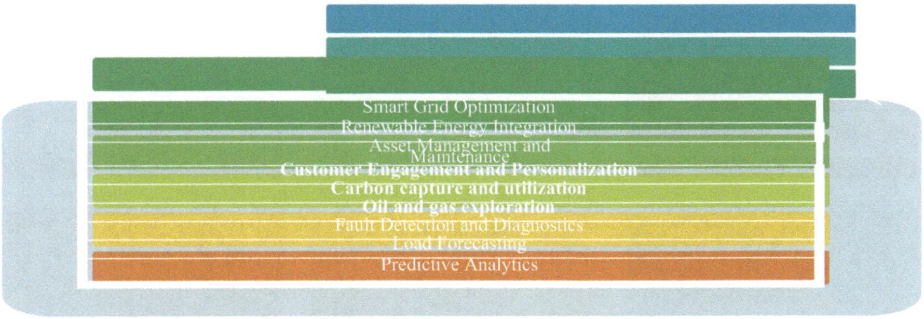

Fig. 2 Role of AI in revolutionizing energy system

optimizing energy consumption. This enhances customer satisfaction and encourages sustainable behavior.

- **Carbon Capture and Utilization:** AI boosts the efficiency of Carbon Capture, Utilization, and Storage (CCUS) operation by improving the extraction of CO_2 from the environment or emission sources. AI-powered systems excel in identifying the effective technique for utilizing captured carbon, whether for industrial applications or secure long-term storage. This technology is pivotal in curbing greenhouse gas emissions and addressing climate change.
- **Oil and Gas Exploration:** The impact of artificial intelligence (AI) on the petroleum and natural production sector is significant. Through meticulous analysis of extensive geological datasets. Furthermore, it evaluates the economic feasibility of these reserves, directing exploration efforts to the most viable opportunities. This not just improves efficiency but also greatly increases the achievement rate of exploration efforts, resulting in reduced resource waste and expenditures. Furthermore, AI plays an important part in the drilling process. AI-powered forecasting algorithms consider a variety of elements, including geological structures, drilling machinery performance, and atmospheric conditions to identify probable hazards and obstacles. This enables drilling professionals to take action on concerns, improve security protocols, and optimize drilling procedures, resulting in more secure and more productive operations in the petroleum and natural gas industry.

Overall, AI plays a multifaceted role in energy management systems, driving efficiency improvements, cost savings, and sustainability across the entire energy value chain.

6 AI Utilization Across Various Energy Sectors

AI is deployed across different energy sectors for activities like predictive maintenance, demand prediction, energy efficiency improvement, grid management, and asset supervision. Within renewable energy, AI enhances the effectiveness of solar

Fig. 3 AI applications in energy system

and wind power generation by employing predictive analytics and resource distribution [21]. In the oil and gas domain, AI is harnessed for tasks such as reservoir oversight, drilling enhancement, and predictive equipment maintenance. In the electric power sector, AI contributes to grid enhancement, load prediction, and demand adjustment. Ultimately, AI substantially enhances operational efficiency, diminishes expenses, and optimizes resource usage throughout the energy sector (Fig. 3).

7 AI Benefits in Smart Power Management

Artificial Intelligence (AI) in smart power management offers several advantages such as improved efficiency, predictive maintenance, accurate demand forecasting, grid optimization, load management, renewable energy integration, and advanced data analytics [30–33]. These benefits lead to a more reliable, resilient, and sustainable energy infrastructure.

- **Efficiency Improvement:** AI system can examine large volume of information from smart grids, sensors, and tools to optimize energy distribution and consumption. This result in more effective resources, reducing waste and costs.
- **Predictive Maintenance**: AI can anticipate equipment breakdowns by studying data trends, allowing for preventive maintenance. This reduces downtime and increases the longevity of the electricity infrastructure.
- **Demand Forecasting:** AI models can accurately forecast energy demand based on historical data, weather patterns, and other variables. This helps utility

companies plan and adjust energy production and distribution accordingly, avoiding shortages or surpluses.

- **Grid Optimization**: AI algorithms can dynamically optimize the operation of power grids in real-time, balancing supply and demand, managing fluctuations in renewable energy sources, and reducing transmission losses.
- **Load Management**: AI-powered systems can optimize energy consumption in buildings, factories, and other facilities by adjusting HVAC systems, lighting, and other equipment based on demand, occupancy, and external factors.
- **Renewable Energy Integration**: AI can help integrate sources of clean energy such as wind and solar energy into the system by forecasting their generation, managing variability, and optimizing their use alongside traditional energy sources.
- **Data Analytics:** AI enables advanced data analytics to uncover insights from energy uses patterns, customer actions, and market trends. This helps utilities and consumers make educated decisions about energy usage and investments.

Overall, the application of AI in smart power management leads to a more reliable, resilient, and sustainable energy infrastructure, benefiting both utility providers and consumers.

8 Key Challenges and Considerations for Implementing Responsible AI in Smart Energy Management

Leveraging Artificial Intelligence (AI) for the management of distributed energy resources in smart grids presents a range of challenges and limitations. These encompass concerns regarding data privacy, security, scalability, and explainability. Moreover, the intermittent nature of renewable energy sources in microgrids introduces complexities for energy management and control, necessitating advanced AI solutions. While AI technologies offer opportunities to optimize energy management in microgrids, they also encounter limitations in addressing variable factors such as wind speed and solar irradiation. Additionally, integrating machine learning algorithms into smart grid management systems faces hurdles including data security, privacy issues, and scalability concerns [34, 35]. Overcoming these obstacles is essential for fully realizing the potential of AI in effectively managing distributed energy resources within smart grids.

- **Data Quality and Accessibility**: Guarantee the access to precise and dependable data for AI algorithms to efficiently optimize energy consumption and distribution. Tackle issues concerning the completeness, accuracy, and consistency of data.
- **Privacy and Security**: Implement strong security measures to safeguard sensitive energy consumption data against unauthorized access or misuse. Ensure

compliance with privacy regulations while mitigating the risks of cyber threats and breaches.

- **Bias and Equity**: Reduce biases within AI algorithms to guarantee equitable treatment across diverse user groups and prevent discriminatory outcomes in energy management decisions.
- **Transparency and Understandability**: Improve the transparency and interpretability of AI-powered energy management systems to foster trust among stakeholders. Users should have the ability to understand the decision-making process.
- **Stakeholder Engagement**: Involve stakeholders, such as energy consumers, utilities, regulators, and policymakers, in both the development and deployment phases of AI solutions. Address their requirements and apprehensions to ensure successful implementation.
- **Regulatory Compliance**: Adhere to applicable regulations and standards governing energy management practices, data privacy, and the use of AI to mitigate legal risks.
- **Energy Efficiency and Sustainability**: Give priority to energy efficiency and sustainability goals in AI-driven energy management strategies. This includes efforts to reduce carbon emissions, encourage the integration of renewable energy sources, and optimize resource utilization.
- **Scalability and Flexibility**: Develop AI systems that can scale and adjust to changing energy demand patterns, advancements in technology, and evolving regulatory requirements to ensure on-going effectiveness.
- **Ethical Considerations**: Consider ethical implications such as environmental impact, social equity, and long-term sustainability when designing and implementing AI solutions for smart energy management.

Continuous Monitoring and Evaluation: Continuously assess the performance and impact of AI-powered energy management systems to identify potential issues, refine algorithms, and improve overall effectiveness while minimizing unintended consequences.

9 Conclusion

AI integration enhances energy management systems by enabling real-time problem detection and resolution for increased efficiency. This leads to cost reduction, optimized energy usage, and improved grid stability, shaping a future of energy management that's more reliable, sustainable, and efficient. In summary, the incorporation of AI into smart grid management marks a significant advancement in energy efficiency and sustainability. Through the utilization of AI technologies like machine learning and data analytics, smart grid systems can actively identify and resolve issues in real time, resulting in improved energy distribution, lowered expenses, and heightened grid reliability. Furthermore, AI facilitates predictive maintenance and asset management, aiding in the prevention of costly downtimes and the extension of critical infrastructure component lifespans. Ultimately, AI-driven smart grid management offers great potential for establishing more robust, dependable, and eco-conscious energy systems capable of meeting the evolving needs of our contemporary society.

References

1. Mladenovikj, V. T. (2020). Machine learning approaches for smart city energy management. *Energy, 1*(1), 1–12.
2. Mahapatra, C., Moharana, A. K., & Leung, V. (2017). Energy management in smart cities based on internet of things: Peak demand reduction and energy savings. *Sensors, 17*(12), 2812–2833.
3. Liu, Y., Yang, C., Jiang, L., Xie, S., & Zhang, Y. (2019). Intelligent edge computing for IoT-based energy management in smart cities. *IEEE Network, 33*(2), 111–117.
4. Masum, A. K. M., Chy, M. K. A., Hasan, M. T., & Sayeed, M. H. (2019). Smart meter with load prediction feature for residential customers in Bangladesh. In *International Conference on Energy and Power Engineering, Bangladesh* (pp. 1–6). IEEE.
5. Ghazal, T. M., Rehman, A. U., Saleem, M., Ahmad, M., Ahmad, S., et al. (2022). Intelligent model to predict early liver disease using machine learning technique. In *International Conference on Business Analytics for Technology and Security, Dubai, United Arab Emirates* (pp. 1–5). IEEE.
6. Luo, J., Wang, Y., & Li, G. (2023). The innovation effect of administrative hierarchy on inter-city connection: The machine learning of twin cities. *Journal of Innovation and Knowledge, 8*(1), 100293–100302.
7. Batool, T., Abbas, S., Alhwaiti, Y., Saleem, M., Ahmad, M., et al. (2021). Intelligent model of ecosystem for smart cities using artificial neural networks. *Intelligent Automation and Soft Computing, 30*(2), 513–525.
8. Sami, F. (2022). Integration of blockchain and edge computing to improve the scalability and latency. *International Journal of Advanced Sciences and Computing, 1*(1), 23–32.
9. Ahmed, A., Khan, M. F. N., Khan, I., Alquhayz, H., Khan, M. A., et al. (2021). A novel framework to determine the impact of time varying load models on wind DG planning. *IEEE Access, 9*(2), 11342–11357.
10. Cheng, B., Wang, M., Zhao, S., Zhai, Z., Zhu, D., et al. (2017). Situation-aware dynamic service coordination in an IoT environment. *IEEE/ACM Transactions on Networking, 25*(4), 2082–2095.
11. Min, C., Pan, Y., Dai, W., Kawsar, I., & Lietal, Z. (2023). Trajectory optimization of an electric vehicle with minimum energy consumption using inverse dynamics model and servo constraints. *Mechanism and Machine Theory, 181*, 105185–105197.
12. Li, X., Wang, F., Razgan, M. A., Awwad, M. E., Abduvaxitovna, Z., et al. (2023). Race to environmental sustainability: Can structural change, economic expansion and natural resource consumption effect environmental sustainability? A novel dynamic ARDL simulations approach. *Resources Policy, 86*(1), 4044–4057. https://doi.org/10.1016/j.resourpol.2023.104044
13. Lu, S., Ban, Y., Zhang, X., Yang, B., Liu, S., et al. (2022). Adaptive control of time delay teleoperation system with uncertain dynamics. *Frontiers in Neuro Robotics, 16*, 928863–928869.
14. Lu, S., Ding, Y., Liu, M., Yin, Z., Yin, L., et al. (2023). Multiscale feature extraction and fusion of image and text in VQA. *International Journal of Computational Intelligence Systems, 16*(1), 45–54.
15. Lu, S., Liu, M., Yin, L., Yin, Z., Liu, X., et al. (2023). The multi-modal fusion in visual question answering: A review of attention mechanisms. *PeerJ Computer Science, 9*, e1400–e1412.
16. Ashraf, S., Ahmed, T., & Saleem, S. (2020). NRSM: Node redeployment shrewd mechanism for wireless sensor network. *Iran Journal of Computer Science, 4*, 171–183.

17. Saleem, Y., Crespi, N., Rehmani, M. H., & Copeland, R. (2019). Internet of things-aided smart grid: Technologies, architectures, applications, prototypes, and future research directions. *IEEE Access, 7*, 62962–63003.
18. Elkazaz, M. H., Hoballah, A., & Azmy, A. M. (2016). Artificial intelligent-based optimization of automated home energy management systems. *International Transactions on Electrical Energy Systems, 26*(9), 2038–2056.
19. Jo, H., & Yoon, Y. I. (2018). Intelligent smart home energy efficiency model using artificial tensorflow engine. *Human-centric Computing and Information Sciences, 8*(1), 1–18.
20. IRENA. (2019). *Innovation landscape brief: Artificial intelligence and big data*. International Renewable Energy Agency.
21. Aguilar, J., Garces-Jimenez, A., R-Moreno, M., & García, R. (2021). A systematic literature review on the use of artificial intelligence in energy self-management in smart buildings. *Renewable and Sustainable Energy Reviews, 151*, 111530. https://doi.org/10.1016/j.rser.2021.111530
22. Amaral, J., Reis, C., & Brandão, R. F. M. (2013). Energy management systems. In *2013 48th International Universities' Power Engineering Conference (UPEC), Dublin, Ireland* (pp. 1–6). IEEE. https://doi.org/10.1109/UPEC.2013.6715015
23. Aman, S., Simmhan, Y., & Prasanna, V. K. (2013). Energy management systems: State of the art and emerging trends. *IEEE Communications Magazine, 51*(1), 114–119.
24. Salehi, V., Mazloomzadeh, A., & Mohammed, O. (2011). Real-time analysis for developed laboratory-based smart micro grid. In *2011 IEEE Power and Energy Society General Meeting* (pp. 1–8). IEEE.
25. Palanisamy, S., Rahiman, Z., & Chenniappan, S. (2023). Introduction to smart power systems. In *Artificial intelligence-based smart power systems* (pp. 1–13). Wiley.
26. Palati, M., Prathap, S. S., & Nagaraju, N. H. (2023). Modeling and analysis of smart power system. In *Artificial intelligence-based smart power systems* (pp. 15–36). Wiley.
27. Uriarte-Gallastegi, N., Arana-Landín, G., Landeta-Manzano, B., & Laskurain-Iturbe, I. (2024). The role of AI in improving environmental sustainability: A focus on energy management. *Energies, 17*(3), 649.
28. Karduri, Rajini. (2019). *The Role of Artificial Intelligence in Optimizing Energy Systems*. 25–35.
29. Musa, A. (2023). The impact of artificial intelligence on energy management: A revolutionary shift in the power industry. *IJCRT International Journal of Creative Research Thoughts, 11*(5), 882–887.
30. Entezari, A., Aslani, A., Zahedi, R., & Noorollahi, Y. (2023). Artificial intelligence and machine learning in energy systems: A bibliographic perspective. *Energy Strategy Reviews, 45*(8), 101017.
31. Zhao, X., Guo, Y., Guo, X., & Li, H. (2023). Artificial intelligence applications and prospects for the smart grid. In *2023 Panda Forum on Power and Energy (PandaFPE)* (pp. 1844–1848). IEEE.
32. Vijayalakshmi, S., Savita, Genish, T., & George, J. P. (2023). The role of artificial intelligence in renewable energy. In *AI-powered IoT in the energy industry: Digital technology and sustainable energy systems* (pp. 253–269). Springer.
33. Vijayalakshmi, S., Savita, & Durgadevi, P. (2023). AI and IoT in improving resilience of smart energy infrastructure. In *AI-powered IoT in the energy industry: Digital technology and sustainable energy systems* (pp. 189–213). Springer International Publishing.
34. Danish, M. S. S. (2023). AI in energy: Overcoming unforeseen obstacles. *AI, 4*(2), 406–425.
35. Ahmad, T., Zhang, D., Huang, C., Zhang, H., Dai, N., Song, Y., & Chen, H. (2021). Artificial intelligence in sustainable energy industry: Status Quo, challenges and opportunities. *Journal of Cleaner Production, 289*, 125834.

Model Selection Strategies for Identifying Effective Energy Storage Systems

Naived George Eapen and Alwin Joseph

1 Introduction

Sustainable and intelligent energy solutions are required in the current era of dynamic energy paradigms, growing energy needs, and environmental concerns. The global energy landscape is transforming profoundly, marked by the ascendancy of sustainable and eco-friendly energy sources. Energy storage systems have emerged as pivotal enablers, bridging intermittent renewable energy resources and evolving energy grid systems. In the contemporary landscape, converging energy intelligence, energy storage systems, and optimised storage solutions are paramount for addressing pressing societal needs [1]. Positioned at the nexus of environmental sustainability and burgeoning energy demands, energy resources' effective harnessing and utilisation have acquired imperativeness. Energy intelligence, rooted in data-driven insights and predictive analysis, emerges as a pivotal driver in this pursuit of managing and storing energy resources and using the resources to sustain the needs of future generations.

Global endeavours towards sustainable energy sourcing, coupled with the intermittent nature of renewable energy, underscore the pivotal role of energy storage systems. These systems serve as linchpins for efficient energy capture and distribution, contributing to grid stability and the seamless assimilation of green energy sources. Consequently, they play a vital part in facilitating the environmental repercussions of energy generation and consumption. Optimised storage systems transcend the ambit of mere energy administration. It encompasses the dynamic orchestration of energy resources to meet the evolving demands of modern society [2]. The evolving urbanisation trends, escalating dependence on electronic

N. G. Eapen (✉) · A. Joseph
CHRIST (Deemed to be University), Pune, Maharashtra, India
e-mail: naived.eapen@christuniversity.in; alwin.joseph@christuniversity.in

S. Vijayalakshmi et al. (eds.), *Smart Power Systems*, Power Systems,
https://doi.org/10.1007/978-3-031-89459-6_11

apparatus, and the quest for cleaner, more efficient energy sourcing underscore the pressing need for sophisticated energy storage and management frameworks [3].

In the face of climate alterations, resource scarcity, and soaring energy expenditures, the synergy of energy intelligence, energy storage systems, and optimised storage methodologies becomes indispensable in pursuing energy security, ecological sustainability, and economic fortitude [4]. The trajectory to a more resilient and ecologically mindful future unfolds within this intricate amalgamation of technology and sustainability. In parallel with this transformation, the confluence of energy intelligence and intelligent grids has ushered in a new era characterised by utilising real-time data analytics, predictive modelling, and optimisation methodologies.

This chapter extensively explores the intricate interplay between energy intelligence and energy storage systems. We analyse aspects such as energy forecasting and analytics, the manifold advantages and applications of energy intelligence, the foundational paradigms of storage system models, and the central concept of optimised storage systems, encompassing elements like live monitoring and predictive analytics, which all attribute to various model selection strategies for the identification of the best model selection strategies. This chapter also delves into optimisation techniques and diverse strategies constituting the energy storage systems' conceptual framework, including simulation tools, design configurations, life cycle analyses, economic valuations, automated load management, and environmental impact assessments. The chapter culminates with exploring efficient models tailored to the meticulous organisation of optimised storage, including optimisation models, demand response strategies, grid synchronisation, dynamic programming paradigms, and predictive maintenance modalities. In a context grappling with the dual imperatives of energy security and environmental preservation, this chapter aims to provide clarity into the evolving terrain of energy storage systems, where astute resolutions harmoniously coalesce with sustainability imperatives, affording a glimpse of a more eco-sensitive, intelligent, and resilient energy future.

1.1 Energy Intelligence and Smart Grids

In the current electrical usage landscape, the rapid proliferation of renewable energy sources necessitates the utilisation of energy storage technologies to mitigate the disparities arising from their intermittent nature. These energy storage technologies are indispensable pillars for stabilising power production and managing energy demand fluctuations [5, 6]. Their operation revolves around accumulating surplus or underutilised energy, which can then be distributed to the grid or consumers. In the forthcoming electric grid framework, energy storage systems are poised to assume the role of primary electricity sources. Researchers and industry experts have diligently worked on various energy storage technologies, intertwining diverse renewable energy resources to engineer robust and adaptable energy storage systems.

The relevance of this context within our study underscores the critical nexus between smart grid paradigms, energy intelligence, and energy storage systems.

The smart grid, underpinned by sophisticated data analytics and real-time decision-making, is pivotal in meeting the rapidly growing energy needs, rating the seamless integration of renewable energy sources. It capitalises on the insights derived from energy intelligence to optimally manage the storage and distribution of renewable energy. Consequently, the interplay between energy storage technologies, smart grids, and energy intelligence forms the cornerstone of our investigation, as it delineates the path to a more sustainable, resilient, and efficient energy ecosystem [7].

The typical applications encompass real-time monitoring and control of energy production and consumption, predictive analytics for load forecasting and grid optimisation, demand response mechanisms, and efficient energy resource management. Integrating energy intelligence with smart grids empowers utilities and consumers to make informed decisions regarding energy usage, reduce costs, and enhance grid resilience [8, 9]. Additionally, the synergy of these technologies facilitates the incorporation of renewable energy sources into the grid, thereby promoting sustainability and environmental stewardship. The applications outlined within this domain are pivotal in modernising the energy infrastructure, enabling a more adaptive, efficient, and eco-conscious energy ecosystem.

The realm of energy intelligence and smart grids serves as the fulcrum for many applications with profound implications for the energy sector. Leveraging real-time data analytics and decision-making, these technologies enable utility companies to manage and optimise their energy resources efficiently, ensuring reliable power supply while minimising costs. Moreover, these insights extend into the domain of energy storage systems, where predictive analytics facilitate optimal energy utilisation and storage, contributing to grid stability and the seamless integration of renewable energy sources [10, 11]. As integral components of this synergy, energy forecasting and analytics offer invaluable insights into load forecasting, grid management, and the orchestration of renewable energy resources. Effectively utilising these technologies is pivotal in ushering in a more innovative, sustainable, and resilient energy future.

1.1.1 Energy Forecasting and Analytics

Energy forecasting and analytics are pivotal components in energy storage systems, encompassing a spectrum of techniques and methodologies that play a crucial role in system optimisation, grid management, and the efficient allocation of energy resources.

- **Load Forecasting**: Load forecasting employs historical consumption data, time-series analysis, and machine learning algorithms to predict future electricity demand accurately. This is instrumental in scheduling energy storage systems for peak demand periods, ensuring an uninterrupted power supply.
- **Renewable Energy Integration**: Predictive analytics are deployed to forecast the intermittency of renewable energy sources such as solar and wind. Advanced forecasting models, like weather data integration and neural networks, enable

energy storage systems to balance and store excess renewable energy for future use.

- **Battery Health Monitoring**: In energy storage systems, analytics are indispensable for monitoring battery health. Techniques such as state of charge (SoC) and state of health (SoH) estimation provide real-time insights into a battery's performance, lifespan, and degradation trends.
- **Grid Frequency Regulation**: Energy storage systems integrated into the grid rely on analytics to respond to fluctuations in grid frequency. Algorithms for predictive control, such as model predictive control (MPC), enable precise energy dispatch to maintain grid stability.
- **Market Price Prediction**: Energy markets often experience price volatility. Predictive analytics forecast market price trends, allowing energy storage operators to optimise energy purchasing and selling strategies.
- **Energy Storage Sizing**: Machine learning models and optimisation algorithms play a critical role in sizing energy storage systems appropriately, ensuring that the system's capacity aligns with the specific energy needs and objectives of a given application.
- **Predictive Maintenance**: Analytics are used for predictive maintenance in energy storage systems. This involves continuously monitoring components and systems and using algorithms to predict maintenance requirements based on performance metrics and failure patterns.

Energy forecasting and analytics provide the technical backbone for energy storage systems, enabling real-time decision-making, efficient energy management, and the seamless integration of renewable energy into the grid [12, 13]. By harnessing the power of data-driven insights and sophisticated algorithms, energy storage systems are poised to play a pivotal role in the energy landscape of the future.

1.1.2 Advantages and Applications of Energy Intelligence

Energy intelligence emerges as a sentinel, facilitating the seamless amalgamation of data-driven insights with the orchestration of energy resources, consequently delineating a future characterised by resilience, efficiency, and sustainability. The planning and organisation of effective resource utilisation have been carried out effectively with the help of intelligent techniques. Various techniques and intelligent applications have been deployed to identify and forecast energy resource needs. Intelligent techniques help to monitor production and serve the customers' needs efficiently [14].

The advantages of energy intelligence manifest in its capacity to optimise energy consumption, thereby maximising grid stability and reducing operational expenditures. By amalgamating historical data and real-time information, energy storage systems can intelligently allocate stored energy, adroitly responding to demand fluctuations, thus minimising wastage while ensuring an uninterrupted power supply [15]. Moreover, energy intelligence empowers utility providers and end-users to proactively manage energy utilisation, culminating in cost reduction and reducing carbon emissions.

Integrating renewable energy sources, energy intelligence plays an instrumental role in forecasting and optimising the intermittent nature of green energy [16]. This dynamic capability allows energy storage systems to store surplus renewable energy during excess generation and release it during peak demand. The applications extend into grid balancing and voltage control, ensuring the perpetuation of a stable and efficient power distribution network.

Energy intelligence fosters predictive maintenance strategies for energy storage systems, guaranteeing longevity and reliability. It leverages sophisticated machine learning algorithms to predict component failures, enabling timely interventions and significant cost savings. Data-driven precision harmoniously converges with sustainability imperatives, resulting in advantages that encompass reducing operational costs, fortifying grid stability, seamless integration of renewable energy sources, and augmentation of system reliability. The smart energy systems represent a fusion of engineering and computational methodologies integral to optimising energy utilisation and grid stability [17]. An in-depth understanding of their underpinnings is vital, as it forms the foundation for discussions in subsequent sections on optimisation techniques and the application of energy intelligence, energy storage systems being a pivotal aspect of it.

1.2 Energy Storage Systems: An Overview

Energy Storage Systems (ESS) constitute pivotal components in contemporary energy infrastructure, serving as reservoirs for electrical energy storage and subsequent disbursement. ESS are instrumental in mitigating the intermittency inherent in renewable energy sources, such as solar and wind, by capturing surplus energy during periods of excess generation and releasing it during high demand or when renewable sources are inactive. Various ESS technologies encompass batteries, pumped hydro storage, flywheels, and more [18, 19]. Some of these are illustrated in Table 1.

Energy storage systems play a crucial role in smart grid frameworks. These systems underpin grid stability, harmonise intermittent renewable energy sources, and ensure the constancy of power supply. Embedded systems and computer-based logic have emerged as instrumental components to enhance their performance further. These technologies facilitate real-time monitoring, management, and optimisation, fine-tuning the energy storage systems' responsiveness to evolving grid dynamics. Advanced control algorithms, informed by predictive models and grid conditions, enable efficient scheduling of charge and discharge cycles. This consequential aspect minimises energy wastage while maximising the integration of renewable energy sources. This paradigm optimally contributes to realising a resilient and efficient smart grid ecosystem.

Table 1 Different types of energy storage systems

Energy storage system	Capacity range	Efficiency (%)	Response time	Lifespan	Applications	Example
Lithium-ion batteries (chemical storage)	kWh to MWh	90–95	Milliseconds	15–20 years	Portable electronics, electric vehicles, grid support	Tesla Powerwall
Pumped hydro storage (mechanical storage)	MWh to GW	70–85	Seconds	50+ years	Grid-scale energy storage, load balancing	Bath County pumped storage
Flywheel energy storage (mechanical storage)	kWh to MWh	85–95	Milliseconds	20+ years	Uninterruptible power supply, grid stabilisation	Beacon power flywheel
Compressed air energy storage (mechanical storage)	MWh to GW	50–70	Seconds	20+ years	Grid energy storage, grid stabilisation	Huntorf CAES
Flow batteries (chemical storage)	kWh to MWh	70–85	Seconds	10–20 years	Grid-scale energy storage, renewable integration	Vanadium redox flow battery
Thermal energy storage (thermal storage)	MWh to GW	70–95	Minutes to hours	20+ years	Solar power generation, heating, and cooling	Molten salt storage tanks

A multifaceted approach is employed in monitoring and managing energy storage systems. Typically, a network of sensors and monitoring devices is integrated into the energy storage infrastructure, collecting real-time data on critical parameters. These parameters encompass state of charge (SoC), state of health (SoH), temperature, voltage, and current, which are pivotal for assessing the performance and health of the storage system. These data points are continually transmitted to a central supervisory control and data acquisition (SCADA) system. The SCADA system, equipped with embedded computer-based logic, processes and analyses this information to make informed decisions regarding energy storage operations. Additionally, advanced algorithms facilitate optimising charge and discharge cycles, considering factors such as grid demand, energy prices, and weather forecasts. Notifications are generated based on these analytics, which can be relayed through automated systems or alerts to grid operators, thereby ensuring energy storage systems' efficient and reliable operation.

Distributing stored energy to various power lines is coordinated through intelligent control systems. Embedded logic governs the routing and timing of energy dispatch, aligning it with the grid's and consumers' specific requirements. In smart grids, real-time demand forecasting and predictive modelling inform the release of stored energy, allowing for precise matching with the fluctuating energy demand. Usage monitoring involves continuous data analysis to ensure efficient energy utilisation and grid stability. Any energy surplus or deficit is detected through real-time SoC and SoH monitoring, prompting the system to adapt and make necessary adjustments. In the event of left-out or excess energy, automated notifications or alerts can be dispatched to relevant stakeholders, including consumers or grid operators, facilitating the timely redistribution or utilisation of surplus energy and minimising wastage within the grid.

1.2.1 Storage System Models

Storage system models are critical for designing, evaluating, and optimising energy storage systems. These models simulate the behaviour of various storage technologies under different conditions, helping to predict performance metrics such as energy density, power density, efficiency, cycle life, and response time. They incorporate various factors, including the physical properties of storage materials, the dynamics of charge and discharge cycles, and the impacts of environmental conditions. Advanced storage system models also integrate economic considerations, such as capital and operational costs, and environmental factors, like carbon footprint and resource availability [20]. These models often employ optimisation algorithms to enhance accuracy and reliability, leveraging mathematical techniques to identify the most cost-effective and efficient storage solutions. Additionally, with the advent of AI and ML, storage system models are increasingly data-driven, utilising large datasets to improve predictions and adapt to changing conditions. These sophisticated models are essential for researchers and engineers as they develop innovative solutions that meet the complex demands of modern energy systems, ensuring sustainability, reliability, and scalability.

1.3 Optimised Storage Systems

Optimised storage systems are crucial for enhancing the efficiency and reliability of energy storage solutions, ensuring they meet the demands of modern energy grids and applications. These systems are designed using advanced optimisation techniques that balance multiple factors such as cost, performance, and environmental impact. Optimisation involves the application of mathematical models and algorithms to determine the best possible configuration and operation of storage systems. This includes selecting the appropriate type of storage technology (e.g., batteries, thermal storage, and flywheels), sizing the system correctly to meet energy demands, and optimising charge and discharge cycles to maximise efficiency and lifespan.

AI and ML play a significant role in optimisation, offering data-driven insights that enhance decision-making. These technologies analyse large datasets to identify patterns and predict system performance under various conditions, allowing for real-time adjustments and proactive maintenance. Optimisation algorithms, such as linear programming, genetic algorithms, and multi-criteria decision analysis (MCDA), help evaluate different scenarios and trade-offs, ensuring that the chosen storage solution delivers optimal performance [20–22].

Optimised storage systems aim to store energy efficiently and integrate seamlessly with renewable energy sources and smart grid technologies. This integration helps balance supply and demand, reduce energy waste, and minimise costs. By optimising energy storage systems, stakeholders can achieve greater sustainability, economic viability, and resilience in the energy infrastructure, ultimately contributing to a more reliable and environmentally friendly future.

1.3.1 Live Monitoring

Live monitoring is essential to modern energy storage systems, providing real-time oversight and control to enhance performance, safety, and efficiency. It involves continuously collecting and analysing data from various sensors and meters embedded within the system [23]. These sensors track critical parameters such as voltage, current, temperature, state of charge, and state of health. By leveraging advanced technologies and data analytics, live monitoring systems enable operators to detect and respond to anomalies promptly, ensuring the reliable and safe operation of the energy storage systems.

Integrating live monitoring with energy management systems allows for dynamic energy storage and distribution optimisation. This real-time feedback loop helps balance supply and demand, predict and prevent potential failures, and optimise the charge and discharge cycles to extend the lifespan of the storage components. Moreover, live monitoring supports predictive maintenance strategies, where data analytics and machine learning algorithms identify trends and patterns that indicate impending issues, allowing for maintenance to be performed proactively rather than reactively.

In addition to operational benefits, live monitoring provides valuable insights for system improvement and development. Data collected can refine models, improve system designs, and enhance the algorithms that control the storage systems. For stakeholders, including researchers, engineers, and policymakers, live monitoring data is crucial for making informed decisions about upgrades, scaling, and integrating new technologies. Live monitoring transforms how energy storage systems are managed, contributing to increased efficiency, reduced operational costs, and enhanced grid stability. It is a vital technology for advancing the reliability and effectiveness of energy storage solutions in an increasingly complex and renewable-driven energy landscape.

1.3.2 Predictive Analytics

Predictive analytics plays a transformative role in managing and optimising energy storage systems. By leveraging historical data, statistical algorithms, and machine learning techniques, predictive analytics enables forecasting future events and conditions, thus facilitating proactive decision-making [24]. In the context of energy storage systems, predictive analytics can anticipate various operational issues and opportunities, from predicting battery life and degradation rates to forecasting energy demand and optimising charge-discharge cycles.

One of the primary applications of predictive analytics is in maintenance and reliability. By analysing patterns in data such as temperature fluctuations, charge cycles, and performance metrics, predictive models can identify signs of potential failures or inefficiencies before they occur. This allows for timely maintenance interventions, reducing downtime, and extending the lifespan of the storage system. For instance, machine learning algorithms can predict when a battery will likely fail based on its usage history and environmental conditions, enabling operators to replace or service it before it causes system disruptions [25].

Predictive analytics also enhances the integration of renewable energy sources. By accurately forecasting energy production from sources like solar and wind, which are inherently variable, predictive models help in planning and optimising the use of stored energy. This ensures a more stable and reliable energy supply, reducing reliance on fossil fuels and improving the overall efficiency of the energy grid. For example, predictive models can forecast periods of high solar generation and schedule energy storage to capture excess energy, which can then be deployed during times of low generation.

Predictive analytics also supports economic optimisation by predicting market conditions and energy prices. Energy storage systems can be strategically charged or discharged based on anticipated price fluctuations, maximising financial returns. This is particularly relevant in energy markets with dynamic pricing structures, where storing energy when prices are low and selling it when prices are high can significantly enhance profitability.

2 Optimisation Techniques for Energy Storage Systems

The optimisation of energy storage systems is fundamental to improving energy efficiency and sustainability. This section focuses on the diverse methodologies and tools developed for enhancing these systems' performance, economic viability, and reliability [26]. The discussion commences with an exposition of simulation tools and techniques. These are critical for forecasting the operational dynamics of energy storage systems under varied scenarios, facilitating informed decision-making in system design and management.

Optimisation of energy storage systems goes beyond just improving parts individually. It involves evaluating how these systems integrate with the energy grid, work with renewable energy, and their impact on the economy and environment. The main goals are improving these systems' operation and reducing costs. This requires a collaborative approach from various fields, including computer science, electrical engineering, economics, and environmental science, aiming to create practical, affordable, and environmentally friendly solutions [27, 28]. The relationship between energy storage systems and the larger energy grid, especially how storage helps stabilise the grid, supports the use of renewable energy, and enables effective demand response strategies, is crucial. This emphasises the need for a unified approach that improves the technical aspects of storage systems and ensures they meet the growing needs and challenges of the global energy situation.

This section sets the stage for a detailed look at specific optimisation techniques and their uses. It stresses the need for collaborative and interdisciplinary efforts to make the energy sector more efficient, sustainable, and resilient. Essential optimisation methods, starting with simulation tools and techniques, design and configuration approaches, life cycle analysis systems, and detailed optimisation systems, including economic evaluation, automatic load management, and environmental impact analysis, are discussed while keeping this idea in context.

2.1 Simulation Tools and Techniques

Simulation models are computational tools that replicate the behaviour of real-world systems under various conditions, allowing researchers and engineers to study and predict the performance of these systems without physical experiments. Simulations enable the detailed analysis of energy storage systems, including how they interact with the power grid, respond to demand fluctuations, and integrate with renewable energy sources [29]. These models allow optimisation strategies to be tested and refined, enhancing energy solutions' efficiency, reliability, and sustainability.

Adapting simulation models to energy optimisation involves creating detailed representations of energy systems, incorporating variables such as energy generation capacities, storage capabilities, consumption patterns, and environmental impacts. These models can simulate the real-time operations of energy storage, the flow of electricity through the grid, and the performance of renewable energy technologies under different scenarios. This capability is crucial for designing more effective energy storage systems, planning grid expansions, and managing the integration of variable renewable energy sources.

Integrating Artificial Intelligence (AI) and machine learning into simulation models brings a new dimension of efficiency and precision to energy optimisation. AI algorithms can analyse vast amounts of simulation data, identifying patterns and insights that may not be immediately obvious to human analysts. This can lead to the discovery of novel optimisation strategies, more accurate demand forecasts, and improved energy management techniques [30]. AI-enhanced simulation models can adapt to changing conditions in real time, offering dynamic optimisation recommendations that balance energy supply and demand, reduce costs, and minimise environmental impacts.

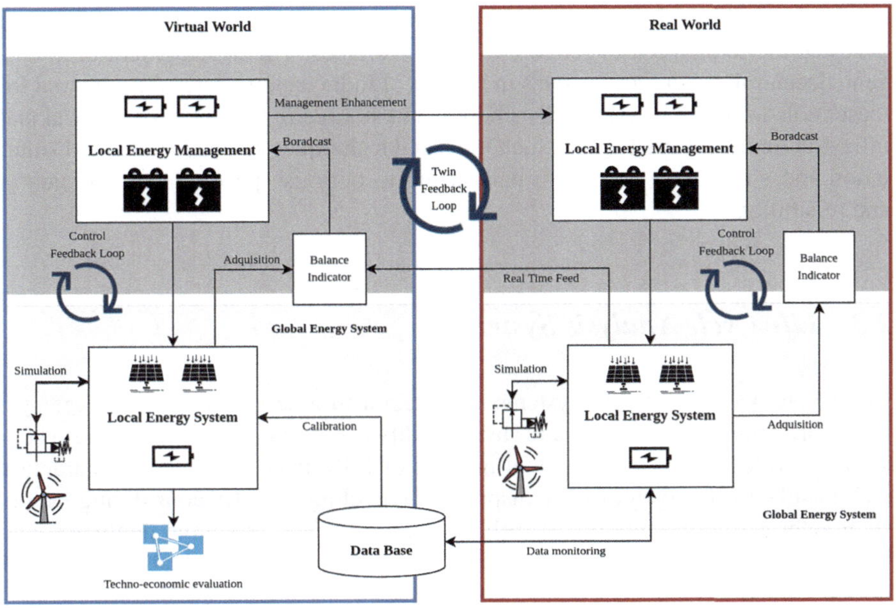

Simulation models offer improved prediction accuracy, cost reduction, and enhanced system design. However, they face challenges such as the need for high-quality data, computational complexity, and accurately replicating real-world conditions, which can limit their effectiveness and applicability in specific scenarios.

2.2 *Design and Configuration Tools and Techniques*

Design and configuration tools are essential in optimising energy storage systems for seamless integration with the energy grid. Critical software tools like CAD and simulation enable visualisation and testing of design concepts. Configuration techniques are essential for aligning systems with specific energy demands and renewable sources. Optimisation algorithms, including linear programming and machine learning models, facilitate optimal system sizing, placement, and operation. Performance modelling tools analyse system performance across diverse operational scenarios [31]. These tools enhance system efficiency but face challenges such as computational complexity and data accuracy requirements. Applying design tools, configuration techniques, and optimisation algorithms in energy storage system optimisation is paramount for achieving optimal system performance and grid integration.

The advent of advanced design and configuration tools has significantly enhanced the development and optimisation of energy storage systems. For instance, deploying advanced battery storage systems in grid stabilisation projects across Germany showcases how precise design and optimisation can enhance grid reliability and facilitate the integration of renewable energy sources. The development of intelligent, decentralised energy systems in regions of India demonstrates the potential for these tools to optimise energy distribution and storage in diverse geographical and infrastructural contexts [3, 32]. Such tools and techniques enable the detailed simulation and adjustment of system parameters to optimise performance, efficiency, and reliability.

2.3 Life Cycle Analysis Systems

Life Cycle Analysis (LCA) systems are integral to assessing energy storage systems' environmental footprint and sustainability across their entire life cycle. These systems evaluate the environmental impacts of ESS, from raw material extraction and manufacturing to end-of-life disposal or recycling [33]. By considering factors such as energy consumption, greenhouse gas emissions, resource depletion, and waste generation at each life cycle stage, LCA systems provide valuable insights for decision-makers, engineers, and policymakers.

Adopting LCA systems in the design and development of storage systems enables stakeholders to make informed decisions that minimise environmental harm and maximise sustainability [33, 34]. For example, by comparing different battery chemistries or storage technologies, LCA systems can identify the most environmentally friendly options and guide research and development efforts towards improving sustainability metrics.

LCA systems support the optimisation of energy storage operations and maintenance practices. By quantifying the environmental impacts of different maintenance strategies, LCA systems help prioritise actions that minimise energy consumption and emissions while maximising system performance and longevity. This proactive approach to environmental management reduces the overall carbon footprint throughout their operational lifespan.

LCA systems facilitate transparency and accountability in the energy storage industry by providing standardised methodologies for assessing and reporting environmental performance [33–36]. This enables stakeholders to evaluate the sustainability claims of different storage systems providers and make informed procurement decisions based on verifiable data.

As the importance of environmental sustainability continues to grow, integrating LCA systems into the energy storage system planning and decision-making processes becomes increasingly critical. By incorporating environmental considerations into the design, operation, and end-of-life management of energy storage systems, LCA systems play a pivotal role in advancing a more sustainable and resilient energy future.

2.4 *Optimisation Systems*

Optimisation systems enhance energy storage systems' efficiency, performance, and cost-effectiveness. These systems employ advanced mathematical models and algorithms to identify the best possible configurations and operational strategies for ESS, considering many variables and constraints [26]. Optimisation systems aim to maximise benefits such as energy efficiency, reliability, and economic returns while minimising costs and environmental impacts.

One of the primary functions of optimisation systems in ESS is to determine the optimal sizing and placement of storage units. This involves analysing the specific energy needs of a location, the characteristics of available storage technologies, and the patterns of energy supply and demand [11, 37]. For example, optimisation models can recommend the most suitable battery size and type for a solar power installation, ensuring that the system can effectively store excess energy during peak production times and release it during periods of high demand.

Optimisation systems also play a vital role in managing the charge and discharge cycles of ESS. By predicting energy production and consumption patterns, these systems can schedule the optimal times for charging and discharging, thereby improving the overall efficiency and lifespan of the storage units [38, 39]. This dynamic management is essential for integrating renewable energy sources characterised by their variability and intermittency. For instance, optimisation algorithms can ensure that a battery is charged during periods of excess wind generation and discharged when wind energy production drops, thus maintaining a stable energy supply.

Along with the technical performance, optimisation systems consider economic factors to enhance the financial viability of ESS. This includes optimising the operational strategies to take advantage of time-of-use pricing, demand response programs, and other market mechanisms [40]. ESS can minimise costs and maximise revenue by strategically timing stored energy, contributing to a more sustainable and profitable energy system.

Environmental considerations are also integral to optimisation systems. These systems can be designed to minimise ESS's carbon footprint and resource use by selecting the most sustainable materials and manufacturing processes and optimising the end-of-life disposal or recycling of storage units. For example, an optimisation system might prioritise using batteries with higher recycling rates and lower environmental impacts over their lifecycle.

2.4.1 Economic Evaluation

Economic evaluation is a critical component in selecting and implementing energy storage systems (ESS). It involves assessing the financial viability and cost-effectiveness of different storage technologies and configurations [41–43]. This evaluation helps stakeholders, including policymakers, investors, and engineers, make informed decisions about the deployment of ESS by considering the initial capital costs and the long-term operational expenses.

A comprehensive economic evaluation typically includes several key metrics:

1. **Capital Costs**: The upfront expenses of purchasing and installing an energy storage system. This includes the cost of the storage technology, infrastructure, and any necessary modifications to the existing energy grid.
2. **Operational and Maintenance Costs:** The ongoing expenses required to keep the ESS functioning effectively. This covers routine maintenance, monitoring, and any repair costs that may arise over the system's lifetime.
3. **Levelized Cost of Storage (LCOS):** A measure of the total cost to store and discharge a unit of energy over the system's lifecycle. LCOS considers all capital and operational expenses, providing a holistic view of the cost-effectiveness of different ESS options.
4. **Revenue Streams and Financial Incentives:** Potential income from the ESS, including payments from grid services, demand response programs, and time-of-use pricing benefits. Additionally, financial incentives such as subsidies, tax credits, and grants can significantly impact the economic evaluation by reducing net costs.
5. **Return on Investment (ROI) and Payback Period:** ROI calculates the profitability of the ESS by comparing the net benefits to the initial investment, while the payback period determines the time required for the ESS to recoup its initial costs through operational savings and revenue generation.
6. **Market and Regulatory Factors:** These include the impact of energy market prices, regulatory frameworks, and policy changes that can affect the economic viability of ESS. For example, favourable policies for renewable energy integration and storage can enhance the economic attractiveness of certain ESS technologies.

Economic evaluation also considers the potential risks and uncertainties associated with ESS investments. These might include fluctuations in energy prices, changes in technology costs, and evolving regulatory landscapes. Scenario analysis and sensitivity analysis are often used to understand how these variables might impact the financial performance of an ESS over time.

2.4.2 Automatic Load Management

Automatic load management (ALM) is a crucial technology for optimising the efficiency and reliability of energy systems, particularly in conjunction with ESS [44, 45]. ALM systems dynamically balance electricity supply and demand by automatically controlling the load on the grid, thereby ensuring optimal energy distribution, reducing peak demand, and enhancing the integration of renewable energy sources.

Key features and benefits of automatic load management include:

1. **Demand Response:** ALM systems enable real-time demand response by automatically adjusting the load based on current grid conditions. During periods of high demand, ALM can reduce or shift non-essential loads to off-peak times, alleviating stress on the grid and reducing the need for expensive peaking power plants.
2. **Energy Efficiency:** By optimising the usage of electrical devices and systems, ALM reduces overall energy consumption. It ensures that energy is used more efficiently, minimising waste and lowering energy costs for consumers and businesses.
3. **Integration with Energy Storage:** ALM systems work synergistically with ESS to maximise their benefits. For instance, ALM can manage when to charge or discharge the storage systems based on grid demand, energy prices, and the availability of renewable energy. This coordination ensures that stored energy is used most effectively, enhancing the stability and resilience of the grid.
4. **Renewable Energy Integration:** ALM facilitates the integration of variable renewable energy sources such as solar and wind by adjusting loads to match the availability of renewable generation. When renewable output is high, ALM can increase the load to use the excess energy; when renewable output is low, it can reduce the load or draw from stored energy.
5. **Cost Savings:** By shifting energy usage to off-peak times when electricity rates are lower, ALM helps consumers save on energy bills. For utilities, ALM reduces the need for investment in additional generation capacity and grid infrastructure, leading to long-term cost savings.
6. **Grid Stability and Reliability:** ALM contributes to grid stability by maintaining a balanced load, preventing overloading, and mitigating the risk of blackouts. It also enhances the reliability of the energy supply by ensuring a continuous and consistent flow of electricity, even during peak demand periods or when there are supply disruptions.
7. **Real-Time Monitoring and Control:** Advanced ALM systems incorporate real-time monitoring and control capabilities, using smart meters and sensors to collect data on energy usage. This data is analysed to make instantaneous adjustments to the load, ensuring optimal performance and efficiency.
8. **User Convenience and Automation:** ALM systems provide automated control of electrical devices and systems, reducing the need for manual intervention. Users can set preferences and priorities, allowing the system to manage energy usage automatically based on these settings.

2.4.3 Environmental Impact Analysis

Environmental Impact Analysis (EIA) is a critical process for assessing the potential environmental effects of energy storage systems (ESS) throughout their lifecycle [46, 47]. This analysis helps stakeholders understand the ecological footprint of different storage technologies, guiding decisions to minimise negative impacts and enhance sustainability.

Key Components of Environmental Impact Analysis:

1. **Lifecycle Assessment (LCA):** LCA is a comprehensive method used in EIA to evaluate the environmental impacts associated with all stages of an ESS's life, from raw material extraction and manufacturing to operation and end-of-life disposal or recycling. It considers energy consumption, greenhouse gas emissions, resource depletion, and waste generation.
2. **Resource Extraction and Material Use:** EIA examines the environmental impacts of extracting and processing raw materials required for ESS, such as lithium, cobalt, and rare earth metals. It assesses the sustainability of these resources and the ecological damage caused by mining activities.
3. **Manufacturing Processes:** The environmental effects of ESS components, including the energy and water used, emissions produced, and waste generated, are analysed. This helps identify cleaner production techniques and more sustainable material choices.
4. **Operational Impacts:** During the operational phase, EIA evaluates the emissions and energy consumption associated with charging and discharging the storage system. This includes the efficiency of the ESS and its integration with renewable energy sources, which can significantly influence its environmental footprint.
5. **End-of-Life Management:** The disposal, recycling, or repurposing of ESS components at the end of their life cycle is a crucial aspect of EIA. It assesses the environmental benefits of recycling materials versus the impacts of disposal in landfills, aiming to maximise resource recovery and minimise waste.
6. **Carbon Footprint:** EIA quantifies the total greenhouse gas emissions associated with the ESS over its entire lifecycle. This includes direct emissions from manufacturing and operation and indirect emissions from associated activities, providing a clear picture of the system's contribution to climate change.
7. **Toxicity and Human Health Risks:** The analysis considers potentially releasing toxic substances during the ESS lifecycle, which can affect human health and ecosystems. This includes examining the impacts of chemicals used in batteries and other storage technologies.
8. **Ecological Impacts:** EIA assesses the broader ecological effects of ESS, such as habitat destruction, biodiversity loss, and soil and water pollution. It evaluates how the deployment of ESS might alter land use patterns and affect local wildlife.

Benefits of Environmental Impact Analysis:

- Informed Decision-Making: EIA provides comprehensive data that helps stakeholders make informed decisions about the deployment and management of ESS, ensuring that environmental considerations are integrated into the planning process.
- Regulatory Compliance: Conducting EIA helps organisations comply with environmental regulations and standards, avoiding legal penalties and promoting sustainable practices.

- Sustainability Improvements: By identifying the most significant environmental impacts, EIA enables the development of strategies to mitigate negative effects and enhance the overall sustainability of ESS.
- Stakeholder Engagement: EIA facilitates transparency and engagement with stakeholders, including communities, policymakers, and investors, fostering trust and support for ESS projects.

3 Efficient Models for Organised Optimised Storage

Efficient energy production, storage, and optimisation are crucial factors to consider in the growing energy demand. Optimisation and planning are required to minimise the loss and maximise the efficiency. Optimisation can be related to production, storage, consumption, and planning to meet future needs. As the energy comes from various sources, it needs to be optimised and organised to tap the maximum potential of each energy source [48–50]. Energy storage is also a challenging task, requiring significant consideration of the environment and sustainability. Moreover, when it comes to energy storage, different considerations need to be addressed, including the type of storage, possible optimisation levels, and the management system. When studying the energy-optimised storage systems and models for identifying effective models, the foremost important part is to address and study the various optimisation models and demand response strategies, which have a crucial role in planning models for energy storage.

3.1 Optimisation Models

Optimisation models are adopted to control and manage energy production, storage, and distribution effectively [48, 51]. The models are designed for optimisation following various strategies, which include the following

- **Physical Storage Optimisation**
- The physical storage of energy systems is a crucial factor to consider. The architecture of multiple storage systems significantly depends on how effectively the system can store the energy. Various storage systems are available to store energy in different forms, from lithium-ion batteries to pumped hydro storage. Some energy sources cannot be stored, so effective planning of systems helps convert the energy from its sources to storage and distribution mechanisms.
- Thermal management is another critical factor that must be considered when we store and generate energy. Various storage mediums generate a significant amount of energy from heat, which needs to be optimised to ensure the stored energy is safe. When identifying the effective way to store the power produced,

many things are to be considered apart from the thermal issues associated with the storage medium.

- The placement and layout of the storage medium are other critical factors in identifying the optimisation for physical storage. Landscape and demographic characteristics are also crucial in determining the effective placement of storage space, which caters to the demand for energy for various stakeholders. The conversion and production of energy from renewable energy sources require geographical advantage, and the distribution of power from such locations to the in-demand locations requires proper transmission mediums that are also optimised for effective transmission.
- Various considerations are essential in determining the effective physical storage system for energy. Identifying such a medium depends on the geographical location, political factors, government investment plans, and many other essential factors. Thus, optimising the physical storage for effective organisation of energy storage systems is challenging.
- **Energy Management System**
- An energy management system (EMS) is associated with distributing, producing, and managing energy sources. This system helps gather data regarding the production, demand, and management of energy supply for different users. These systems help analyse and support the decision-making process based on the data and the historical information from various data points related to the production and consumption of energy from different sources. The energy from multiple sources has been effectively managed, planned, and distributed with the support of these systems.
- Predictive analytics is an essential factor in the EMS that supports the decision-makers with the energy demand, optimises charging/discharging cycles in storage mediums, reduces the impact of heavy energy consumption, and manages the grids effectively. The analytics are based on historical data collected from energy demand and usage patterns, which helps management prepare for energy demands. Along with historical data, various other factors are essential to energy needs. These external factors not part of the regular patterns also have to be considered, addressed, and prepared for energy needs without affecting the systems in place and to minimise the tension. AI and ML also help in this process.
- Reducing the strain of energy in the grid helps minimise the loss of power, which is energy loss that cannot be eliminated but can be organised to optimise and minimise. The demand forecast helps in the management of energy loss and production.

3.1.1 Load Scheduling Models

Energy demands change over time and are highly dependent on various factors. Identifying the exact load requirements significantly impacts energy storage and transmission. Multiple models for determining the load scheduling are available to predict the load and energy requirements [52, 53]. These mathematical models are aimed to optimise the generation and dispatch of electricity resources to meet the forecasted demand with careful consideration of various technical and economic factors.

- **Unit Commitment Model**
- This model helps decide which power generation units to work to meet the energy demand. The model helps determine whether the power generation units must be on/off, how long they should work, what days they need to work, and so on. The main objective of this model is to minimise the total operational cost while effectively meeting the expected demand, minimising downtime, and maximising the plant's lifetime.
- **Economic Dispatch**
- The economic dispatch model helps to determine the optimal power output for each committed generating unit at a specific point in time. The main application of this model is to minimise the power generation cost while meeting the real-time demand and optimising the distribution capacities.
- **Optimal Power Flow**
- The optimal power flow model has been designed to calculate the optimal voltage and power flowing across the transmission network, which helps in efficient and secure operation. This model considers power losses on lines, generation unit capacities, storage capacity, and various security constraints.
- **Stochastic Programming**
- The stochastic programming model helps to manage the uncertainty in the power forecasts. The forecast can be applied to production, consumption, and storage requirements. It is a probabilistic model that finds robust solutions that can adapt to real-world conditions, accommodating various factors, including weather and other factors.
- **Model Predictive Control**
- This model uses real-time data and short-term forecasts to update the schedules and continuously optimise energy generation and storage. This model suits a dynamic and responsive approach and efficiently manages the power grid when scheduling loads.

Various models are applicable for multiple purposes, and identifying and deploying effective models for accurate predictions and forecasts are required. Various applications require different approaches to determine the appropriate model. The models have to be selected based on three main factors:

1. **Planning Horizon:** Choose the model based on the term for which you are planning to rely on the model. Models like unit consumption are ideal for long-term planning, while economic dispatch and optimal power flow models are suitable for real-time planning and decision-making.
2. **Level of Detail:** Models like unit consumption are appropriate for identifying the energy load commitment-based directions, while the economic dispatch model helps plan power distribution during a committed period.
3. **Uncertainty:** Energy production, consumption, and demand have uncertainties. Stochastic programming models are models that can accommodate uncertainty. Using these models helps to tackle the uncertainties that arise due to the demand, supply, and storage of energy.

Production, storage, and consumption are significant activities that are involved in energy usage and production. Identification and taping of required energy in various forms are associated with production, which is strictly driven by demand. Storing energy in different forms and using it when needed is challenging. Multiple parameters must be closely monitored to determine production, consumption, and storage.

3.1.2 Grid Synchronisation Models

An easy means of energy transmission and storage is using grid systems can be identifyed by making use of the following to ensure the management of grids effectively for power distribution and storage

- Swing equation model
- Phase-locked loop model
- Droop control model
- IEEE Standard Model
- Detailed machine model

3.1.3 Dynamic Programming Approaches for Storage Optimisation

Dynamic programming (DP) is a powerful technique well-suited for optimising storage systems, mainly when dealing with energy storage. Here's how DP can be applied to storage optimisation:

3.1.4 Genetic Algorithms and Heuristic Optimisation

Genetic algorithms (GAs) are a heuristic optimisation technique inspired by natural selection. Both GAs and heuristic optimisation aim to find solutions to complex problems that are good enough, especially when the ideal solution is difficult or time-consuming to see directly.

3.2 Demand Response Strategies

Meeting the energy requirements is one of the major factors that helps in planning the required energy storage means. The higher the demand, the less time is necessary to keep the energy stored, but if the production is not happening in the time of need, storage is a crucial factor that must be considered [54]. Energy storage system planning and identification requires considering various factors, and the systems can only be designed for such applications where the energy needs and requirements are correctly identified. The solution for deployed problems will only work in fixed use cases and may not be suitable for other applications.

3.2.1 Automated Strategies for Smart Grid Systems

Automation is the central part of making the grid smart; the production, distribution, and storage of energy are managed by the grid systems in place. The need for identification and the planning of appropriate action plans have to be quickly addressed to meet the demand. Hence, the deployment of automated systems plays a huge role in the planning and creation of such automated systems that keep monitoring the existing systems and usage and forecasting and meeting the unexpected need for energy in the grid systems. Various automation tests can be employed to check whether the systems are up and running [55, 56]. Keeping the sensors and integrating the sensor network to gather data help to attain automated systems for the smart systems. AI–ML technologies can also gather data from heterogeneous sensors and sensory points, which helps build a fully functional automated system for tracking and maintaining such smart systems.

3.2.2 Grid Balancing and Stability

Grid balancing and grid's stability are maintained by monitoring the load and the planned dispatch. The grid usage has to be optimised in such a way that the needs of the users are met and the demand is satisfied without managing the grid or optimising the load that has been shared. Balancing the load and maintaining stability are achieved with constant upgrades in the monitoring strategies [57–59]. The grid balancing systems that require time-series data of the components' demand, supply, distribution, and maintenance history all go upon an input to a multimodal prediction model that can predict the load requirements in the smart grid and energy systems. Based on the normal predictions, the systems can be planned and prepared so that keeping a buffer of resources will help meet unexpected demands and situations without affecting the stakeholders.

3.2.3 Energy Storage Sizing and Integration

The storage system plan heavily depends on various factors, including the potential need for energy in various forms. The power generated from various sources must be saved in the battery systems or grids. These storage requirements, when planned appropriately, the size for the storage mediums can be identified and then integrated with the supplying grids. The energy storage requirements come with respect to the unexpected rise in need or energy usage for some abnormal activities. The storage plans and sizing requirements can only be forecasted using various mathematical and statistical modelling methods when an abnormal activity is detected in the connected endpoints. This can be an outage of some of the components in the smart grid components, an outage in some part of the distribution channel or the storage system, or a rise in the usage and potential customers [27, 60, 61]. A smart system is planned and prepared to address this incident without manual intervention. The storage sizing requirements of a region vary from others. Hence, the data analysis helps in planning the required hardware infrastructure requirements for storage sizing integrations to the grid and the possibility of expansion of storage mediums.

3.2.4 Predictive Maintenance of Systems

The systems in place require constant maintenance and wear and tear checks for continuous running to meet the expectations of various stakeholders. Maintaining the various components of the sophisticated planned and integrated systems requires data input from various sources. Apart from the regular maintenance schedules of the machines in an annual or other timeframe, the systems are to be made smarter with the help of interconnected sensors and enabled for constant monitoring of the various devices in the systems [62, 63]. The connected systems and constant monitoring of these systems enable us to efficiently identify the maintenance schedules and plan for the continuous running of the systems to meet the requirements and serve the demand of energy needs. There are various case studies that we can refer to see the implementation of various systems to constantly track and monitor the systems in the grid and for storage of energy resources for future needs.

4 Emerging Trends and Future Directions

The evolving landscape of energy storage, a convergence of cutting-edge technologies, sustainability-driven initiatives, uncharted research frontiers, and their profound implications are shaping the present and future of energy storage systems. In this section, a detailed study on dynamic facets of energy storage, exploring the promise of emerging technologies such as advanced batteries and artificial intelligence, delving into the sustainable paradigms that underpin green energy initiatives, charting the course for future research endeavours and identifying research gaps, and, most importantly, understanding the wide-reaching implications of these trends [64, 65]. From revolutionising energy storage efficiency to aligning with global sustainability goals, these developments propel us towards a future where energy storage systems are at the forefront of an efficient, sustainable, and resilient energy landscape.

4.1 Emerging Technologies in Energy Storage

As the field of energy storage continues to evolve, emerging technologies are at the forefront of innovation, offering the promise of more efficient, sustainable, and resilient energy storage systems. These technologies are crucial in addressing the increasing demands of modern energy systems, including integrating renewable energy sources, transportation electrification, and the need for reliable grid solutions [66, 67]. In this subsection, we delve into the most promising emerging technologies in energy storage and their implications for the future of optimised storage systems.

4.1.1 Advanced Battery Technologies

One of the most prominent areas of innovation in energy storage is the development of advanced battery technologies. Solid-state batteries, for instance, are poised to revolutionise the industry. These batteries replace traditional liquid or gel electrolytes with solid materials, offering higher energy density, longer cycle life, and enhanced safety. Such advancements are of particular significance in applications like electric vehicles (EVs) and grid-scale storage, where energy density and safety are paramount. Lithium–sulfur batteries, with their potential for even higher energy density, and flow batteries, offering scalability and long cycle life, are also gaining momentum [68]. Integrating these advanced battery technologies into optimised storage systems can transform energy storage efficiency and sustainability.

4.1.2 Machine Learning and Artificial Intelligence (AI)

Machine learning and AI have emerged as powerful tools for enhancing the performance of energy storage systems. These technologies leverage historical data to predict energy demand patterns and optimise the real-time operation of storage systems. Predictive analytics, driven by machine learning, enable better capacity planning, while real-time optimisation adjusts charging and discharging rates based on dynamic factors such as electricity prices and grid conditions. Machine learning and AI can significantly improve energy storage's economic and operational performance of energy storage, making them integral to future optimised storage systems.

4.1.3 Integration of Renewable Energy Sources

The increasing adoption of renewable energy sources, such as solar and wind power, underscores the importance of energy storage in balancing supply and demand. Grid-scale energy storage solutions are vital for storing excess energy generated during periods of high renewable output and releasing it during low generation, ensuring grid stability and a reliable power supply. Innovations in integrating renewable sources with storage, including advanced control systems and hybrid storage solutions, will play a pivotal role in the future of optimised storage systems, facilitating the transition to cleaner and more sustainable energy grids.

4.1.4 Circular Economy and Recycling

Sustainability and recycling have gained prominence with the growing demand for batteries and energy storage components. The principles of the circular economy, emphasising responsible sourcing, eco-friendly manufacturing, and end-of-life recycling, are guiding the industry [68, 69]. Emerging trends in recycling processes aim to repurpose and recycle materials from old batteries, reducing waste and environmental impact. Integrating sustainable practices into optimised storage systems aligns with global efforts to reduce the carbon footprint and minimise resource depletion.

These emerging technologies and trends in the development and optimisation of energy storage systems are paramount for meeting the challenges and opportunities of the future. By leveraging advanced batteries, harnessing the power of machine learning and AI, embracing the integration of renewable energy, and adopting sustainable and recycling practices, we are paving the way for more efficient, eco-friendly, and resilient energy storage solutions [68–71]. These emerging technologies represent a transformative force in the field of energy storage, offering the potential to shape the energy landscape for years to come.

4.2 Sustainability and Green Energy Initiatives

In an era marked by heightened environmental awareness and a collective commitment to reduce carbon footprints, sustainability and green energy initiatives are pivotal considerations in the evolution of energy storage systems. As society increasingly embraces the transition to cleaner and more sustainable energy sources, energy storage is crucial in facilitating this transition [72]. In this subsection, we explore the burgeoning trends and future directions in sustainability and green energy initiatives as they relate to optimised storage systems for energy.

4.2.1 Sustainable Battery Design

Central to sustainability in energy storage is the development of batteries with minimal environmental impact. Sustainable battery design encompasses several fundamental principles, including responsible sourcing of materials, eco-friendly manufacturing processes, and end-of-life recycling. Manufacturers are exploring methods to reduce the carbon footprint of battery production by sourcing materials ethically, implementing energy-efficient manufacturing techniques, and reducing harmful emissions [72, 73]. Moreover, adopting eco-friendly chemistries and materials in battery construction is a promising avenue for minimising environmental harm. The objective is to create batteries that are not only energy-efficient but also produced and recycled with the least possible impact on the planet.

4.2.2 Second-Life Batteries

An emerging trend in sustainability is the concept of second-life batteries. These batteries have served their primary function but still retain a considerable amount of their initial capacity. Instead of discarding these batteries, they are repurposed for secondary applications, such as stationary energy storage in homes or businesses. Second-life batteries can provide cost-effective energy storage solutions while reducing waste and requiring new battery production. Using second-life batteries represents a sustainable approach to prolonging the life and value of battery components.

4.2.3 Grid Decarbonisation

As the world shifts towards decarbonisation, green energy initiatives are increasingly focused on reducing the carbon footprint of energy systems. Optimised storage systems for energy play a pivotal role in this transition by enabling the efficient integration of renewable energy sources, such as solar and wind power, into the grid. Advanced models are being developed to balance and optimise renewables and conventional energy sources, reducing reliance on fossil fuels. Grid decarbonisation efforts emphasise the need for more effective storage systems to handle intermittent renewables, ultimately leading to greener and more sustainable energy grids.

4.2.4 Green Certifications and Standards

Green energy initiatives are backed by certifications and standards that guide the design and operation of energy storage systems. Various organisations have established green building and energy storage certification programs, including the U.S. Green Building Council (USGBC) and the Leadership in Energy and Environmental Design (LEED). Compliance with these standards ensures that energy storage systems adhere to stringent sustainability criteria, including energy efficiency, responsible material sourcing, and environmentally friendly operation. Adopting these certifications reflects a commitment to sustainability and provides consumers and stakeholders with assurance regarding the environmental credentials of energy storage systems.

Incorporating sustainability and green energy initiatives into the development and operation of optimised storage systems is essential for addressing the growing global imperative to reduce greenhouse gas emissions and minimise the environmental impact of energy technologies [74, 75]. By emphasising responsible battery design, exploring the potential of second-life batteries, contributing to grid decarbonisation, and adhering to green certifications and standards, energy storage systems are poised to play a pivotal role in transitioning to a cleaner, more sustainable energy future. These initiatives underscore the significance of striking a balance between energy efficiency, economic viability, and environmental responsibility in the quest for optimised storage systems for energy.

4.3 Research Gaps and Future Research Directions

While advancements in energy storage systems and technologies are making significant strides, several critical research gaps and uncharted frontiers warrant exploration. Identifying these gaps serves as a compass guiding future research endeavours, enabling the development of more effective models for computing optimised storage systems for energy.

4.3.1 Longevity and Degradation Modelling

One of the foremost research gaps is longevity and degradation modelling in energy storage systems. Understanding and accurately predicting degradation patterns are crucial as these systems are exposed to continuous charge and discharge cycles. Future research should focus on developing comprehensive degradation models encompassing various storage technologies, including advanced batteries and flow systems. This will not only extend the lifespan of storage systems but also enhance the economic viability of energy storage.

4.3.2 Sustainability Metrics and Assessment

Sustainability is an increasingly critical aspect of energy storage technology. Yet, there is a need for more standardised metrics and assessment tools to evaluate the environmental impact of storage solutions. Future research should delve into the development of comprehensive sustainability assessment frameworks and eco-labelling systems. Such tools enable stakeholders to make informed choices when selecting energy storage systems that align with sustainability and green energy initiatives.

4.3.3 Grid Integration and Control Models

The efficient integration of energy storage systems into the grid remains an active research area. Future endeavours should concentrate on refining grid integration and control models to maximise the utilisation of renewable energy sources and minimise grid disruptions. Advanced machine learning and AI models that can predict and optimise energy storage operations in real time will be instrumental in achieving this goal.

4.3.4 Techno-Economic Optimisation

The development of models for techno-economic optimisation is paramount. These models should consider more than just the technical performance of energy storage systems and their economic feasibility. Future research should provide a holistic approach to optimising storage systems, accounting for various factors, including initial investment, operating costs, and the value of services provided to the grid.

4.3.5 Safety and Reliability Enhancement

Safety and reliability in energy storage systems are non-negotiable. There remains room for research in enhancing the safety mechanisms and reliability of advanced battery technologies, particularly solid-state and lithium–sulfur batteries. This includes developing fail-safe mechanisms, thermal management strategies, and methods for early fault detection to mitigate safety risks.

4.3.6 Electrified Transportation Infrastructure

As the electrification of transportation gains momentum, research needs to be expanded into developing infrastructure and charging solutions that can support large-scale EV adoption. This includes exploring advanced charging technologies, grid compatibility, and establishing charging standards that ensure convenient and accessible charging for a growing number of EVs.

4.3.7 Interdisciplinary Collaborations

Future research directions should encourage interdisciplinary collaborations. The energy storage requires expertise in various domains, from materials science and engineering to economics and policy. Cross-disciplinary collaborations can facilitate the development of holistic models considering the multifaceted aspects of energy storage optimisation.

Identifying gaps and future research directions is fundamental to advancing energy storage systems. Addressing these gaps will propel the development of more efficient, sustainable, and resilient energy storage solutions. Through interdisciplinary collaborations and innovative research endeavours, the energy storage community can contribute to a future where optimised storage systems are central in transitioning to cleaner, more reliable, viable energy ecosystems.

4.4 Implications for Energy Storage Systems

The evolving landscape of energy storage systems and the dynamic trends and research directions discussed in the previous sections carry profound implications for the present and future of energy storage. Understanding these implications is vital for policymakers, industry stakeholders, and researchers as they navigate the path towards more efficient, sustainable, and economically viable energy storage solutions.

4.4.1 Enhanced Grid Stability

Integrating advanced models for grid integration and control models, driven by machine learning and AI, will lead to enhanced grid stability. Energy storage systems will play a pivotal role in buffering the intermittency of renewable energy sources, providing grid operators with the flexibility to balance supply and demand, reduce grid disruptions, and improve the overall reliability of energy supply.

4.4.2 Sustainability and Green Initiatives

Incorporating sustainability metrics and eco-labelling for energy storage systems will promote a greater focus on eco-friendly designs and practices. This will reduce the environmental footprint of storage systems and align the industry with global sustainability goals, making energy storage an integral component of green energy initiatives.

4.4.3 Economic Viability

Techno-economic optimisation models will have profound implications for the economic viability of energy storage systems. Considering technical and financial factors, these models will guide investments towards storage solutions that enhance grid performance and provide a favourable return on investment. This will encourage more widespread adoption of energy storage technology.

4.4.4 Energy Decentralisation

The development of models for distributed energy storage systems models will facilitate energy decentralisation, allowing storage solutions to be deployed closer to consumption points. This will reduce transmission and distribution losses, improve grid resilience, and provide localised energy sources during grid disruptions. The implications include greater energy security and reduced dependence on centralised energy generation.

4.4.5 Electrified Transportation

As the electrification of transportation expands, there will be implications for the development of charging infrastructure and grid compatibility. Energy storage systems will be integral to ensuring that the grid can support the charging needs of many electric vehicles. This will foster a more sustainable and convenient transportation ecosystem.

4.4.6 Safety and Reliability

Enhancements in safety and reliability mechanisms, particularly for advanced battery technologies, will boost consumer and industry confidence in the safety of energy storage systems. This will accelerate the adoption of advanced solid-state and lithium–sulfur batteries in applications such as electric vehicles and grid-scale storage.

4.4.7 Transition to Cleaner Energy Systems

The overarching implication is that optimised storage systems are pivotal in transitioning to cleaner and more sustainable energy systems. By maximising the efficiency and reliability of energy storage, we can leverage the full potential of renewable energy sources, reduce greenhouse gas emissions, and create more resilient and adaptive energy ecosystems.

The implications of the evolving trends and research directions in energy storage are profound. They encompass enhanced grid stability, a focus on sustainability and green initiatives, economic viability, energy decentralisation, support for electrified transportation, improved safety and reliability, and, most importantly, the accelerated transition to cleaner and more sustainable energy systems [64, 76–80]. These implications underscore the pivotal role of energy storage in reshaping the energy landscape and the importance of continued research and innovation in the field. As we embrace these implications, we pave the way for a future where optimised energy storage systems are central to achieving a more efficient, sustainable, and reliable energy ecosystem.

5 Conclusion

While studying the various models for computing optimised storage systems for energy, we have traversed a landscape marked by innovation, sustainability, evolving research, and far-reaching implications. From emerging technologies in energy storage, including advanced battery systems and AI-driven optimisation, to the dedication to sustainability and green energy initiatives, the journey has revealed the transformative potential of energy storage systems. The study could identify research gaps that invite further inquiry, guiding the way to a more sophisticated understanding of these systems and the potential avenues for growth.

The implications of these advancements extend beyond the realm of energy storage alone. They transcend into the very fabric of our modern energy landscape. Enhanced grid stability, economic viability, energy decentralisation, and the facilitation of electrified transportation are only a few of the profound shifts we anticipate. Our commitment to sustainability is reflected in the design and operation of energy storage systems, aligning them with the global imperative to reduce our environmental footprint. The optimised energy storage systems of the future are not merely a technological achievement but a crucial enabler of a sustainable, efficient, and resilient energy ecosystem. Their potential is limited only by our collective commitment to research, innovation, and the embrace of forward-thinking models that guide our path.

The evolving energy landscape continues to beckon us forward, and the journey towards optimised energy storage is ongoing. It is a journey propelled by advanced technology, sustainability principles, interdisciplinary collaborations, and a steadfast commitment to a cleaner, greener future. By navigating this landscape with foresight, we stand on the cusp of an energy paradigm where storage systems serve as the cornerstone of a brighter, more sustainable energy future, and we have the tools, insights, and vision to make it a reality.

References

1. Eapen, N. G., Harsha, K. G., & Kesan, A. (2023). Energy intelligence: The smart grid perspective. In *AI-powered IoT in the energy industry* (pp. 35–74). Springer. https://doi.org/10.1007/978-3-031-15044-9_3
2. Koirala, B., van Oost, E., & van der Windt, H. J. (2021). *New pathways for community energy and storage*. Mdpi AG. https://books.google.com/books/about/New_Pathways_for_Community_Energy_and_St.html?hl=&id=cs7mzgEACAAJ
3. Garcia, J. (2020). *Analysis and design of hybrid energy storage systems*. MDPI. https://books.google.com/books/about/Analysis_and_Design_of_Hybrid_Energy_Sto.html?hl=&id=_VjdDwAAQBAJ
4. Taha Demirkan, M., & Attia, A. (2019). *Energy storage devices*. Books on Demand. https://books.google.com/books/about/Energy_Storage_Devices.html?hl=&id=SHb-DwAAQBAJ
5. Huang, X., Lin, Y., Ruan, X., Li, J., & Cheng, N. (2023). Smart grid energy scheduling based on improved dynamic programming algorithm and LSTM. *PeerJ Computer Science, 9*, e1482. https://doi.org/10.7717/peerj-cs.1482
6. Zobaa, A. F., & Vaccaro, A. (2014). *Computational intelligence applications in smart grids: Enabling methodologies for proactive and self-organizing power systems*. World Scientific. https://play.google.com/store/books/details?id=Wdm3CgAAQBAJ
7. Vinoth Kumar, B., Sivakumar, P., Rajan Singaravel, M. M., & Vijayakumar, K. (2020). *Intelligent paradigms for smart grid and renewable energy systems*. Springer Nature. https://play.google.com/store/books/details?id=tTMPEAAAQBAJ
8. Gabbar, H. (2016). *Smart energy grid engineering*. Academic Press. https://play.google.com/store/books/details?id=VEq0CwAAQBAJ
9. Gerasopoulos, S. I., Manousakis, N. M., & Psomopoulos, C. S. (2024). A novel methodology for the estimation of failure behavior of "fair" smart meters and analysis of their circular economy chain. *Environmental Science and Pollution Research International, 31*(12), 17533–17545. https://doi.org/10.1007/s11356-022-24784-x
10. Ma, C.-T., & Yao, B.-H. (2023). Design and implementation of an integrated control scheme for GaN-based multiple power converters. *Micromachines, 14*(4), 833. https://doi.org/10.3390/mi14040833
11. Mathur, A., Kumari, R., Meena, V. P., Singh, V. P., Azar, A. T., & Hameed, I. A. (2024). Data-driven optimization for microgrid control under distributed energy resource variability. *Scientific Reports, 14*(1), 10806. https://doi.org/10.1038/s41598-024-58767-4
12. Galanis, G. (2020). *Renewable energy resource assessment and forecasting*. MDPI. https://books.google.com/books/about/Renewable_Energy_Resource_Assessment_and.html?hl=&id=NZoLEAAAQBAJ
13. Malik, H., Yadav, A. K., & Alotaibi, M. A. (2023). *Intelligent data analytics for solar energy prediction and forecasting: Advances in resource assessment and PV systems optimization*. Elsevier. https://books.google.com/books/about/Intelligent_Data_Analytics_for_Solar_Ene.html?hl=&id=MPa6zwEACAAJ
14. Nagpal, N., Alhelou, H. H., Siano, P., Padmanaban, S., & Lakshmi, D. (2023). *Applications of big data and artificial intelligence in smart energy systems: Vol. 2. Energy planning, operations, control and market perspectives*. CRC Press. https://play.google.com/store/books/details?id=Q5DZEAAAQBAJ
15. Shaw, R. N., Ghosh, A., Mekhilef, S., & Balas, V. E. (2022). *Applications of AI and IOT in renewable energy*. Academic Press. https://play.google.com/store/books/details?id=x5xBEAAAQBAJ
16. Vyas, A. K., Balamurugan, S., Hiran, K. K., & Dhiman, H. S. (2022). *Artificial intelligence for renewable energy systems*. John Wiley & Sons. https://books.google.com/books/about/Artificial_Intelligence_for_Renewable_En.html?hl=&id=LcV6EAAAQBAJ
17. Doukas, H., Marinakis, V., & Sarmas, E. (2024). *Machine learning applications for intelligent energy management: Invited chapters from experts on the energy field*. Springer.

https://books.google.com/books/about/Machine_Learning_Applications_for_Intell.html?hl=&id=6As60AEACAAJ
18. Cabeza, L. F. (2014). *Advances in thermal energy storage systems: Methods and applications*. Elsevier. https://play.google.com/store/books/details?id=_590AwAAQBAJ
19. U.S. Department of Energy, Division of Energy Storage Systems. (1979). *Energy storage systems: Program overview*. https://play.google.com/store/books/details?id=1KTgHeSy0scC
20. Rufer, A. (2017). *Energy storage: Systems and components*. CRC Press. https://books.google.com/books/about/Energy_Storage.html?hl=&id=CjU8DwAAQBAJ
21. Kim, Y., Pedram, M., Wang, Y., & Chang, N. (2013). Computer-aided design and optimization of hybrid energy storage systems. https://books.google.com/books/about/Computer_Aided_Design_and_Optimization_o.html?hl=&id=XTHYxQEACAAJ
22. Recioui, A., & Bentarzi, H. (2020). *Optimizing and measuring smart grid operation and control*. IGI Global. https://play.google.com/store/books/details?id=4rgIEAAAQBAJ
23. Fabri, J. (2018). *Automatic storage optimization (classic reprint)*. 1kg Limited. https://books.google.com/books/about/Automatic_Storage_Optimization_Classic_R.html?hl=&id=MeVpswEACAAJ
24. Quintero, D., Ancel, T., Cassie, G., Ceron, R., Darwish, A., Felix, G., He, J. J., Keshavamurthy, B., Makineedi, S., Nikalje, G., Pal, S., Salie, Z., Tiwary, A., & Redbooks, I. B. M. (2013). *Workload optimized systems: Tuning POWER7 for analytics*. IBM Redbooks. https://play.google.com/store/books/details?id=-PLHAgAAQBAJ
25. Kumar, V., & Ram, M. (2021). *Predictive analytics: Modeling and optimization*. CRC Press. https://play.google.com/store/books/details?id=rSUMEAAAQBAJ
26. Elsayed, M. E. A., Abdo, S., Attia, A. A. A., Attia, E.-A., & Abd Elrahman, M. A. (2022). Parametric optimisation for the design of gravity energy storage system using Taguchi method. *Scientific Reports, 12*(1), 19648. https://doi.org/10.1038/s41598-022-20514-y
27. Hu, Z. (2020). *Energy storage for power system planning and operation*. Wiley. https://play.google.com/store/books/details?id=2a7MDwAAQBAJ
28. Mathew, V. K., Hotta, T. K., Ali, H. M., & Sundaram, S. (2022). *Energy storage systems: Optimization and applications*. Springer Nature. https://play.google.com/store/books/details?id=_LOSEAAAQBAJ
29. Antunes, C. H. (2019). *Optimisation models and methods in energy systems*. MDPI. https://books.google.com/books/about/Optimisation_Models_and_Methods_in_Energ.html?hl=&id=ToanDwAAQBAJ
30. Mohammadi-Ivatloo, B., Shotorbani, A. M., & Anvari-Moghaddam, A. (2021). *Energy storage in energy markets: Uncertainties, modelling, analysis and optimization*. Academic Press. https://play.google.com/store/books/details?id=AjYXEAAAQBAJ
31. Kim, Y., & Chang, N. (2016). *Design and management of energy-efficient hybrid electrical energy storage systems*. Springer. https://books.google.com/books/about/Design_and_Management_of_Energy_Efficien.html?hl=&id=7J9oDQEACAAJ
32. Schmiegel, A. U. (2023). *Energy storage systems: System design and storage technologies*. Oxford University Press. https://books.google.com/books/about/Energy_Storage_Systems.html?hl=&id=rES7EAAAQBAJ
33. Martínez, E., Jiménez, E., Blanco, J., & Pérez, M. (2016). *Life cycle analysis of renewable energy systems: Environmental issues*. Springer. https://books.google.com/books/about/Life_Cycle_Analysis_of_Renewable_Energy.html?hl=&id=F8TssgEACAAJ
34. Passarini, F., & Ciacci, L. (2021). *Life cycle assessment (LCA) of environmental and energy systems*. MDPI. https://books.google.com/books/about/Life_Cycle_Assessment_LCA_of_Environment.html?hl=&id=c94mEAAAQBAJ
35. Basosi, R., Cellura, M., Longo, S., & Parisi, M. L. (2018). *Life cycle assessment of energy systems and sustainable energy technologies: The Italian experience*. Springer. https://play.google.com/store/books/details?id=jXB_DwAAQBAJ

36. Miguel, G. S., & Alvarez, S. (2021). *Life cycle assessment of energy systems*. MDPI. https://books.google.com/books/about/Life_Cycle_Assessment_of_Energy_Systems.html?hl=&id=GzspEAAAQBAJ
37. Dinçer, I., Rosen, M. A., & Ahmadi, P. (2017). *Optimization of energy systems*. Wiley. https://books.google.com/books/about/Optimization_of_Energy_Systems.html?hl=&id=UD_CDgAAQBAJ
38. Assad, M., & Rosen, M. A. (2021). *Design and performance optimization of renewable energy systems*. Academic Press. https://play.google.com/store/books/details?id=ODoEEAAAQBAJ
39. Sieniutycz, S., & Jezowski, J. (2013). *Energy optimization in process systems and fuel cells*. Newnes. https://play.google.com/store/books/details?id=wMX-e6Sk1ZAC
40. Nagel, J. (2018). *Optimization of energy supply systems: Modelling, programming and analysis*. Springer. https://play.google.com/store/books/details?id=89hwDwAAQBAJ
41. Decision Sciences Corporation. (1973). *Economic evaluation of total energy guidelines*. https://play.google.com/store/books/details?id=fMGrJIxM_9cC
42. Oelert, G., Auer, F., & Pertz, K. (1987). *Economic issues of renewable energy systems: A guide to project planning*. Deutsche Gesellschaft für Technische Zusammenarbeit (GTZ). https://books.google.com/books/about/Economic_Issues_of_Renewable_Energy_Syst.html?hl=&id=5fJbAAAACAAJ
43. Short, W., et al. (2005). *A manual for the economic evaluation of energy efficiency and renewable energy technologies*. University Press of the Pacific. https://books.google.com/books/about/A_Manual_for_the_Economic_Evaluation_of.html?hl=&id=IIMiHAAACAAJ
44. Handschin, E., & Petroianu, A. (2012). *Energy management systems: Operation and control of electric energy transmission systems*. Springer. https://play.google.com/store/books/details?id=lcbuCAAAQBAJ
45. Intelligent Energy Systems. (1989). Load management experiment: Feasibility study. https://books.google.com/books/about/Load_Management_Experiment.html?hl=&id=MQfHAAAACAAJ
46. Hills, P. J., & Ramani, K. V. (1990). *Energy systems and the environment: Approaches to impact assessment in Asian developing countries*. Asian and Pacific Development Centre. https://books.google.com/books/about/Energy_Systems_and_the_Environment.html?hl=&id=Gse5AAAAIAAJ
47. Ren, J. (2021). *Energy systems evaluation (Vol. 1): Sustainability assessment*. Springer. https://play.google.com/store/books/details?id=fzsxEAAAQBAJ
48. Demirbaga, Ü. (n.d.). *Big data analytics: Theory, techniques, platforms, and applications*. Springer. https://books.google.com/books/about/Big_Data_Analytics.html?hl=&id=mmcGEQAAQBAJ
49. Ladwig, G. (2014). *Efficient optimization and processing of queries over text-rich graph-structured data*. KIT Scientific Publishing. https://books.google.com/books/about/Efficient_Optimization_and_Processing_of.html?hl=&id=flOumc5lmk8C
50. Vincent, B. T., Baddeley, R. J., Troscianko, T., & Gilchrist, I. D. (2005). Is the early visual system optimised to be energy efficient? *Network, 16*(2–3), 175–190. https://doi.org/10.1080/09548980500290047
51. Kallrath, J., Pardalos, P. M., Rebennack, S., & Scheidt, M. (2008). *Optimization in the energy industry*. Springer. https://play.google.com/store/books/details?id=eC0voHnTDaQC
52. An, D., Xi, H., Zhang, H., Yang, J., & Chai, L. (2023). *Future electricity system based on energy internet: Energy storage system design, optimal scheduling, security, attack model and countermeasures*. Frontiers Media SA. https://books.google.com/books/about/Future_Electricity_System_Based_on_Energ.html?hl=&id=Gl_iEAAAQBAJ
53. Dolores Esteban, M., López-Gutiérrez, J.-S., & Negro, V. (2021). *Renewable energies for sustainable development*. MDPI. https://books.google.com/books/about/Renewable_Energies_for_Sustainable_Devel.html?hl=&id=gE0-EAAAQBAJ
54. Wang, R., Qin, D., Hu, W., Chang, F., & Wang, J. (2023). *Advanced cooperative control and optimization strategies for integrated energy systems*. Frontiers Media

SA. https://books.google.com/books/about/Advanced_cooperative_control_and_optimiz.html?hl=&id=nBKwEAAAQBAJ
55. Das, S. K., Islam, M. R., & Xu, W. (2022). *Advances in control techniques for smart grid applications*. Springer. https://play.google.com/store/books/details?id=P99mEAAAQBAJ
56. Sulaiman, A., Nagu, B., Kaur, G., Karuppaiah, P., Alshahrani, H., Reshan, M. S. A., AlYami, S., & Shaikh, A. (2023). Artificial intelligence-based secured power grid protocol for smart city. *Sensors, 23*(19), 8016. https://doi.org/10.3390/s23198016
57. Dixon, A. (2019). *Modern aspects of power system frequency stability and control*. Academic Press. https://play.google.com/store/books/details?id=9niWDwAAQBAJ
58. Golpîra, H., Román-Messina, A., & Bevrani, H. (2021). *Renewable integrated power system stability and control*. Wiley. https://play.google.com/store/books/details?id=_FMmEAAAQBAJ
59. Nuthalapati, S. (2021). *Use of voltage stability assessment and transient stability assessment tools in grid operations*. Springer. https://play.google.com/store/books/details?id=o5YwEAAAQBAJ
60. Selim, A., El-Shimy, M., Amer, G., Ihoume, I., Masrur, H., & Guerrero, J. M. (2024). Hybrid off-grid energy systems optimal sizing with integrated hydrogen storage based on deterministic balance approach. *Scientific Reports, 14*(1), 6888. https://doi.org/10.1038/s41598-024-55631-3
61. Zobaa, A. F., Ribeiro, P. F., Abdel Aleem, S. H., & Afifi, S. N. (2018). *Energy storage at different voltage levels: Technology, integration, and market aspects*. Energy Engineering. https://books.google.com/books/about/Energy_Storage_at_Different_Voltage_Leve.html?hl=&id=O-aEDwAAQBAJ
62. Hughes, J. J., Lagunas-Robles, G., & Campbell, P. (2024). The role of conflict in the formation and maintenance of variant sex chromosome systems in mammals. *The Journal of Heredity, 115*(6), 601–624. https://doi.org/10.1093/jhered/esae031
63. Lughofer, E., & Sayed-Mouchaweh, M. (2019). *Predictive maintenance in dynamic systems: Advanced methods, decision support tools and real-world applications*. Springer. https://play.google.com/store/books/details?id=lfCKDwAAQBAJ
64. Prabhansu, & Kumar, N. (2022). *Emerging trends in energy storage systems and industrial applications*. Elsevier. https://play.google.com/store/books/details?id=2RxoEAAAQBAJ
65. Khalid, M. (2023). *Sustainable energy systems: Emerging technologies and practices in renewable energy storage*. Mdpi AG. https://books.google.com/books/about/Sustainable_Energy_Systems.html?hl=&id=uqxA0AEACAAJ
66. Dhundhara, S., Verma, Y. P., & Kumar, A. (2023). *Energy storage technologies in grid modernization*. Wiley. https://books.google.com/books/about/Energy_Storage_Technologies_in_Grid_Mode.html?hl=&id=9EHJEAAAQBAJ
67. Kishan, D., Kannan, R., Dastagiri Reddy, B., & Prabhakaran, P. (2023). *Power electronics for electric vehicles and energy storage: Emerging technologies and developments*. CRC Press. https://play.google.com/store/books/details?id=5nm2EAAAQBAJ
68. Anvari-Moghaddam, A., Mohammadi-ivatloo, B., & Mohammadi, F. (2021). *Emerging technologies for the energy systems of the future*. MDPI. https://books.google.com/books/about/Emerging_Technologies_for_the_Energy_Sys.html?hl=&id=4e8_EAAAQBAJ
69. Sircar, A., Tripathi, G., Bist, N., Shakil, K. A., & Sathiyanarayanan, M. (2022). *Emerging technologies for sustainable and smart energy*. CRC Press. https://play.google.com/store/books/details?id=_mJxEAAAQBAJ
70. Graditi, G., & Di Somma, M. (2022). *Technologies for integrated energy systems and networks*. Wiley. https://books.google.com/books/about/Technologies_for_Integrated_Energy_Syste.html?hl=&id=f816EAAAQBAJ
71. Salkuti, S. R. (2024). *Energy and environmental aspects of emerging technologies for smart grid*. Springer. https://books.google.com/books/about/Energy_and_Environmental_Aspects_of_Emer.html?hl=&id=9ypgzwEACAAJ
72. Dhayal, K. S., Agrawal, S., Agrawal, R., Kumar, A., & Giri, A. K. (2024). Green energy innovation initiatives for environmental sustainability: Current state and future research directions.

Environmental Science and Pollution Research International, 31(22), 31752–31770. https://doi.org/10.1007/s11356-024-33286-x
73. Rezić, I., & Meštrović, E. (2024). Challenges of green transition in polymer production: Applications in zero energy innovations and hydrogen storage. *Polymers, 16*(10), 1310. https://doi.org/10.3390/polym16101310
74. Dinçer, H., & Yüksel, S. (2022). *Sustainability in energy business and finance: Approaches and developments in the energy market*. Springer. https://play.google.com/store/books/details?id=1ExgEAAAQBAJ
75. Yang, P. (2019). *Cases on green energy and sustainable development*. IGI Global. https://play.google.com/store/books/details?id=LwyjDwAAQBAJ
76. Ahmad, M., Nawaz, T., Hussain, I., Meharban, F., Chen, X., Khan, S. A., Iqbal, S., Rosaiah, P., Ansari, M. Z., Zoubi, W. A., & Zhang, K. (2024). Evolution of metal tellurides for energy storage/conversion: From synthesis to applications. *Small, 20*(28), e2310099. https://doi.org/10.1002/smll.202310099
77. Kilkis, B., & Kakaç, S. (2012). *Energy storage systems*. Springer. https://play.google.com/store/books/details?id=NCryCAAAQBAJ
78. Márquez, J. M. A., Manzano, F. S., & Luengo, J. R. (2023). *Energy storage systems: Fundamentals, classification and a technical comparative*. Springer. https://play.google.com/store/books/details?id=ZK7QEAAAQBAJ
79. Sahoo, U. (2021). *Energy storage*. Wiley. https://books.google.com/books/about/Energy_Storage.html?hl=&id=PWI9EAAAQBAJ
80. Ter-Gazarian, A. G. (1994). *Energy storage for power systems*. IET. https://books.google.com/books/about/Energy_Storage_for_Power_Systems.html?hl=&id=5VMotgCfJmAC

Machine Learning Algorithms for Identification of Equipment Overloading

B. Meena Preethi, S. Gokul, M. Parameshvar, and B. Dharshini

1 Novelty

This is the first open paper highlighting the detection of overloaded vehicles by the comparison of Gross Vehicle Weight (GVW) and current weight of the vehicle. The You Only Look Once (YOLO) object detection architecture is used to vehicle detection, and its details were taken. Bending plates are used to detect the weight of each axle, and GVW is calculated as sum of all axles. The calibration of bending plates is done by the latest genetic algorithm which will mind on rectifying minute errors in weight detection because of some external factors.

2 Introduction

Overloaded vehicles exceed prescribed weight limits, compromising safety and road infrastructure. Excessive loads strain braking systems, impair stability, and increase accident risks. Road damage ensues, impacting maintenance costs. Regulatory compliance is crucial to ensure safe transport, preserve infrastructure, and safeguard public well-being on the roads. Overloaded vehicle is a threat for the following reasons

- **Road Safety**: Overloaded trucks present a substantial threat to road safety as the excess weight compromises crucial aspects of vehicle performance. The additional mass results in prolonged stopping distances, challenging maneuverability, and an elevated risk of accidents. Braking systems, strained by the heavier

B. M. Preethi (✉) · S. Gokul · M. Parameshvar · B. Dharshini
Sri Krishna Arts and Science College, Coimbatore, Tamil Nadu, India
e-mail: meenapreethib@skasc.ac.in

S. Vijayalakshmi et al. (eds.), *Smart Power Systems*, Power Systems,
https://doi.org/10.1007/978-3-031-89459-6_12

load, may exhibit reduced effectiveness, particularly in emergency scenarios. Handling becomes more challenging, raising the likelihood of collisions and road incidents. Beyond the immediate safety concerns, overloaded trucks also contribute to increased wear on components like tires and suspension, escalating the potential for mechanical failures that further jeopardize road safety for both truck drivers and other road users.

- **Infrastructure Preservation**: Roads and bridges are meticulously designed to accommodate specific weight limits, ensuring structural integrity and longevity. Overloaded trucks, surpassing these limits, exert excessive pressure on road surfaces, accelerating wear and tear. The consequential damage manifests as potholes, cracks, and structural compromises. Such deterioration not only necessitates more frequent and costly maintenance but also diminishes the overall lifespan of the infrastructure. Higher repair expenses strain public resources and disrupt transportation networks, underscoring the imperative of enforcing weight regulations to preserve the integrity of critical infrastructure, reduce maintenance costs, and sustain the long-term functionality of roads and bridges for the benefit of society.
- **Traffic Congestion**: Overloaded trucks, constrained by reduced maneuverability and slower speeds, become contributors to traffic congestion, impeding the efficient flow of vehicles. The diminished ability to navigate smoothly through traffic creates bottlenecks, leading to delays and disruptions in the transportation network. Traffic congestion, exacerbated by overloaded trucks, has broader economic implications, hindering overall productivity and contributing to increased transportation-related delays. Efficient traffic flow is integral to the timely movement of goods and people, and the presence of overloaded trucks can compromise this essential aspect of transportation management. Striking a balance by adhering to weight regulations is crucial to maintaining an effective and productive transportation system.
- **Environmental Impact**: Overloaded trucks, characterized by lower fuel efficiency, substantially escalate carbon emissions and environmental impact. The excessive weight demands more energy to propel the vehicle, resulting in increased fuel consumption and greater greenhouse gas emissions. Adhering to prescribed weight limits is pivotal, as it allows trucks to operate optimally, enhancing fuel efficiency and subsequently mitigating their carbon footprint. This commitment to weight regulations is integral to fostering environmental sustainability, as it not only minimizes the ecological impact of transportation activities but also aligns with broader efforts to promote greener practices and reduce the carbon intensity associated with the movement of goods on our roadways.
- **Public Safety**: The heightened risk of accidents associated with overloaded trucks extends beyond infrastructure damage, encompassing a significant threat to public safety. The increased weight adversely impacts vehicle dynamics, lengthening stopping distances and compromising maneuverability, raising the likelihood of collisions. Prevention of overloading is paramount in safeguarding road users and minimizing the potential for accidents. Authorities play a pivotal

role in enforcing weight regulations, ensuring that trucks adhere to prescribed limits. This commitment contributes to creating safer roadways, protecting not only the integrity of the transportation infrastructure but also the well-being of all individuals who share the road—pedestrians, cyclists, and other motorists alike.

The U.S. Bureau of Public Roads initiated the development of Weigh-in-Motion (WIM) systems in the early 1950s to monitor truck weights. Weigh-in-Motion (WIM) is a technology used to measure the weight of vehicles while they are in motion, without the need for them to stop at a weigh station. The traditional method of weighing vehicles involves stationary scales at weigh stations, where trucks and other vehicles are required to stop for inspection and weighing. Weigh-in-motion systems, on the other hand, allow for continuous weight monitoring as vehicles travel at normal speeds along the road. Modern weigh-in-motion (WIM) systems generally comprise three components aimed at: (1) recording vehicle weight, (2) capturing additional vehicle data like speed, axle count, type, and spacing, and (3) gathering, processing, transmitting, and analyzing vehicle data. Weigh-in-motion systems find applications in various sectors, including transportation management, toll collection, and enforcement of weight regulations. They contribute to road safety, infrastructure durability, and the overall efficiency of transportation networks. As technology continues to advance, weigh-in-motion systems are likely to play an increasingly important role in the optimization of transportation systems worldwide.

Weigh-in-motion (WIM) systems operate by strategically placing sensors, such as strain gauges or load cells, on the road surface to detect and measure the dynamic forces exerted by moving vehicles. As a vehicle passes over these sensors, the forces are converted into electrical signals, which are then processed by a central unit using advanced algorithms to calculate the vehicle's weight. The system considers factors like axle spacing and vehicle speed to ensure accuracy. Real-time data transmission capabilities enable authorities to monitor vehicle weights remotely and enforce weight regulations efficiently, contributing to improved traffic flow, road safety, and infrastructure management. Regular calibration and integration with broader transportation systems are crucial aspects of WIM technology to maintain reliability and effectiveness.

2.1 Factors Affecting WIM Data

Several factors can impact the quality of data supplied by Weigh-in-Motion (WIM) systems.

- **Roadway Factors**: The impact of vehicular and roadway factors is multifaceted, encompassing variables such as vehicle speed, gross vehicle weight, axle configuration, and load distribution. The interplay of these elements, coupled with external factors like road roughness, truck speed, suspension type, and tire pressure, results in dynamic variations in the load applied to a pavement surface by a truck. This instantaneous variability can have significant implications for pavement integrity, influencing wear and tear. A comprehensive understanding of these factors is crucial for effective road design, maintenance, and management, ensuring the durability and safety of the transportation infrastructure under the diverse and dynamic conditions imposed by vehicular traffic.

- **Climatic Conditions**: Climatic conditions, including temperature fluctuations, significantly influence the performance of Weigh-in-Motion (WIM) scales. Manufacturers acknowledge the impact of temperature on their equipment and incorporate measures to address these variations. Pavement temperature alterations can affect stiffness, disrupting the conventional "opposite and equal reaction" principle. Scale accuracy is particularly sensitive to temperature changes. Variables such as temperature, wind, snow, ice, and moisture introduce complexities in data interpretation. Recognizing the dynamic nature of climatic factors is imperative in refining WIM technology, ensuring reliable and accurate weight measurements for vehicles in diverse weather conditions and Calibration.
- **Equipment factors**: Equipment factors pose inherent challenges in Weigh-in-Motion (WIM) technology. The intricate interplay between sophisticated electronic components and the constantly evolving environment introduces an element of unpredictability. This unpredictability can lead to systematic errors that, to a certain extent, are unavoidable. The sensitivity of electronic equipment to environmental changes requires ongoing technological advancements to minimize potential inaccuracies. Factors such as temperature fluctuations, electrical interference, and wear on sensors contribute to the complexity of achieving absolute precision. Acknowledging and addressing these challenges are integral to refining WIM systems, ensuring their reliability and accuracy in diverse and dynamic operating conditions.

3 Traditional Methods to Detect Overloaded Vehicles

In the realm of transportation management, traditional methods for identifying overloaded vehicles primarily involve weigh stations. These stations strategically positioned along highways require trucks to divert and drive onto scales for precise weight verification. Law enforcement agencies play a pivotal role in ensuring compliance with weight limits and promoting road safety. They often employ portable scales, allowing them to conduct weight assessments at various locations. Visual inspections of vehicles also contribute to detection efforts, as officers look for signs of excessive load through observation of the vehicle's physical condition and alignment. Furthermore, documentation checks, such as scrutinizing permits and cargo manifests, aid in confirming compliance with weight regulations. While these traditional methods have been effective, advancements in technology, such as

Weigh-in-Motion systems and smart sensors, are increasingly complementing or replacing these practices to enhance efficiency and accuracy in identifying overloaded vehicles. WIM sensors offer various advantages and are chosen based on factors such as accuracy requirements, installation ease, and environmental considerations. Weigh-in-Motion (WIM) sensors are devices designed to measure the weight of vehicles while they are in motion. There are several types of WIM sensors, each utilizing different technologies for weight measurement. Some common types include:

3.1 Sensor 1: Strain Gauge

Strain gauge sensors play a critical role in the realm of Weigh-in-Motion (WIM) systems, where their application involves the measurement of deformation or strain induced by the weight of passing vehicles. At the core of their functionality lies the piezoresistive effect, a principle wherein the electrical resistance of specific materials undergoes alterations in response to applied mechanical stress.

In the operational context of WIM, strain gauge sensors are strategically positioned either within the road surface itself or on supportive structures such as bridges or weighbridge platforms. As a vehicle traverses over the sensor location, the inherent weight of the vehicle leads to deformation in the road surface or the supporting structure. This deformation, in turn, triggers a change in the resistance of the embedded strain gauges.

The composition of a strain gauge sensor typically involves one or more thin wires or strips of material with high resistance, securely affixed to the surface under scrutiny. When the material undergoes deformation due to the applied load, the resistance experiences a proportional change. Subsequently, this alteration in resistance is meticulously measured and then converted into an electrical signal.

One of the primary advantages of employing strain gauge sensors in WIM systems is their inherent sensitivity to even minute changes in load. This characteristic makes them particularly well-suited for accurately measuring the weight distribution of individual axles or the aggregate weight of the entire vehicle. Additionally, strain gauge sensors offer a cost-effective solution, and their adaptability allows for installation in various locations, including existing road surfaces, facilitating traffic monitoring without substantial disruption.

However, it is essential to acknowledge certain considerations in the utilization of strain gauge sensors. Calibration becomes a crucial aspect of their deployment, ensuring precise measurements align with the expected values. Moreover, environmental factors, notably temperature fluctuations, can influence the accuracy of strain gauge sensors. To counteract this, thorough temperature compensation mechanisms are implemented to maintain the reliability and precision of the measurements.

3.1.1 Advantages of Strain Gauge

Strain gauge sensors offer several advantages when utilized in Weigh-in-Motion (WIM) systems, contributing to their effectiveness in accurately measuring the weights of vehicles on roadways.

Sensitivity to Small Changes: Due to their exceptional sensitivity, strain gauge sensors can detect even the slightest alterations in deformation or strain. This heightened sensitivity translates into the ability to make precise measurements, facilitating accurate assessments of load distribution across individual axles and determining the overall weight of a vehicle with a high degree of accuracy. The capability to capture minute changes in strain positions strain gauge sensors as valuable tools in Weigh-in-Motion (WIM) systems, ensuring reliable and detailed data for effective traffic monitoring, regulatory compliance, and infrastructure management.

Accuracy: Strain gauge sensors exhibit a commendable responsiveness to mechanical stress, yielding precise weight measurements. This accuracy plays a pivotal role in compliance monitoring, facilitating the enforcement of weight regulations essential for road safety. Ensuring the integrity of road infrastructure is paramount, and the dependable accuracy of strain gauge sensors aids in preventing structural damage caused by overloaded vehicles. By upholding stringent weight limits, these sensors contribute significantly to maintaining road safety standards, preserving the longevity of roads and bridges, and enhancing overall transportation infrastructure management.

Adaptability and Versatility: Possessing notable versatility, strain gauge sensors can be seamlessly integrated into diverse locations such as existing road surfaces or various support structures. Their adaptable nature enables flexible deployment, rendering them suitable for a broad spectrum of Weigh-in-Motion (WIM) scenarios. This flexibility in installation locations ensures that strain gauge sensors can be efficiently utilized in a range of settings, contributing to their widespread applicability in monitoring traffic, enforcing weight regulations, and optimizing transportation infrastructure management across varying geographical and operational conditions.

3.1.2 Disadvantages of Strain Gauge

While strain gauge sensors have notable advantages in Weigh-in-Motion (WIM) systems, they also come with certain disadvantages that should be considered in the context of specific applications. Some notable disadvantages include:

Susceptibility to Environmental Factors: Strain gauge sensors, while effective, are susceptible to environmental influences like temperature fluctuations and humidity. These external factors have the potential to affect the accuracy of measurements, necessitating meticulous calibration and compensation mechanisms. To mitigate the impact of environmental variations on readings, calibration processes are carefully conducted to ensure the sensors maintain accuracy under diverse conditions. The sensitivity of strain gauge sensors to environmental changes underscores the importance of ongoing monitoring and calibration practices, emphasizing the need for a comprehensive approach to maintain the reliability and precision of Weigh-in-Motion (WIM) systems in real-world operating environments.

Calibration Challenges: Attaining and sustaining precise calibration for strain gauge sensors constitutes a nuanced and intricate process. The meticulous calibration required is essential to guarantee accurate weight measurements in Weigh-in-Motion (WIM) systems. This process is not only time-consuming but also demands ongoing attention, as periodic adjustments may be necessary to account for factors such as sensor wear or environmental changes. The complexity of calibration underscores the importance of thorough and regular maintenance practices, ensuring the continued reliability and accuracy of strain gauge sensors in the dynamic and demanding operational contexts of traffic monitoring and regulatory compliance.

Maintenance Requirements: Although strain gauge sensors are designed for durability, periodic maintenance remains essential. Even with their robust construction, these sensors may necessitate inspections, repairs, or replacements over time. This ongoing maintenance is crucial for ensuring the proper functioning and sustained accuracy of Weigh-in-Motion (WIM) systems. While the durability of strain gauge sensors minimizes the frequency of maintenance needs, organizations must allocate resources for occasional upkeep to uphold the reliability of these sensors. This consideration highlights the importance of a proactive approach to maintenance, balancing the initial investment in durable technology with the long-term benefits of reliable and accurate weight measurements.

3.2 Sensor 2: Load Cell

The load cell sensor, a cornerstone in the intricate landscape of Weigh-in-Motion (WIM) systems, assumes a pivotal role as a meticulously designed instrument dedicated to unraveling the weight dynamics of vehicles traversing specific sensor locations. Its fundamental operational principle revolves around a transformative process—converting applied force or load into an electrical signal, thereby facilitating meticulous and accurate weight measurements.

As the bedrock of WIM technology, load cells exhibit a rich array of technological implementations, with strain gauge load cells emerging as a preeminent and widely adopted variant. These load cells distinguish themselves through the integration of strain gauges onto a precisely engineered metal structure. The orchestrated interplay unfolds when an external force, epitomized by the weight of a passing vehicle, impacts the load cell. This force induces a nuanced deformation in the metal structure, triggering a corresponding alteration in the electrical resistance of the embedded strain gauges. The crux of this transformative process lies in the subsequent conversion of these alterations into an electrical signal, meticulously calibrated to faithfully mirror the applied force. This resulting signal becomes an authentic representation of the weight borne by the vehicle. Various load cell types include compression load cells, tension load cells, bending beam load cells, and shear beam load cells.

3.2.1 Advantages of Load Cell

High Accuracy: Load cell sensors, distinguished for their exceptional precision, stand as stalwarts in delivering elevated levels of accuracy when measuring weights. This characteristic proves indispensable in applications where precision holds paramount significance, notably in the realms of regulatory compliance assessments and infrastructure evaluations. The reliability of load cell sensors in providing precise weight measurements establishes them as pivotal instruments, ensuring data integrity in critical scenarios. Whether applied in traffic monitoring or infrastructure management, the high precision of load cell sensors remains a cornerstone for informed decision-making and the overall efficacy of weigh-in-motion systems.

Non-intrusive Installation: Load cell sensors exhibit unparalleled adaptability, fostering seamless integration into diverse Weigh-in-Motion (WIM) scenarios. Their versatility is evident in both embedded installations within the road surface and surface-mounted configurations. This non-intrusive nature ensures effortless deployment without causing disruptions to road surfaces or impeding the smooth flow of traffic. Whether strategically positioned beneath the road or affixed on the surface, load cell sensors exemplify a harmonious balance between precision and ease of installation. This flexibility positions them as instrumental components, effectively contributing to the efficiency and reliability of WIM systems across a spectrum of operational environments.

Low Maintenance: Meticulously engineered with a focus on longevity, load cells boast minimal maintenance requirements owing to their robust construction and inherent durability. This sturdy design is pivotal in enhancing reliability and diminishing the necessity for frequent attention and upkeep. Load cells enduring nature not only ensures sustained accuracy in weight measurements but also minimizes operational disruptions. This attribute is particularly advantageous in weigh-in-motion systems, where the dependable performance of load cells translates into prolonged periods of functionality, reinforcing their role as dependable components with lasting efficacy in various applications.

3.2.2 Disadvantages of Load Cell

While load cells are integral components in Weigh-in-Motion (WIM) systems, they are not without certain disadvantages. Here are some notable drawbacks associated with the use of load cells in WIM:

Potential for Overloading: Each load cell is designed with specific weight limits, and surpassing these thresholds poses challenges for accurate measurements. In instances of vehicles exceeding the load cell's designated capacity, there exists a notable risk of overloading. This scenario not only jeopardizes the load cell's structural integrity but also introduces the potential for damage, subsequently influencing the precision of weight measurements. Vigilance in adhering to load cell specifications becomes imperative to avoid such instances, ensuring the sustained accuracy and reliability of weigh-in-motion systems, especially in environments with varying vehicle sizes and weights.

Cost: The acquisition and installation of load cells, particularly high-precision variants, entail an initial investment that can be relatively substantial. Beyond the initial costs, ongoing maintenance and calibration activities further contribute to the overall cost of ownership. While high-precision load cells offer unparalleled accuracy, the associated expenses necessitate careful consideration in budgetary planning. Organizations must weigh the long-term benefits of precise weight measurements against the initial and recurring costs, ensuring a judicious investment that aligns with operational requirements and financial considerations in weigh-in-motion systems and related applications.

Susceptibility to Wear and Tear: Load cells, whether embedded in road surfaces or exposed to environmental elements, are susceptible to wear and tear as a consequence of continuous use. Prolonged exposure to heavy vehicle loads and diverse environmental conditions can result in gradual degradation, impacting the overall lifespan of load cells. This wear and tear highlight the importance of periodic inspections and maintenance to ensure sustained functionality. Addressing these factors becomes pivotal in enhancing the longevity of load cells, maintaining their accuracy, and optimizing their performance in weigh-in-motion systems, where reliability over an extended operational period is paramount.

3.3 Sensor 3: Piezoelectric

In the dynamic landscape of WIM technology, while strain gauge sensors continue to be a stalwart choice, advancements have led to the emergence of alternative sensor types. These include piezoelectric sensors, which harness the piezoelectric effect for weight measurement, and load cell sensors, directly measuring force or weight. Piezoelectric sensors, serving as a pivotal component in Weigh-in-Motion (WIM) systems, provide an innovative alternative to traditional load cell sensors. At the core of their functionality lies the piezoelectric effect, a phenomenon wherein certain materials produce an electric charge when subjected to mechanical stress. In the WIM context, piezoelectric sensors find strategic placement embedded within the road surface, ushering in a paradigm shift in weight measurement technology.

The operational intricacies unfold when a vehicle traverses over a piezoelectric sensor. The sensor, in response to the vehicle's weight, undergoes deformation, instigating the generation of an electric charge within the piezoelectric material. This charge's magnitude correlates directly with the applied force or the weight of the passing vehicle. The ensuing electric signal is meticulously captured and subsequently converted into a weight measurement, offering real-time insights into the vehicular load.

3.3.1 Advantages of Piezoelectric Sensor

Piezoelectric sensors offer several advantages in WIM applications:

Cost-Effective: Piezoelectric sensors present a compelling proposition in specific Weigh-in-Motion (WIM) applications, as they can be more cost-effective to both install and maintain. This cost efficiency enhances their attractiveness as a viable option, particularly in scenarios where budget considerations are pivotal. The streamlined installation process, coupled with lower maintenance requirements, positions piezoelectric sensors as a pragmatic and economical choice, aligning with the fiscal constraints and operational needs of certain WIM applications.

Real-time Monitoring: The inherent capability of piezoelectric sensors to deliver real-time weight measurements introduces a dynamic dimension to vehicle monitoring as they traverse over sensor locations. This real-time functionality ensures swift and accurate insights into the weight variations of vehicles in motion. Such immediacy in weight data is invaluable in applications where timely information on load distribution and traffic patterns is crucial. Piezoelectric sensors, by offering this real-time monitoring feature, enhance the responsiveness and efficiency of weigh-in-motion systems, contributing to informed decision-making in traffic management, regulatory compliance, and infrastructure planning.

Durability: Leveraging the robust nature of piezoelectric materials, these sensors exhibit remarkable durability, showcasing resilience in the face of heavy loads and diverse environmental conditions. Engineered to withstand the rigors of continuous use and external factors, piezoelectric sensors maintain their structural integrity, ensuring prolonged functionality in challenging scenarios. This durability is paramount in weigh-in-motion applications, where the sensors are subjected to varying weights, weather conditions, and vehicular traffic.

The inherent sturdiness of piezoelectric sensors contributes to their reliability, affirming their role as dependable instruments in providing enduring and accurate weight measurements over extended periods.

3.3.2 Disadvantages of Piezoelectric Sensor

Despite these advantages, it's important to note some limitations:

Sensitivity to Temperature Changes: While piezoelectric sensors offer numerous advantages, their sensitivity to temperature variations can impact weight measurement accuracy. Fluctuations in temperature may introduce deviations in the output signals of these sensors. To mitigate this sensitivity, meticulous calibration procedures and temperature compensation mechanisms become imperative. Calibration ensures that the sensors consistently deliver accurate measurements, accounting for the influence of temperature changes. Implementing temperature compensation further refines the precision of piezoelectric sensors, aligning them with the stringent demands of weigh-in-motion applications. This nuanced approach becomes crucial to maintaining reliable performance, especially in environments characterized by varying temperature conditions.

Dynamic Range: Piezoelectric sensors, while offering versatility, may exhibit a limited dynamic range when contrasted with specific load cell technologies. This limitation manifests in their capacity to accurately measure loads at the extremes of the weight spectrum—both very light and very heavy loads. The dynamic range constraints underscore the importance of a nuanced selection process, ensuring that piezoelectric sensors are deployed in scenarios aligning with their optimal operational range. Despite this limitation, the distinctive advantages of piezoelectric sensors, including real-time monitoring and non-intrusive installation, position them as valuable components in weigh-in-motion systems tailored to specific weight measurement requirements.

Calibration Requirements: As with all sensor technologies, precise calibration is essential for piezoelectric sensors to ensure accuracy and reliability in weight measurements. This meticulous calibration process becomes paramount, accounting for potential variations in sensor performance over time. Calibration aligns the sensor's output with a known standard, correcting for any drift or changes in sensitivity. This attention to calibration safeguards the accuracy of weight measurements, providing confidence in the sensor's performance throughout its operational lifespan. The careful calibration regimen contributes to the sustained precision of piezoelectric sensors, reinforcing their effectiveness in delivering consistent and dependable weight data in weigh-in-motion applications.

4 Review of Traditional Methods

Weigh-in-motion (WIM) systems are integral to modern transportation infrastructure, providing real-time weight measurements of moving vehicles. Central to the functionality of these systems are strain gauge and load cell sensors, each playing a crucial role in ensuring accurate and reliable weight data. In this exploration, we delve into the operational principles, applications, and advancements of strain gauge and load cell sensors in WIM. Additionally, we scrutinize the emergence of piezoelectric sensors as a cost-effective alternative, considering their benefits and inherent limitations.

4.1 Strain Gauge Sensors

- **Operational Mechanism**
- Strain gauge sensors operate on the piezoresistive effect, a phenomenon where the electrical resistance of certain materials changes in response to applied mechanical stress. In the context of WIM, these sensors, typically composed of thin, high-resistance wires or strips, are strategically installed on the road surface or support structures like bridges. When a vehicle traverses over the sensor

location, its weight induces deformation in the road surface or structure. This deformation, in turn, alters the electrical resistance of the embedded strain gauges.

- **Signal Conversion and Accuracy**
- The change in resistance is meticulously measured and converted into an electrical signal. This signal is proportional to the weight of the passing vehicle, allowing for precise weight measurements. Strain gauge sensors exhibit exceptional sensitivity, enabling the differentiation of individual axle loads and providing an accurate depiction of the overall vehicle weight. This data proves invaluable in various applications, including traffic monitoring, weight enforcement, and ensuring the structural integrity of roads.
- **Applications in Weigh-in-Motion**

- Strain gauge sensors serve as the backbone of weigh-in-motion systems, offering real-time insights into the dynamic weights of moving vehicles. Their versatility allows for deployment in diverse scenarios, ranging from highways to bridges, contributing to the efficiency and safety of transportation systems.

4.2 Load Cell Sensors

- **Role in Weigh-in-Motion**

- Load cell sensors stand as indispensable components in WIM systems, known for their ability to deliver accuracy, durability, and adaptability. These sensors are strategically placed on the road surface or within support structures, operating on the principle of converting a force or load into an electrical signal.
- **Precision in Weight Measurement**
- The load cell consists of strain gauges bonded to a metal structure. As a vehicle passes over the sensor location, the weight of the vehicle causes deformation in the metal structure, leading to a change in the electrical resistance of the strain gauges. This change is converted into an electrical signal, reflecting the weight of the vehicle. Load cell sensors excel in providing precise weight measurements, contributing significantly to traffic monitoring, regulatory compliance enforcement, and infrastructure management.
- **Technological Advancements**

- With ongoing technological advancements, load cells continue to evolve, playing a pivotal role in enhancing the efficiency and reliability of weigh-in-motion applications. Their adaptability makes them suitable for various environments and scenarios, ensuring consistent and accurate weight measurements.

4.3 *Piezoelectric Sensors*

- **Real-Time Weight Measurements**
- Piezoelectric sensors present a cost-effective alternative in WIM applications, offering real-time weight measurements with non-intrusive installation. Embedded within the road surface, these sensors generate electric charges in response to mechanical stress when a vehicle passes over them.
- **Considerations and Limitations**
- While piezoelectric sensors bring cost-effectiveness and real-time monitoring to the forefront, careful consideration is essential due to their sensitivity to temperature changes and dynamic range limitations. Calibration and informed application approaches become crucial to mitigate these considerations, ensuring accurate and reliable performance in diverse traffic and environmental conditions.
- **Future Prospects**
- As technology advances, piezoelectric sensors carve a niche in the WIM landscape. Their contributions extend beyond weight monitoring, influencing the trajectory of intelligent transportation systems toward enhanced sustainability and operational excellence.

Strain gauge, load cell, and piezoelectric sensors play pivotal roles in the functionality of weigh-in-motion systems. Strain gauge sensors, with their precision and sensitivity, provide accurate weight measurements. Load cell sensors contribute accuracy, durability, and adaptability to diverse scenarios. Piezoelectric sensors, as cost-effective alternatives, offer real-time monitoring, though careful consideration of their limitations is essential. Together, these sensors form the technological backbone of WIM, contributing to safer, more efficient transportation systems and paving the way for future innovations in intelligent transportation infrastructure.

5 Modern Methods to Detect Overloaded Vehicles

5.1 *Stage 1: Integrating Bending Plates and Deep Learning Techniques*

Bending plate Weigh-in-Motion (WIM) sensors are a form of technology that is used to measure a vehicle's weight and other properties while it is moving, usually as it passes over an instrument that is embedded in the road surface and is specifically intended for that purpose.

5.1.1 Bending Plates

WIM Sensor Components

Bending Plates: The bending plate is the fundamental part of a bending plate WIM sensor. These plates are meant to be strong and long-lasting, and they are usually made of steel or composite materials. They are positioned flush with the road surface and frequently have a rectangular or square shape.

Sensors: Strain gauges and various kinds of sensors are integrated into the bending plates. When a car drives over the plates, these sensors are in charge of determining how much the plates bend or deform. The sensors measure the stress or strain that the vehicle's weight causes the bending plate to undergo when it flexes.

Data Processing Unit: A control system or data processing unit receives the information gathered by the sensors. This unit is in charge of real-time data processing and parameter determination, including wheelbase, axle loads, vehicle weight, and vehicle categorization.

Working Principles of Bending Plate WIM Sensors

The working principles of bending plate WIM sensors involve the interaction between the vehicle's weight and the bending plates:

Vehicle Passage: The weight of the vehicle is dispersed over the bending plates as it passes over them.

Plate Bending: When a vehicle is loaded, the bending plates, which are pliable, distort, or bend somewhat. The weight of the car's tires pressing against the plates is what's causing this bending.

Sensor Measurements: The amount of this deformation or strain is determined by the sensors or strain gauges that are implanted in the bending plates. In reaction to the bending, the strain gauges produce electrical signals that are proportionate to the strain encountered.

Data Collection: The data processing unit receives the electrical signals that are gathered from the strain gauges.

Real-Time Analysis: The signals are examined in real time by the data processing unit to determine a number of parameters, including the wheelbase, total weight of the vehicle, and the weights on each axle. If integrated, vehicle classification is also carried out by examining the sequence and timing of the axles' passage over the sensors.

Data Output: A variety of applications, such as toll collection, traffic management, and weight enforcement, can make use of the computed data, which includes vehicle weight and classification.

Installation and Flexibility

Bending plate scales can be permanently mounted within the road surface or designed to be movable, depending on the specific requirements of the toll gate or application. The flexibility in installation makes them adaptable to various road types and locations.

Precise Weight Measurement

One of the key advantages of bending plate WIM systems is their ability to accurately measure the weight of a vehicle. They achieve this by adding up the individual weights of each axle to determine the gross weight of the entire vehicle. This approach ensures that even variations in load distribution are taken into account, resulting in highly precise weight measurements.

Axle-Spacing Detection Technology

To further enhance the capabilities of WIM systems, many incorporate axle-spacing detection technology. When a vehicle passes over the sensor, the system not only measures the weight but also automatically calculates the number of axles the vehicle possesses. This feature is particularly useful for toll gate operations, as it aids in determining the appropriate fee based on the number of axles and the overall weight of the vehicle.

Real-Time Data Collection

A significant advantage of WIM systems, including bending plates, is their ability to collect data in real time as vehicles pass through the toll gate. This is a remarkable feat, as the entire measurement and calculation process occur seamlessly while the vehicle is in motion. It eliminates the need for vehicles to come to a halt, contributing to smoother traffic flow and reducing delays at the toll gate.

Applications of Bending Plate WIM Systems

Bending plate WIM systems find widespread usage in a variety of scenarios, including but not limited to:

1. Pre-sorting Systems: Pre-sorting systems employ WIM technology on weigh station ramps to collect data about vehicles passing through. This data is vital for sorting vehicles based on weight classes and ensuring that heavy vehicles are directed to the appropriate lanes for further inspection or fee assessment.

2. Industrial and Military Weighing: In industrial and military contexts, where precise weight measurements are crucial, bending plate WIM systems are frequently employed. They help determine the weight of vehicles and their loads, ensuring compliance with safety and regulatory standards.
3. Dynamic Weight Measurement: Bending plates are capable of measuring dynamic weights, which is especially valuable for vehicles in motion. This real-time data collection allows for quick and accurate weight measurements without requiring vehicles to stop or slow down significantly.

5.1.2 YOLO

YOLO, an acronym for "You Only Look Once," is an advanced real-time object detection system that has garnered considerable acclaim within the realms of computer vision and deep learning. YOLO was first introduced in the paper "You Only Look Once: Unified, Real-Time Object Detection" by Joseph Redmon, Santosh Divvala, Ross Girshick, and Ali Farhadi in 2016. It has since seen several iterations and versions, with YOLOv4 and YOLOv5 being some of the most recent developments.

The YOLO Detection Process

The YOLO object detection process can be broken down into several key steps:

Input Image: The process begins with an input image that the model needs to analyze for objects. Image Preprocessing: The input image is preprocessed to meet the network's requirements, including resizing and normalizing pixel values.

Passing Through the Network: The preprocessed image passes through the YOLO neural network. The network consists of various convolutional and fully connected layers.

Output Processing: YOLO divides the image into a grid of cells. For each cell, the model predicts bounding boxes, objectness scores, and class probabilities. Each bounding box consists of the predicted coordinates (x, y, width, height), an objectness score (indicating the likelihood of an object being present within the box), and class probabilities (predicting the object's class). Non-Maximum Suppression (NMS): After obtaining multiple bounding box predictions, YOLO employs non-maximum suppression to filter out redundant and overlapping boxes, retaining only the most confident ones.

Object Classification: YOLO assigns class labels to the remaining bounding boxes based on the highest class probability.

Final Detection Output: The output of the YOLO model consists of the bounding boxes around detected objects, their associated class labels, and the corresponding confidence scores.

Working of YOLO

The proposed approach is a comprehensive three-step process designed to enhance vehicle identification and recognition. Each step plays a crucial role in achieving accurate results, particularly in complex scenarios. Let's delve deeper into each of these steps.

Step 1: Vehicle Identification Through License Plate and Vehicle Tag Detection

The first and foundational step in this approach involves identifying a vehicle using license plate and vehicle tag detection. This step is essential as it provides the initial link to the vehicle's identity. Vehicle identification is not limited to just license plates and vehicle tags; it also encompasses various textual information and even vehicle lights, which bear distinctive characteristics of the vehicle. License plates, while serving as the primary identifier, can be obscured by dust or other environmental factors. Therefore, it is important to consider alternative visual cues in the form of textural information and vehicle lights to enhance the reliability of the identification process.

In this step, the system employs image recognition technology to locate and extract information from license plates and vehicle tags. This information serves as the foundation for subsequent classification and recognition tasks.

Step 2: Machine Learning Algorithm for Vehicle Brand and Model Classification

Following the initial identification, the second step involves the application of machine learning algorithms to classify the vehicle's brand and model. This process comprises two distinct phases: feature extraction and vehicle classification.

Feature extraction involves isolating and quantifying relevant features from the acquired data. These features can include patterns, colors, shapes, and other visual elements that distinguish one vehicle brand and model from another. The extracted features serve as input data for the subsequent classification task.

Vehicle classification, the second phase of this step, employs machine learning algorithms to categorize the vehicle based on the extracted features. The algorithms are trained on a dataset that contains a diverse range of vehicle images, making it capable of recognizing different brands and models with a high degree of accuracy.

Step 3: Vehicle Pattern Recognition Using Deep Learning Algorithms

The final step in this approach focuses on recognizing vehicle patterns using deep learning algorithms. Deep learning, a subset of machine learning, is particularly well-suited for complex pattern recognition tasks. In this context, it merges the extracted features and conducts recognition after analyzing vehicle features from various viewpoints and angles.

Deep learning algorithms, such as YOLO, are employed to create a robust system that excels in identifying vehicle brands and models, even in challenging scenarios. These algorithms are trained on a vast and diverse dataset, enabling them to generalize patterns effectively and adapt to different conditions. By considering the extracted features from multiple perspectives, the system becomes highly effective at recognizing vehicles under various real-world conditions.

The three-step approach outlined above represents a comprehensive and sophisticated method for vehicle identification and recognition. It leverages advanced technologies, including image recognition, machine learning, and deep learning, to enhance accuracy and reliability in identifying vehicles, including their brands and models. This approach is especially valuable in scenarios where license plates may be obscured, and it offers a robust solution for a wide range of applications, including security, surveillance, and traffic management.

5.1.3 Integrating Bending Plates and Deep Learning Techniques

Toll gates play a pivotal role in managing traffic flow, collecting fees, and ensuring road safety. Modern toll gates are equipped with a plethora of technologies, including multiple cameras, to streamline operations and enhance security. These cameras are strategically positioned to capture both front and side images of vehicles passing through. The collected images serve various purposes, from vehicle identification to load capacity determination, and contribute to improving overall toll gate efficiency. In this comprehensive exploration, we will delve into the multifaceted roles of these cameras, the significance of front and side images, and how advanced technologies like Weigh-in-Motion (WIM) systems, such as bending plates, are integrated into toll gate operations.

Multiple cameras at toll gates are essential components of the toll collection process. They serve several crucial functions, primarily focused on vehicle identification, security, and monitoring. The front and side images captured by these cameras are instrumental in achieving these objectives.

In the first step of using these photos for recognition, the frontal picture (front view) plays a pivotal role. This frontal image is the key element in identifying the vehicle. Specifically, it is used to recognize the model of the truck. The reason for this is that the frontal photograph provides a clear and unobstructed view of the vehicle, which is crucial for accurate model recognition. It enables the system to analyze the distinctive features and characteristics of the truck model, such as its shape, size, and other visual attributes.

While the frontal image is the primary source of information for vehicle recognition, the side picture is also harnessed to enhance the precision of frontal view identification. The side image offers additional viewpoints and details, which can augment the recognition process. By providing an alternative perspective of the vehicle, it offers complementary information that can be crucial in certain situations.

Moreover, the side image serves the purpose of dividing the truck into multiple segments. This segmentation is valuable for improving the accuracy of model recognition. By breaking down the vehicle into distinct sections and analyzing each of them separately, the system gains a more comprehensive understanding of the truck's features. This level of granularity is particularly useful for complex or unique vehicle models that may have distinguishing characteristics not readily visible in a single frontal image.

The integration of side images into the recognition process is a sophisticated approach to improving accuracy. It ensures that the system can identify trucks with a higher degree of precision, even in challenging conditions.

The combination of both frontal and side photos enables a thorough analysis of the vehicle, offering a holistic view of its attributes. This thorough analysis is pivotal for recognizing the finer details that may not be apparent from a single perspective. For example, certain trucks may have specific body traits, such as unique logos, designs, or modifications. These distinctive traits can be better identified and categorized through the examination of both frontal and side images. The result is a more precise identification process that can classify trucks based on their specific characteristics.

In addition to vehicle identification, toll gates often play a role in assessing a truck's load capacity, which is critical for safety and fee calculation. To determine a truck's load capacity, information about weight measurement is acquired, typically using Weigh-in-Motion (WIM) systems.

WIM systems, including bending plate systems, play a pivotal role in measuring a vehicle's weight as it passes through the toll gate. Bending plate WIM systems involve specialized plates with sensors located on the underside. These plates are strategically placed within the road surface to ensure accurate data collection.

As a vehicle traverses over the bending plate, the sensors detect and log the strain experienced by the plate using strain gauges. This strain data is collected and processed by the system to calculate the dynamic load of the vehicle. Bending plate WIM systems are renowned for their accuracy and efficiency, making them a preferred choice in various applications (Fig. 1).

5.2 *Stage 2: Calibrating Weight Sensors*

Calibration of Weigh-in-Motion (WIM) systems involves measuring and adjusting the system parameters when a vehicle is stationary on the sensors. This calibration is essential for accurately estimating the static weight of vehicles, as opposed to dynamic weights measured when vehicles are in motion. Here are the steps for static calibration of a WIM system:

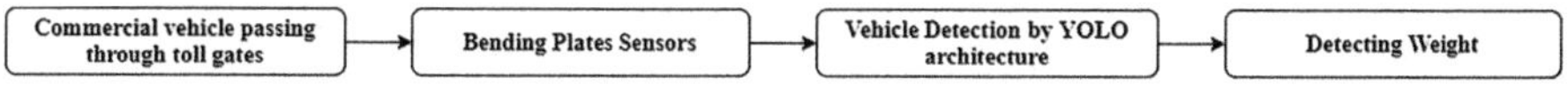

Fig. 1 Working of bending plates

Choose a suitable site for static calibration. Ensure that the site is level, stable, and free from vibrations. It should represent typical conditions where the WIM system will be deployed. Place known and certified calibration weights on the sensors or load cells of the WIM system. These weights should cover the entire range of expected loads and represent the loads that the WIM system will encounter during regular operation. Record the actual weights of the calibration loads accurately. This information will be used as a reference for comparing the WIM system's measurements. Capture the readings produced by the WIM system as it measures the static weights of the calibration loads. Ensure that the data collection process is thorough and that readings are stable. Compare the actual weights of the calibration loads with the WIM system's measurements. Identify any discrepancies or errors in the readings. This analysis helps determine how closely the WIM system aligns with the true weights.

If discrepancies are found, adjust the calibration factors within the WIM system. Calibration factors may include correction factors for individual sensors, load cell sensitivity, or other parameters. The goal is to align the WIM system's measurements with the true weights. Repeat the static calibration process multiple times with different calibration weights to ensure consistency and accuracy across the entire weight range. This iterative process helps fine-tune the calibration factors.

5.2.1 Automatic Calibration of Bending Plates

Weigh-in-Motion (WIM) systems are used to measure the weight of vehicles while they are in motion, typically on highways or bridges. The automatic calibration of WIM systems involves adjusting and optimizing the system parameters to ensure accurate and reliable weight measurements. Several algorithms and techniques may be employed for this purpose, depending on the specific WIM system and its components.

One machine learning algorithm commonly used for automatic calibration of vehicle-related systems is the Genetic Algorithm (GA). Genetic algorithms are a type of optimization algorithm inspired by the process of natural selection. They are particularly useful for fine-tuning parameters to achieve optimal performance in complex systems.

In the context of vehicle calibration, genetic algorithms can be employed to automatically adjust parameters in a systematic way, mimicking the process of evolution. The algorithm creates a set of parameter values, evaluates their performance against a defined objective or fitness function (e.g., accuracy in vehicle weight estimation), and then evolves the population over several iterations to converge towards an optimal solution.

Here is a general guide on how GAs can be applied to the automatic calibration of WIM:

- **Define the Calibration Problem:** Clearly define the calibration problem, including the parameters that need adjustment in the WIM system. These parameters might include sensor calibration factors, filtering parameters, or other settings affecting weight estimation accuracy.

- **Objective Function:** Develop an objective function (fitness function) that quantifies the performance of the WIM system. This function should reflect the calibration goals, such as minimizing the difference between estimated and actual weights or optimizing other relevant performance metrics.
- **Parameter Encoding:** Represent potential solutions as individuals in a population, with each individual corresponding to a set of parameter values for the WIM system. Design an encoding scheme to represent these parameter values.
- **Initialization:** Create an initial population of potential solutions, comprising sets of parameters, either randomly or using a designated initialization method.
- **Fitness Evaluation:** Assess the fitness of each individual within the population by employing the specified objective function. This involves running the WIM system with the parameters represented by each individual and measuring how well it performs.
- **Selection**: Select individuals for reproduction based on their fitness. Individuals with higher fitness have a higher chance of being selected, simulating the natural selection process.
- **Crossover:** Apply crossover (recombination) to pairs of selected individuals to create new individuals (offspring). This involves exchanging genetic material (parameter values) between the selected parents.
- **Mutation:** Introduce small random changes (mutation) to some of the parameters in the population to maintain diversity and explore the solution space more thoroughly.
- **Evaluate Offspring:** Evaluate the fitness of the newly created offspring using the objective function.
- **Replacement:** Replace the least fit individuals in the current population with the newly created offspring, forming the next generation.
- **Termination Criteria:** Iterate through the steps of selection, crossover, mutation, and replacement for multiple generations or until a termination criterion is satisfied. Termination criteria may involve reaching a predetermined number of generations, achieving a satisfactory fitness level, or adhering to a predefined computational budget.
- **Extract Optimal Parameters:** Once the genetic algorithm converges, extract the parameter values from the individual with the highest fitness, as these represent the optimized calibration settings for the WIM system.
- **Apply Optimized Parameters:** Apply the optimized parameters to the WIM system for practical use (Fig. 2).

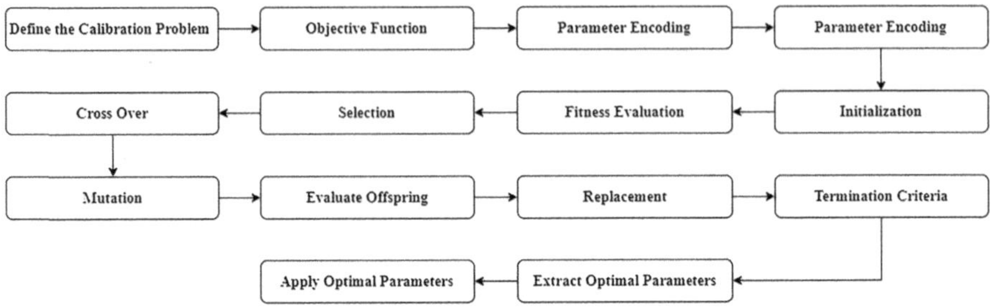

Fig. 2 Working of genetic algorithm

6 Comparison Between Traditional and Modern Methodologies

Bending plates, load cells, strain gauges, and piezoelectric sensors are all technologies used in Weigh-in-Motion (WIM) systems, each with its own set of advantages and disadvantages. The choice of which technology to use depends on various factors, and there isn't a one-size-fits-all solution. Below are some potential advantages of bending plates over load cells, strain gauges, and piezoelectric sensors in certain situations.

6.1 Simplicity and Durability

Bending plates, uncomplicated structures bending under vehicular loads, offer durability through their inherent simplicity, minimizing potential points of failure. The reduction in components contributes to increased resilience, translating into lower maintenance demands when contrasted with more intricate systems. This simplicity is a notable advantage, as it not only enhances reliability but also makes bending plates a cost-effective and practical choice for weigh-in-motion applications. This straightforward design, combined with ease of installation and suitability for both static and dynamic measurements, positions bending plates as an efficient solution for accurately measuring vehicle weights on roadways while minimizing the need for extensive upkeep.

6.2 Cost-Effectiveness

Bending plates stand out as a cost-effective solution within the realm of sensor technologies. Their affordability becomes evident in both manufacturing and installation, presenting a compelling choice for scenarios where financial considerations are paramount. This cost-effectiveness not only addresses budget constraints effectively but also enhances the overall feasibility of implementing weigh-in-motion systems. As a result, bending plates offer an economical yet reliable means of accurately measuring vehicle weights, making them particularly advantageous in applications where cost considerations play a pivotal role in decision-making processes.

6.3 Suitability for Static and Dynamic Measurements

The versatility of bending plates is manifest in their ability to measure both static and dynamic loads with precision. This adaptability renders them well-suited for applications demanding a comprehensive range of measurements, accommodating diverse vehicle weights and speeds. Whether capturing the stationary weight of parked vehicles or effectively gauging the dynamic loads imposed by vehicles in motion, bending plates prove to be a versatile solution. This capability to seamlessly handle various scenarios enhances the applicability of bending plates in weigh-in-motion systems, providing an efficient means to gather accurate weight data across a spectrum of operational conditions.

6.4 Ease of Installation

The installation simplicity of bending plates becomes apparent when contrasted with specific load cell configurations. Their straightforward integration involves embedding them into the road surface, a process that is notably more accessible, particularly when retrofitting existing infrastructure. This ease of installation is a valuable characteristic, streamlining the deployment of weigh-in-motion systems. The ability to seamlessly incorporate bending plates into roadway surfaces reduces installation complexities, making them an efficient choice for projects where simplicity and adaptability are paramount. This advantage contributes to their appeal in applications where rapid, uncomplicated installation is a crucial requirement for successful implementation.

6.5 Low Maintenance

The straightforward design and robust nature of bending plates often translate into reduced maintenance needs when compared to alternative technologies. This inherent simplicity proves advantageous in scenarios where minimizing downtime and associated maintenance costs takes precedence. The durable construction of bending plates contributes to their reliability over time, requiring less frequent attention and troubleshooting. This characteristic is particularly beneficial for weigh-in-motion systems deployed in critical applications where operational continuity and cost-effective maintenance are paramount considerations, ensuring that the system can consistently and dependably deliver accurate weight measurements without compromising efficiency.

7 Results and Discussion

Metrics provide a comprehensive assessment of the performance of image detection models. The choice of metrics depends on the specific requirements and goals of the task at hand.

7.1 *Accuracy*

Model accuracy is determined by metrics that incorporate true positives, true negatives, false positives, and false negatives.

$$\text{Accuracy} = \frac{\text{Number of Correct Predictions}}{\text{Total number of predictions}}$$

For binary classifications, the model's accuracy can be calculated using metrics such as true positives, true negatives, false positives, and false negatives.

$$\text{Accuracy} = \frac{\text{True Negatives} + \text{True Positive}}{\text{True Positive} + \text{False Positive} + \text{True Negative} + \text{False Negative}}$$

7.2 *Precision*

Precision assesses the precision of positive predictions by determining the ratio of true positives to the sum of true positives and false positives.

$$\text{Precision} = \frac{\text{True Positives} + \text{False Positives}}{\text{True Positives}}$$

8 Conclusion

As technology evolves, the integration of sophisticated sensor arrays and smart data processing further refines the capabilities of WIM systems, contributing to enhanced accuracy, efficiency, and reliability in monitoring vehicular weights on roadways. Bending plates, with their proven efficacy and adaptability, remain integral components in the intricate tapestry of weigh-in-motion technology, continually advancing the field of traffic monitoring and infrastructure management.

9 Future Enhancement

GNSS, or global navigation satellite system, can be employed for detecting and tracking vehicles. GNSS systems, which include GPS (Global Positioning System), provide accurate positioning information that can be leveraged for various applications related to vehicle detection. One of the primary applications of GNSS in vehicle detection is real-time tracking. GNSS-enabled devices in vehicles can receive signals from multiple satellites, allowing them to calculate their precise location on Earth. This information is invaluable for fleet management, logistics optimization, and emergency response. Fleet managers can monitor the movement of vehicles, assess their current location, and plan efficient routes in real time. GNSS is employed in road usage and toll collection systems. GNSS is integral to road usage and toll collection systems. By equipping vehicles with GNSS receivers, authorities can track their movement through specific points on the road. This tracking allows for the calculation of toll charges based on the distance traveled or the utilization of particular road segments. The integration of GNSS in toll collection systems has streamlined the process, enabling virtual toll booths and reducing the need for physical infrastructure.

Through the integration of GNSS with bending plates, we can eliminate the need for traditional physical toll booths and transition to a system of virtual toll booths.

AI-Based Methods and Algorithms for Fault Detection in the Transmission Lines

W. Jaisingh, P. Durgadevi, and R. K. Kavitha

1 Introduction

Within the dynamic realm of power systems, the consistent and effective functioning of transmission lines assumes utmost significance. This chapter commences an investigation into revolutionary developments in the field of defect detection for these vital components, capitalizing on the paradigm-shifting capabilities of artificial intelligence (AI). Given the capacity of transmission line faults to cause power supply disruptions and jeopardize the stability of electrical infrastructures, an alternative approach to fault detection is required. The introductory section offers a perceptive synopsis of the factors that contribute to transmission line faults and their repercussions, establishing the context for the critical need to implement advanced methodologies that can effectively manage the complexities and ever-changing characteristics of such faults [1]. Acknowledging the deficiencies of traditional methodologies with regard to precision and speed of reaction, the subsequent section of the chapter centers on the potential of artificial intelligence to fundamentally transform the field of fault detection. With an emphasis on the capabilities of artificial intelligence (AI) to process extensive datasets, adjust to changing circumstances, and detect subtle fault patterns, this chapter provides an overview of diverse AI-based techniques. These techniques encompass deep learning approaches like convolutional and recurrent neural networks, in addition to machine learning

W. Jaisingh (✉)
School of Information Science, Presidency University, Bangalore, Karnataka, India

P. Durgadevi
School of Computer Science and Engineering, Presidency University, Bengaluru, Karnataka, India

R. K. Kavitha
Kumaraguru College of Technology, Coimbatore, India
e-mail: kavitha.rk.mca@kct.ac.in

S. Vijayalakshmi et al. (eds.), *Smart Power Systems*, Power Systems,
https://doi.org/10.1007/978-3-031-89459-6_13

algorithms like support vector machines and neural networks. The discourse further encompasses the utilization of sophisticated optimization algorithms and the amalgamation of artificial intelligence (AI) with other intelligent systems, including the Internet of Things, in order to leverage real-time data and enhance the capabilities of fault detection. After an in-depth examination of practical implementations and case studies, the chapter concludes with a proactive discourse regarding the future possibilities and emerging patterns at the convergence of artificial intelligence and fault detection. This provides a comprehensive resource for practitioners, industry experts, and researchers who are interested in advancing the domain towards transmission line systems that are more intelligent and resilient.

1.1 A Brief Synopsis of Transmission Line Defects

The introductory segment of this chapter provides a comprehensive examination of transmission line faults, elucidating their genesis and possible ramifications within the framework of power systems. Critical elements of electrical infrastructures, transmission lines are vulnerable to a wide range of faults that have the potential to interrupt the uninterrupted flow of power, thereby compromising the overall stability and dependability of the system. The discourse undertakes an exhaustive analysis of the various factors that contribute to transmission line failures, including equipment malfunctions and environmental influences [2]. Moreover, it delves into the complex ramifications that ensue from defects, placing particular emphasis on the critical necessity for prompt and precise detection techniques in order to alleviate disturbances and bolster the robustness of power systems. The aforementioned fundamental comprehension of transmission line faults provides an indispensable context for the following investigation into cutting-edge AI-driven techniques and algorithms that aim to resolve the intrinsic difficulties linked to fault identification in these ever-changing and vital elements of the electrical infrastructure.

1.2 Importance of Sophisticated Fault Detection Methods

In the subsequent segment of this chapter, the necessity for sophisticated defect detection methodologies in the domain of transmission lines is emphasized. In light of the critical function that transmission lines perform in facilitating the smooth flow of electricity throughout electrical infrastructures, this chapter explicates the drawbacks of conventional fault detection techniques with regard to their precision and timely response. Given the ever-changing intricacies and fluid characteristics of transmission line faults, there exists an urgent requirement for state-of-the-art methodologies that can efficiently improve the functionality of power systems [3]. This segment underscores the importance of implementing sophisticated fault detection techniques in order to promptly detect and rectify faults, as well as to enhance the overall dependability and stability of electrical infrastructures. With the escalating electricity demand and the growing interconnectivity of grid infrastructure, there is an amplified need for inventive resolutions. This has led to the subsequent investigation of artificial intelligence (AI) algorithms and methods as a paradigm-shifting strategy to tackle these obstacles.

1.3 Function of Artificial Intelligence in the Detection of Faults

The subsequent segment of this chapter explores the critical significance of artificial intelligence (AI) in transforming transmission line defect detection. Recognizing the limitations of conventional approaches, this segment emphasizes the potential of artificial intelligence to revolutionize fault detection through its capacity to process vast quantities of data, adjust to dynamic environments, and identify complex fault patterns. The narrative underscores the suitability of AI technologies, such as machine learning algorithms and deep learning approaches, for effectively managing the intricacies associated with transmission line faults. By acquiring knowledge from past data, these artificial intelligence methodologies possess the capacity to provide enhanced precision and effectiveness in defect detection procedures. This section establishes the foundation for an extensive examination of diverse AI-based algorithms and methods in the following sections, emphasizing the transformative potential of AI in bolstering the robustness of transmission line systems and improving the capabilities of fault detection systems in the face of ever-changing challenges.

2 Conventional Failure Detection Techniques

This segment establishes the foundation by conducting an analysis of the current state of conventional defect detection techniques utilized in the domain of transmission lines. In the past, transmission lines, which are vital elements of power systems, have predominantly utilized traditional methods to detect and rectify faults. This preliminary investigation establishes the foundation for a comparative examination involving emergent AI-driven methodologies, underscoring the imperative for inventive solutions to surmount the intrinsic constraints of conventional approaches.

2.1 Constraints of Conventional Approaches

Expanding upon the groundwork laid in the preceding subsection, this one conducts a discerning assessment of the drawbacks associated with conventional fault detection techniques as they pertain to transmission lines. Despite their historical efficacy, these approaches exhibit deficiencies when confronted with the complex and ever-changing characteristics of transmission line faults, such as diminished precision and prolonged response durations [4]. It is essential to comprehend these constraints in order to provide justification for the investigation and implementation of more sophisticated and adaptable defect detection methods.

2.2 Synopsis of Traditional Methods

This section presents a thorough examination of the conventional methods that have historically been utilized to detect faults in transmission lines. Examining the fundamentals of everything from rule-based systems to signal processing techniques, this chapter covers a wide range of topics. This analysis establishes the foundation for a more profound comprehension of the divergence between conventional and AI-driven approaches, emphasizing the necessity for a framework for defect detection that is more sophisticated and adaptable.

2.3 Obstacles Facing Conventional Fault Detection

This section explores the complexities associated with defect detection using traditional methods, shedding light on the inherent difficulties that these well-established techniques present. These obstacles, which may arise from difficulties in discerning fault patterns, constraints in the ability to adapt to dynamic environments, or inefficiencies in managing extensive datasets, emphasize the need for a fundamental transformation in approaches to fault detection. This critical analysis establishes a foundation for the following investigation into artificial intelligence as a revolutionary remedy that can overcome these obstacles and improve the effectiveness of defect detection in transmission lines.

3 Fault Detection Algorithms Utilizing Machine Learning

This segment commences a comprehensive examination of the utilization of machine learning (ML) algorithms in the context of defect detection in transmission lines, acknowledging the revolutionary possibilities that these methodologies hold.

This section establishes the foundation for a thorough examination of diverse ML techniques in the following subsections by introducing the overarching function of machine learning in leveraging historical data to improve the accuracy of fault detection, prior to delving into the detailed examination of specific algorithms.

3.1 Algorithms for Machine Learning: An Overview

This subsection presents an introductory overview of the wide range of machine learning algorithms utilized in transmission line defect detection, thereby establishing a fundamental understanding [5]. This text explicates the underlying principle by which these algorithms acquire knowledge from past data, emphasizing their ability to recognize patterns and generate precise forecasts. This introductory section lays the foundation for a comprehensive examination of particular machine learning algorithms that are customized to address the intricacies of fault detection in electrical infrastructure.

3.2 Machines for Support Vectors

With respect to a particular category of machine learning algorithms, this subsection explores the utilization of Support Vector Machines (SVMs) in the context of fault detection. The text provides an overview of the underlying principles that govern SVMs, with a particular focus on their capacity to identify defect patterns within a high-dimensional space. Through a thorough analysis of the merits and drawbacks of SVMs within this particular framework, the chapter makes a valuable contribution to the overall comprehension of their effectiveness in improving the precision of defect detection systems.

3.3 Random Forests and Decision Trees

In order to broaden the range of machine learning algorithms available, this section delves into the utilization of random forests, an ensemble form of decision trees, for the purpose of defect detection. The text clarifies the hierarchical decision-making processes utilized by these algorithms, demonstrating their effectiveness in distinguishing intricate fault scenarios. This section assesses the efficacy of decision trees and random forests in the context of transmission line faults, in addition to examining their operational principles.

3.4 Deep Learning and Neural Networks

This subsection explores the domain of deep learning and neural networks, emphasizing their capability to capture complex defect patterns. This text presents a comprehensive examination of the exceptional performance exhibited by advanced algorithms such as convolutional neural networks (CNNs) and recurrent neural networks (RNNs) in the domains of automated feature extraction and temporal dependency modeling [6]. The discourse highlights their potential to surpass conventional approaches through their ability to adjust to various fault scenarios, thereby emphasizing their significance in the progression of fault detection capabilities.

3.5 Considerations and Obstacles in the Training of ML Models

This subsection concludes the segment on machine learning algorithms by discussing the inherent difficulties and factors to be taken into account when training such models for the purpose of defect detection in transmission lines. The chapter offers a comprehensive analysis of the practical factors that researchers and practitioners must take into account when implementing fault detection systems based on machine learning. These factors range from complexities related to the computational demands and issues surrounding data availability. This observation establishes the foundation for a more intricate comprehension of the intricacies involved in the integration of machine learning algorithms into practical implementations.

4 Methods Utilizing Deep Learning for Fault Detection

This segment explores the domain of deep learning as a cutting-edge framework for transmission line fault detection. Acknowledging the intricate and ever-changing characteristics of fault patterns, the chapter delves into particular deep learning methodologies that have demonstrated potential in augmenting fault detection functionalities. This section explores the utilization of advanced techniques such as convolutional neural networks (CNNs), which are skilled at automatically extracting features, and recurrent neural networks (RNNs), which are adept at capturing temporal dependencies. It highlights the potential of these cutting-edge methods to outperform conventional approaches when it comes to identifying complex fault scenarios.

4.1 CNNs: Convolutional Neural Networks

This subsection presents a comprehensive examination of Convolutional Neural Networks (CNNs) as they pertain to the identification of faults. The discourse describes how CNNs demonstrate their expertise in signal and image processing by automatically deriving hierarchical features from data. Convolutional neural networks (CNNs), which are uniquely designed to analyze the spatial attributes of fault patterns, have exhibited exceptional efficacy in detecting nuanced irregularities present in transmission lines [7]. This segment not only provides a comprehensive explanation of the operational principles of CNNs, but also illustrates concrete examples and situations in which the accuracy of defect detection has been substantially enhanced by these networks.

4.2 RNNs: Recurrent Neural Networks

This subsection delves into an additional aspect of deep learning by examining the utilization of Recurrent Neural Networks (RNNs) in the context of transmission line fault detection. The chapter explores the significance of temporal dependencies in defect patterns and explains how recurrent neural networks (RNNs), due to their recurrent connections, are highly suitable for modeling sequential data. The discourse delineates cases in which recurrent neural networks (RNNs) have exhibited efficacy in forecasting and pinpointing defects as they occur gradually, thereby augmenting the holistic comprehension of how deep learning can be utilized to improve the temporal dimensions of fault detection.

4.3 Implementations and Benefits of Deep Learning

By transcending specific methodologies, this segment of the chapter offers a more comprehensive outlook on the benefits and practical implementations of deep learning in the domain of fault detection [8]. This statement clarifies the manner in which the effectiveness of deep learning models is enhanced by their inherent capabilities, which comprise automated feature extraction, learning of hierarchical representations, and adaptability to various fault scenarios. Through the illustration of a variety of applications spanning various industries and domains, this segment emphasizes the adaptability of deep learning in tackling the intricate obstacles linked to transmission line defect detection.

4.4 Obstacles in Training and Optimization in Deep Learning

In summary, this subsection concludes the examination of deep learning methodologies by discussing the inherent difficulties associated with training and optimizing these sophisticated models for the purpose of defect detection. The chapter evaluates the practical considerations and challenges that practitioners and researchers face when implementing defect detection systems based on deep learning [9]. These include the computational demands, the requirement for large labeled datasets, and the complexities of hyperparameter tuning. This revelation enhances the collective comprehension of the intricacies associated with optimizing deep learning models for practical implementations in the field of electrical grid dependability.

5 Algorithms for Optimizing Fault Detection

This segment delves into the incorporation of optimization algorithms into transmission line fault detection models, emphasizing their function in improving and augmenting the overall performance of such systems. Embedded with intelligence, optimization algorithms (e.g., particle swarm optimization and genetic algorithms) refine fault detection models, select pertinent features, and ultimately enhance the overall efficacy and precision of the fault detection procedure. The chapter explores the operational principles underlying these optimization techniques and evaluates their particular implementations as they pertain to the reliability of electrical grids.

5.1 Genetic Procedures

With a specific emphasis on genetic algorithms as a fundamental optimization technique, this subsection conducts an exhaustive analysis of how they are implemented in defect detection models. Genetic algorithms, which draw inspiration from the workings of evolution and natural selection, possess remarkable proficiency in traversing intricate solution spaces and discerning optimal configurations [10, 11]. The discourse delves into the manner in which these algorithms facilitate the development of defect detection models by optimizing feature sets and parameters in order to enhance the system's overall precision and adaptability. Success tales and real-world illustrations serve to underscore the effectiveness of genetic algorithms when it comes to the development of optimal fault detection solutions.

5.2 Optimization of Particle Swarms

Particle swarm optimization (PSO) is introduced as an alternative integral optimization algorithm in defect detection models in this section. PSO, which is motivated by natural social behavior, entails the cooperative motion of particles within a solution space with the objective of identifying optimal solutions. The chapter provides a detailed explanation of how PSO is utilized to optimize features, weights, and parameters in fault detection models, thereby improving their precision in identifying and categorizing defects. The discourse covers the merits of PSO as a method for tackling the difficulties presented by the dynamic and intricate characteristics of transmission line faults.

5.3 Function of Optimization in Models for Fault Detection

This subsection presents a comprehensive overview of the fundamental function that optimization algorithms perform in defect detection models. This underscores the manner in which these algorithms enhance the resilience and flexibility of the models through the optimization of parameters, the identification of pertinent features, and the refinement of the overall architecture [12, 13]. The discourse underscores the reciprocal association between fault detection algorithms and optimization techniques, demonstrating how their amalgamation enhances the precision and effectiveness of fault identification and mitigation for transmission line defects.

5.4 Enhancing Performance and Efficiency Via Optimization

This subsection concludes the examination of optimization algorithms by emphasizing the concrete effects that optimization has on the performance and efficiency of defect detection models. Through the methodical resolution of obstacles including feature selection, model parameterization, and system-wide optimization, the defect detection procedure is enhanced through the integration of genetic algorithms and particle swarm optimization. The chapter presents an analysis of how optimization impacts fault detection solutions for transmission lines by reducing response times, increasing accuracy, and bolstering overall resilience in the face of evolving fault scenarios. These findings provide a strong argument for the incorporation of intelligent techniques such as optimization in the quest for more efficient fault detection solutions.

6 The Incorporation of Intelligent Systems

This segment explores the ways in which fault detection systems utilizing artificial intelligence (AI) can be effectively integrated with other intelligent systems. It emphasizes the critical importance of collaborative technologies in enhancing the reliability of transmission lines. The chapter delves into two critical aspects of this integration—data analytics platforms and the Internet of Things (IoT)—both of which offer distinct functionalities that augment fault detection procedures. Through the seamless integration of AI algorithms, IoT device data insights in real time, and robust analytics platforms, this segment sheds light on the possibility of a more comprehensive and adaptable fault detection methodology for electrical grids.

6.1 Fault Detection and the Internet of Things (IoT)

This subsection explores the mutually beneficial correlation that exists between defect detection in transmission lines and the Internet of Things (IoT). IoT devices, which are networked and outfitted with sensors, provide a continuous flow of data in real time. This data can be utilized to identify and address faults with an unparalleled degree of precision and velocity [14]. The conversation examines the ways in which the combination of Internet of Things (IoT) technology and AI-driven fault detection creates a flexible and effective system that can consistently assess the condition of transmission lines and promptly detect irregularities. The integration in question significantly enhances the resilience of power systems, as evidenced by its tangible implementations and notable achievements.

6.2 Platforms for Data Analytics and Fault Detection

This subsection examines the function of data analytics platforms in enhancing the capabilities of defect detection. The provision of robust analytics platforms facilitates the processing and interpretation of enormous quantities of data produced by transmission lines, thereby empowering decision-makers to be more informed. This chapter analyzes the ways in which these platforms enable the incorporation of AI algorithms, providing valuable perspectives on fault patterns and the behavior of the system. By capitalizing on sophisticated analytics, fault detection models can enhance their comprehension of past patterns, thereby making valuable contributions to the development of more precise prognostications and proactive maintenance strategies.

6.3 Capitalizing on Real-Time Data to Enhance Fault Detection

This subsection, which concludes the section on intelligent system integration, examines the overarching concept of utilizing real-time data to improve transmission line fault detection. By integrating AI-powered fault detection with data analytics platforms and real-time data streams from IoT devices, a dynamic and adaptable response to evolving fault scenarios is made possible. The chapter highlights the manner in which this synergy improves the precision and effectiveness of defect detection systems, thereby decreasing response times and ultimately strengthening the robustness of power systems. By adopting real-time data as a fundamental component, the incorporation of intelligent systems becomes a revolutionary approach to further develop fault detection in the context of electrical infrastructures.

7 Implementation Difficulties and Factors to be Considered

This segment addresses the pragmatic obstacles and factors that must be taken into account when integrating defect detection systems for transmission lines that rely on artificial intelligence (AI). The efficacy of these sophisticated technologies relies on the successful resolution of intricate challenges pertaining to ethical considerations, data management, model interpretability, and computational resources. Through a thorough analysis and resolution of these obstacles, the chapter imparts significant knowledge for practitioners, industry experts, and researchers engaged in the deployment of AI-driven fault detection solutions. This contributes to a more comprehensive comprehension of the pragmatic factors that impact the achievement of successful implementation.

7.1 Quality and Availability of Data

This subsection emphasizes the critical significance of data in the processes of training and validating artificial intelligence models designed for fault detection. This paper explores the complexities related to data quality and availability, with a particular focus on the criticality of having comprehensive, diverse, and representative datasets in order to guarantee the resilience of defect detection models. The discourse delves into approaches to surmount constraints in data, including data augmentation and synthesis [15]. Additionally, it emphasizes the criticality of data quality in bolstering the dependability and applicability of artificial intelligence models as they pertain to transmission line defect detection.

7.2 Interpretability and Explanation of the Model

Regarding the interpretability of AI models, this segment discusses the essential requirement to comprehend and elucidate the reasoning behind the decisions generated by fault detection algorithms. The ability to interpret models is critical in order to establish confidence among stakeholders and guarantee that AI-driven systems can be effectively utilized in real-life situations. In this chapter, methods and strategies for improving the interpretability of models are examined, providing insight into the complexities of such models. Enhancing transparency and accountability can be achieved in the implementation of AI-based defect detection models by recognizing and addressing challenges associated with interpretability.

7.3 Constraints on Resources and Computational Requirements

This subsection examines the resource limitations and computational requirements that are inherent in the implementation of AI-driven fault detection systems. As the complexity of these models rises, critical factors to be taken into account include computational capacity, memory demands, and scalability of deployment. This chapter explores approaches for optimizing and customizing artificial intelligence models to address the limitations of real-world conditions. Its goal is to ensure that defect detection systems continue to be functional, efficient, and viable for integration into operational environments that require different levels of computational resources.

7.4 Privacy and Ethical Considerations

In the concluding portion of this segment, the ethical and privacy implications inherent in the deployment of AI-powered defect detection are examined. Due to the fact that these systems contribute to critical infrastructure and interact with sensitive data, privacy, and ethical considerations become indispensable. The chapter discusses issues pertaining to the possession of data, the need for consent, and the potential ramifications of AI decisions on human lives. Through a critical analysis of the ethical aspects associated with AI-driven fault detection, this chapter promotes a conscientious and transparent methodology that nurtures a harmonious coexistence between groundbreaking advancements and the ethical deliberations that are indispensable for the reliable implementation of such systems in the domain of transmission line uptime.

8 Case Studies and Applications in the Real World

This segment provides an in-depth examination of practical implementations via case studies, with a particular emphasis on the integration of AI-powered defect detection within a transmission line network. The case study provides an in-depth analysis of the network's challenges, showcasing how sophisticated AI algorithms were strategically chosen and implemented to tackle these intricacies. By conducting an in-depth analysis of the integration with intelligent systems, such as data analytics platforms and IoT devices, the case study demonstrates the manner in which real-time data enhances the adaptability and precision of the fault detection mechanism. Case Study 1 presents a concrete illustration of effective implementation that imparts significant knowledge regarding the revolutionary effects of artificial intelligence in bolstering the dependability and robustness of transmission line networks. This concrete example functions as a significant point of reference for practitioners and scholars in the field, providing concrete proof of the effectiveness and practicality of defect detection methodologies based on artificial intelligence.

8.1 Case Study 1: Transmission Line Network Fault Detection Utilizing AI

A vital transmission line network is situated within an expansive metropolitan region, playing a pivotal role in providing electricity to millions of inhabitants and sustaining numerous industries. Nevertheless, the network encounters the enduring obstacle of effectively identifying defects that have the potential to disrupt the provision of power and jeopardize the overall stability of the electrical grid.

In the past, the identification of faults was accomplished through the utilization of rule-based systems and signal processing techniques. However, the inadequate precision and response time of these approaches incited the investigation of more sophisticated AI-driven alternatives. The transmission line network operators reach the consensus to adopt AI-based fault detection in order to augment their capacity to promptly detect and alleviate defects, thereby enhancing the electrical grid's dependability and robustness.

Machine learning algorithms, such as support vector machines, decision trees, and neural networks, are integrated into the AI system [16]. The algorithms undergo training using historical data that covers a wide range of fault scenarios. This allows them to acquire knowledge of complex patterns and attributes that are linked to various types of faults. The adaptability of these algorithms is essential for detecting defects in a timely and precise manner, thereby ensuring greater responsiveness and accuracy.

In order to confront the difficulties presented by various fault patterns, the network administrators implement deep learning methodologies, specifically convolutional neural networks (CNNs) and recurrent neural networks (RNNs). Convolutional neural networks (CNNs) demonstrate exceptional proficiency in the automated extraction of features from spatial data, whereas RNNs capture temporal dependencies within fault patterns. The implementation of this dual-layered strategy improves the system's capacity to identify nuanced irregularities, thereby enhancing the precision of defect detection.

Concurrently, the incorporation of intelligent systems assumes a critical function. Sensor-equipped IoT devices are strategically positioned along transmission lines in order to furnish up-to-the-minute information on a multitude of parameters, including but not limited to temperature, voltage, and current. As data sources, these Internet of Things devices continuously supply the AI algorithms with information. By analyzing this real-time data, data analytics platforms are able to extricate insightful information regarding the condition of the transmission lines.

Having been updated with real-time data, the defect detection algorithms of the AI system now function in a dynamic environment. In the event of a malfunction, the system expeditiously examines the incoming data, discerns the nature of the defect, and initiates an automated course of action, which may involve redirecting power or notifying maintenance teams. By integrating AI with IoT devices and data analytics platforms in a seamless manner, the transmission line network experiences a substantial reduction in response times, outage is minimized, and overall resilience is enhanced.

By conducting routine assessments and making necessary adjustments, the AI-driven fault detection system consistently adjusts to changing fault scenarios, showcasing its capacity to acquire knowledge and enhance itself in the long run. This transmission line network's utilization of artificial intelligence (AI) demonstrates the transformative potential of AI in defect detection and the dependability of critical infrastructure, thereby serving as a model for other urban areas confronting comparable difficulties.

8.2 Case Study 2: Deep Learning-Based Real-Time Fault Detection

The electricity distribution system in an expansive smart city is outfitted with sophisticated sensors and smart grid technology in order to guarantee a dependable and efficient supply of power. Rapid urbanization and the intricacy of the interconnected electrical infrastructure, on the other hand, make it difficult to detect and mitigate faults that have the potential to disrupt the power distribution.

In response to this matter, the utility company opts to deploy real-time defect detection through the utilization of deep learning, capitalizing on the functionalities of sophisticated neural networks. Convolutional neural networks (CNNs) and recurrent neural networks (RNNs) are the focal points of the deep learning system due to their respective proficiency in processing spatial and temporal data [17, 18].

Suppose a defect occurs in the distribution network, such as a short circuit or equipment malfunction, and anomalies in voltage, current, and other pertinent parameters are detected by the sensors distributed throughout the grid. The deep learning model receives the real-time data feeds from these sensors. The model has undergone pre-training using an extensive dataset that encompasses a wide range of fault scenarios.

The deep learning model's CNN component demonstrates exceptional proficiency in the analysis of spatial data. By detecting anomalies and discernible patterns in the current and voltage distributions across the power grid, it precisely determines the fault's location and characteristics. Concurrently, the RNN element captures temporal dependencies, enabling the system to conduct an analysis of the temporal progression of defects.

The deep learning model consistently enhances its comprehension of the typical operational conditions of the grid as it processes the incoming real-time data. In the event that a fault is identified, the system initiates an expeditious reaction, which may involve actions such as isolating the defective component, rerouting power via alternate routes, or promptly dispatching maintenance personnel to rectify the situation.

By incorporating real-time data and deep learning, the fault detection system is not only rendered precise but also capable of adjusting to changing fault patterns. The system's capacity to acquire knowledge from and adjust to fresh information empowers it to sustain a considerable degree of precision, even within ever-changing urban settings characterized by substantial variations in electrical demand and fault scenarios.

By employing deep learning for real-time fault detection, the city's power supply experiences reduced downtime and the detrimental effects of faults are mitigated. Moreover, this system enhances the overall resilience of the electrical infrastructure. The system's proactive and adaptive characteristics guarantee prompt resolution of prospective problems, thereby demonstrating the profound influence of deep learning in augmenting the dependability and effectiveness of real-time fault detection in the electricity distribution system of a smart city.

8.3 Success Stories and Lessons Learned Capability to Adjust to Changing Environments

Insight Gained: The scenarios underscore the significance of designing defect detection systems that possess the capability to adjust to the ever-changing and dynamic characteristics of urban settings. Transmission line networks and smart city systems encounter diverse fault patterns and operational circumstances, which call for artificial intelligence models capable of ongoing learning and adaptation.

Achievement: The deep learning model, which underwent real-time updates and was trained on a heterogeneous dataset, exhibited exceptional flexibility in both situations. The system's capacity to develop and adjust in response to shifting malfunction scenarios led to enhanced precision and decreased response durations.

Consensus with Intelligent Systems for Insights in Real Time
Key Takeaway: The fault detection process is improved through the integration of intelligent systems, including data analytics platforms and Internet of Things (IoT) devices, which offer up-to-the-minute insights into the infrastructure's condition. By means of this integration, the defect detection system is able to function in a constantly changing environment utilizing the most up-to-date data.

Achievement: The IoT devices were instrumental in both scenarios by providing malfunction detection systems with real-time data, which enabled them to make prompt and well-informed decisions. The integration of AI algorithms with intelligent systems resulted in a substantial enhancement in the overall effectiveness and promptness of fault detection.

Integration of Deep Learning and Machine Learning for Comprehensive Solutions
A comprehensive solution to defect detection can be obtained through the integration of machine learning and deep learning methodologies. Machine learning algorithms, including neural networks, support vector machines, and decision trees, establish a robust framework, whereas deep learning models, such as recurrent neural networks (RNNs) and convolutional neural networks (CNNs), demonstrate exceptional proficiency in capturing intricate temporal and spatial relationships.

Success Story: The presented scenarios effectively illustrate the implementation of deep learning and machine learning methodologies. Accurate predictions are facilitated by machine learning algorithms, whereas temporal dependency modeling and feature extraction are improved by deep learning models, culminating in a defect detection system that is both resilient and precise.

Ethical Considerations and Implementation Transparency
Incorporating artificial intelligence (AI) into critical infrastructure necessitates meticulous deliberation of privacy apprehensions, ethical ramifications, and transparent decision-making procedures. This guarantees that the implementation of AI-driven fault detection systems is conscientious, reliable, and in accordance with ethical principles.

Success Story: An ethical implementation of AI-based fault detection systems would encompass transparent communication regarding data usage, privacy protections, and ethical considerations although not explicitly referenced in the scenarios. In this context, success tales consist of preserving public confidence and ensuring the responsible deployment of AI.

Ongoing Assessment and Improvement
Key Takeaway: Ongoing assessment and improvement of artificial intelligence models are critical for sustaining peak performance. Periodic evaluations of the system's precision, flexibility, and effectiveness facilitate the detection of potential areas that require enhancement and optimization.

Success Story: The scenarios underscore the importance of conducting regular evaluations, which allow for the adaptation of AI models to changing fault characteristics and operational circumstances. The iterative nature of this process is a significant factor in maintaining success and continuously enhancing fault detection capabilities. In brief, the insights gained and instances of achievement from these situations underscore the criticality of flexibility, incorporation with intelligent systems, a hybrid approach involving machine learning and deep learning, ethical deliberations, and ongoing assessment for the effective deployment of AI-driven fault detection systems. These insights offer significant guidance for practitioners, researchers, and industry experts who are interested in the implementation and enhancement of these systems in practical situations.

9 Prospects for the Future and Emerging Trends

With an eye towards the future, the developments and emerging patterns in defect detection for transmission lines utilizing artificial intelligence are positioned to significantly influence the reliability of electrical grids. By incorporating sophisticated artificial intelligence algorithms alongside emerging technologies like edge computing, 5G networks, and distributed sensing systems, fault detection solutions could potentially become even more intelligent and resilient. Real-time processing at the point of data generation can be enabled by edge computing, thereby augmenting the velocity and effectiveness of defect detection. The introduction of 5G networks facilitates accelerated and more dependable device-to-device communication, which contributes to the responsiveness and integration of AI-based systems. In addition, the integration of distributed sensing systems facilitates a more holistic comprehension of the condition of the power grid. The convergence of these technologies, in conjunction with AI methodologies, is anticipated to introduce an era characterized by defect detection that is more precise, flexible, and robust than ever before. This will guarantee the ongoing dependability of transmission lines amidst constantly changing challenges. It is highly recommended that researchers, practitioners, and industry professionals investigate and capitalize on these nascent trends in order to propel the field forward and make valuable contributions to the continuous evolution of electrical grid management.

9.1 Edge Computing and Fog Computing Integration

The convergence of edge computing and fog computing with AI-based fault detection systems signifies a significant progression in the domain of transmission line reliability. Decentralized data processing in close proximity to the data source, edge computing improves defect detection by facilitating real-time analysis at the location where the data is generated. The close proximity reduces latency, which enables prompt decision-making in the event of defects. As an expansion of edge computing, fog computing enhances functionality through the implementation of a hierarchical framework that extends from the periphery to the cloud. By utilizing cloud resources for more complex analyses and optimizing the distribution of computational duties, this integration guarantees that critical fault detection processes take place in closer proximity to the source. The integration of edge and fog computing not only enhances the velocity and agility of fault detection in transmission lines but also optimizes the allocation of resources, signifying a paradigmatic transition towards electrical grid management that is more streamlined, decentralized, and adaptable.

9.2 Leveraging 5G Networks to Accomplish Fault Detection

The utilization of 5G networks signifies an innovative pathway towards augmenting fault detection in transmission lines. The introduction of 5G technology has brought about network capacity expansion, reduced latency, and unprecedented data transmission speeds, thereby transforming the communication infrastructure of electrical grids. This progression enables uninterrupted connectivity and rapid data exchange in real time between sensors and AI- powered defect detection systems that are dispersed along transmission lines. Due to the high-speed and low-latency attributes of 5G, fault detection algorithms are able to promptly receive, process, and respond to data, which facilitates the identification of faults in a precise and timely manner. Moreover, the robustness and dependability of 5G networks guarantee uninterrupted communication amidst turbulent and changing surroundings, thereby enhancing the overall efficacy of fault detection systems. Therefore, the incorporation of 5G networks becomes a crucial facilitator, initiating an unprecedented period of effectiveness, promptness, and enhanced dependability with regard to the detection of faults in the electrical grid.

9.3 Collaboration and Distributed Sensing Systems for Fault Detection

The incorporation of distributed sensing systems into collaborative defect detection for transmission lines signifies a fundamental transformation. By utilizing a network of sensors strategically positioned throughout the electrical infrastructure, distributed sensing is capable of gathering an extensive range of data pertaining to variables including temperature, voltage, and current. The abundance of up-to-date data plays a crucial role in constructing a comprehensive assessment of the grid's condition. These sensing systems operate in conjunction with AI-based defect detection algorithms, providing them with a wide range of complex data inputs. The integration of sensor data improves the accuracy and flexibility of fault detection models, thereby fostering a collective intelligence fostered by the interaction between distributed sensing and fault detection. Through the integration of data from various sensors, the system acquires a more comprehensive comprehension of fault patterns and operational circumstances. This, in turn, enhances the precision and effectiveness of the collaborative identification and resolution of transmission line faults. By integrating distributed sensing systems, a collaborative and adaptive approach to defect detection is established, thereby enhancing the resilience and responsiveness of the electrical grid infrastructure.

9.4 Developments in Algorithms and Techniques within AI

Ongoing progress in artificial intelligence methodologies and algorithms serves as a fundamental pillar for the forthcoming era of transmission line fault detection. The continuous development of machine learning and deep learning techniques, in conjunction with novel algorithmic strategies, holds the potential to enhance the precision and flexibility of fault detection systems even further. Various techniques, including transfer learning, reinforcement learning, and ensemble methods, provide opportunities to augment the learning capabilities of artificial intelligence models. This allows them to draw conclusions from a wide range of datasets and adjust to changing malfunction scenarios. Furthermore, the investigation of Explainable AI (XAI) guarantees not only precise prognostications but also open and transparent decision-making, thereby mitigating apprehensions pertaining to the interpretability of models. As scholars explore state-of-the-art advancements, the interaction between artificial intelligence (AI) developments and the complexities of fault detection in transmission lines is positioned to reveal unparalleled levels of accuracy, resilience, and intelligence in ensuring the dependability of electrical infrastructures. The ongoing advancements in artificial intelligence methodologies validate the fluid and continually growing characteristics of fault detection functionalities in the domain of transmission lines.

10 Conclusion

10.1 Synopsis of Transmission Line Fault Detection Utilizing AI

In summary, the investigation into fault detection in transmission lines utilizing artificial intelligence highlights a paradigm shift in the direction of improving the dependability and robustness of electrical infrastructure. This chapter presented a thorough examination of the difficulties presented by transmission line defects, with a particular focus on the shortcomings of conventional fault detection techniques and the imperative for sophisticated resolutions. Artificial intelligence (AI) has emerged as a prominent driver of innovation, primarily due to the prominence of machine learning and deep learning algorithms. The fault detection applications of numerous AI techniques, including support vector machines, decision trees, convolutional neural networks (CNNs), and recurrent neural networks (RNNs), were exhaustively evaluated. The feasibility of incorporating AI into intelligent systems, including data analytics platforms and Internet of Things (IoT) devices, was illustrated through the utilization of real-time data to strengthen fault detection capabilities.

In addition, the examination of optimization algorithms, ethical implications, and real-world case studies demonstrated the complexities and achievements associated with the deployment of AI-driven fault detection systems. Gaining knowledge from situations requiring practitioners and researchers to adapt to dynamic environments, integrate with intelligent systems, and utilize a combination of machine learning and deep learning are all significant contributions.

Anticipating the future, the amalgamation of artificial intelligence (AI) and nascent technologies such as edge computing, 5G networks, and distributed sensing systems presents tremendous potential for more sophisticated and prompt defect detection solutions. The ongoing progress in artificial intelligence methodologies reinforces the trend towards increased accuracy and flexibility. With the ongoing electrification of societies, the knowledge and perspectives presented in this chapter strive to motivate further advancements, cultivating a future in which fault detection systems powered by artificial intelligence are pivotal in safeguarding the dependability of transmission lines and, consequently, the stability of our ever-changing electrical infrastructure.

10.2 Essential Conclusions and Suggestions

An examination of transmission line defect detection utilizing artificial intelligence reveals a number of significant findings and suggestions that are applicable to practitioners, researchers, and industry experts:

A Variety of AI Methods Employed in Holistic Solutions:

Incorporate a wide range of artificial intelligence methodologies, such as deep learning strategies (CNNs, RNNs) and machine learning algorithms (support vector machines, decision trees), in order to develop comprehensive fault detection systems that can effectively manage a variety of fault patterns and operational circumstances.The amalgamation of intelligent systems to obtain real-time insights:

Capitalize on real-time data by utilizing the integration of AI-driven fault detection systems with intelligent systems, including IoT devices and data analytics platforms. The integration of these components improves the overall efficacy, responsiveness, and adaptability of fault detection processes.Ethical Considerations and Implementation Transparency:

Ethical considerations should take precedence over the implementation of AI-based systems. Maintain transparency regarding privacy protections, data usage, and decision-making processes in order to foster confidence and guarantee responsible implementation.Ongoing Assessment and Improvement:

Establish a cycle of ongoing refinement and evaluation for AI models. Consistently evaluate their efficacy, flexibility, and performance to pinpoint areas that require enhancement and guarantee ongoing triumph in fault detection. Developments in AI Methods:

Remain informed about the latest developments in artificial intelligence methodologies and algorithms. Incorporate advancements such as transfer learning, reinforcement learning, ensemble methods, and Explainable AI (XAI) in order to address interpretability concerns and further refine defect detection capabilities.5G Networks, Edge Computing, and Distributed Sensing System Integration:

This paper investigates the potential for AI-based defect detection systems to be integrated with emerging technologies such as distributed sensing systems, edge computing, and 5G networks. Leverage these developments to improve the velocity, effectiveness, and flexibility of fault detection procedures.Detection of Collaborative Faults via Distributed Sensing:

Leverage the collective effort in defect detection through the integration of artificial intelligence systems and distributed sensing networks. The integration of insights from numerous sensors in collaboration improves the precision and effectiveness of fault detection by fostering a more nuanced comprehension of fault patterns.Lessons in Practical Implementation:

Gain valuable insights from the pragmatic case studies that have been presented within this chapter. Gain knowledge of the obstacles encountered, calculated decisions executed, and results attained in tangible situations in order to provide direction and information for the pragmatic deployment of AI-powered defect detection.

By applying these insights and suggestions to subsequent undertakings, one can enhance the continuous development and prosperity of AI-driven defect detection in transmission lines, thereby promoting the establishment of a robust and discerning electrical grid infrastructure.

References

1. Kanwal, S., & Jiriwibhakorn, S. (2023). Artificial intelligence based faults identification, classification, and localization techniques in transmission lines-A review. *IEEE Latin America Transactions, 21*(12), 1291–1305. Retrieved from https://latamt.ieeer9.org/index.php/transactions/article/view/8269
2. Leh, N.A., Zain, F.M., Muhammad, Z., Hamid, S.A., & Rosli, A.D. (2020). Fault Detection Method Using ANN for Power Transmission Line. 2020 10th IEEE International Conference on Control System, Computing and Engineering (ICCSCE), 79–84.
3. Rezapour, H., Jamali, S., & Bahmanyar, A. (2023). Review on artificial intelligence-based fault location methods in power distribution networks. *Energies, 16*, 4636. https://doi.org/10.3390/en16124636

4. Wang, Y., Li, Y., & Li, Y. (2021). Fast and accurate fault detection and classification in transmission lines using extreme learning machine. *IEEE Access, 9*, 123042–123052.
5. Rafique, F., Ling, F., & Mai, R. (2021). End to end machine learning for fault detection and classification in power transmission lines‖. *Electric Power Systems Research, 199*, 107430. https://doi.org/10.1016/j.epsr.2021.107430
6. Belay, M. A., Blakseth, S. S., Rasheed, A., & Salvo Rossi, P. (2023). Unsupervised anomaly detection for IoT-based multivariate time series: Existing solutions, performance analysis and future directions. *Sensors, 23*, 2844. https://doi.org/10.3390/s23052844
7. Tama, B. A., Vania, M., Lee, S., et al. (2023). Recent advances in the application of deep learning for fault diagnosis of rotating machinery using vibration signals. *Artificial Intelligence Review, 56*, 4667–4709. https://doi.org/10.1007/s10462-022-10293-3
8. Qiu, S., Cui, X., Ping, Z., Shan, N., Li, Z., Bao, X., & Xu, X. (2023). Deep learning techniques in intelligent fault diagnosis and prognosis for industrial systems: A review. *Sensors, 23*(3), 1305.
9. Hwang, T., Ha, J. M., & Youn, B. D. (2023). Robust deep learning-based fault detection of planetary gearbox using enhanced health data map under domain shift problem. *Journal of Computational Design and Engineering, 10*(4), 1677–1693.
10. Daudi Mnyanghwalo, Herald Kundaeli, Ellen Kalinga & Ndyetabura Hamisi I James Lam (2020) Deep learning approaches for fault detection and classifications in the electrical secondary distribution network: Methods comparison and recurrent neural network accuracy comparison, Cogent Engineering, 7:1, DOI: https://doi.org/10.1080/23311916.2020.1857500.
11. Zhang, R., Zhang, Y., Wu, Z., & Guo, Y. (2022). Deep learning-based fault classification in power systems considering disturbances and uncertainties. *Automation and Robotics, 10*(5), 1096–1106.
12. Guo, M., Guo, C., Zhang, C., Zhang, X., & Liu, Y. (2023). Intelligent fault detection algorithm based on Hi/H∞ optimization and a cascaded neural networks. *Ocean Engineering, 272*, 113835. ISSN 0029-8018.
13. Xu, H., Ren, P., Wu, Z., Guo, Y., & Wang, F. (2023). Transmission line fault detection threshold optimization based on improved grey wolf optimization algorithm. *Electric Power Components and Systems*, 1–10.
14. Wu, H., Zhou, J., Liu, X., & Huang, X. (2023). An improved multi-objective chaotic particle swarm optimization algorithm for transmission line fault detection threshold optimization. *CSEE Journal of Power and Energy, 9*(3), 351–361.
15. Wang, S., Fan, W., Wu, Z., Wang, Z., & Guo, Y. (2022). Multi-objective optimization of fault detection thresholds in microgrids considering cost and accuracy. *Journal of Modern Power Systems and Clean Energy, 10*(5), 886–898.
16. Leh, M. A., & Zain, M. F. M. (2022). Fault detection on transmission lines using artificial neural network. *IEEE Latin America Transactions, 20*(1), 1–7.
17. Wang, F., Wang, Z., Wang, Y., & Wu, Z. (2023). A comprehensive review of deep learning applications for fault detection in power systems. *CSEE Journal of Power and Energy, 9*(3), 339–350.
18. Li, S., & Pan, Y. (2017). Fault diagnosis method of power transformer based on deep learning algorithm. *IOP Conference Series: Materials Science and Engineering, 245*(5), 052029.

Blockchain Technology and Smart Power Systems: Transforming the Modernization of Power Systems

Palanichamy Naveen, T. Akilan, M. Maragatharajan, K. Mohanasundaram, and H. James Deva Koresh

1 Introduction

The growing demand for electricity, coupled with the need for efficient and cleaner energy sources, has prompted significant advancements in the power sector in the twenty-first century. The integration of modern technologies such as Artificial Intelligence (AI), the Internet of Things (IoT), and blockchain is revolutionizing power systems, leading to the development of smart power systems. These systems leverage intelligent algorithms, real-time data analysis, and decentralized transaction systems to enhance power distribution, energy management, and overall grid efficiency.

This article explores the role of blockchain technology in the modernization of power systems, focusing on the concept of smart power systems. It discusses the applications of AI and IoT in power distribution, power quality control, energy management, and the significant impact of blockchain technology in creating secure and transparent smart power systems. Through practical case studies and real-world examples, this article aims to provide valuable insights for students, researchers, and professionals in the electrical engineering, power systems, and renewable energy fields.

P. Naveen (✉)
Department of Electronics and Communication Engineering,
Dr. N.G.P. Institute of Technology, Coimbatore, India

K. Mohanasundaram · H. J. D. Koresh
Department of Electrical and Electronics Engineering, KPR Institute of Engineering and Technology, Coimbatore, India

T. Akilan
Department of Computer Science Engineering, Amity University, Kolkata, India

M. Maragatharajan
Department of Computer Science Engineering, VIT Bhopal, Bhopal, Madhya Pradesh, India

S. Vijayalakshmi et al. (eds.), *Smart Power Systems*, Power Systems,
https://doi.org/10.1007/978-3-031-89459-6_14

Blockchain technology is rapidly transforming the modernization of power systems by offering unique solutions to enhance decentralization, security, and efficiency. This paper explores the integration of blockchain technology in smart power systems and its impact on various aspects of the power industry. The paper begins with an introduction to blockchain technology, highlighting its fundamental concepts and features. It then delves into the significance of decentralization and security in smart power systems and how blockchain technology addresses these challenges.

The role of blockchain in power system transactions is extensively discussed, emphasizing its ability to enable secure, transparent, and automated energy trading. The use of smart contracts in energy trading is explored, showcasing how these self-executing agreements streamline transactions, eliminate intermediaries, and improve transparency. Grid data management and privacy preservation are also examined, focusing on the secure handling of sensitive data and the protection of consumer privacy rights.

In the subsequent section, the paper presents case studies and real-world examples that demonstrate the practical applications of blockchain in smart power systems. It highlights blockchain-based peer-to-peer energy trading platforms, showcasing their effectiveness in promoting decentralized energy markets and enabling direct interactions between producers and consumers. The integration of renewable energy sources with blockchain is also explored, illustrating how blockchain technology facilitates the traceability and verification of renewable energy generation.

Furthermore, the paper explores the use of blockchain in decentralized energy management systems and smart grids for power grid security. It showcases how blockchain technology enhances grid resilience, optimizes energy management, and mitigates cybersecurity risks in power systems.

The challenges and future perspectives of blockchain in smart power systems are discussed in the subsequent section. Scalability and integration challenges are addressed, along with regulatory and policy considerations that need to be addressed for the widespread adoption of blockchain technology in the power industry. The importance of collaborative efforts and standardization is emphasized to facilitate interoperability and ensure seamless integration of blockchain solutions. Finally, the paper concludes with an overview of future trends in smart power systems, highlighting the potential impact of blockchain technology on the power industry.

This paper provides valuable insights into the transformative potential of blockchain technology in modernizing power systems. It sheds light on the benefits, challenges, and real-world applications of blockchain in smart power systems and paves the way for a more efficient, secure, and decentralized energy future. The overall architecture of the system is depicted in Figure 1.

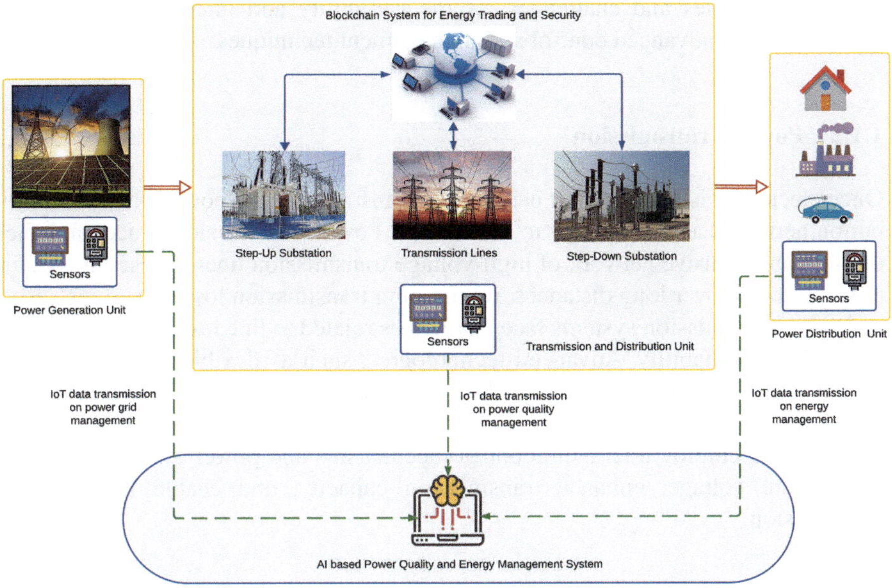

Fig. 1 AI based power quality and energy management system

1.1 Smart Power Systems and the Need for Modernization

Power systems play a vital role in generating, transmitting, and distributing electrical energy to meet the growing demand for electricity. These systems encompass a complex network of power generation sources, transmission lines, distribution networks, and consumer connections. Understanding the fundamental components and functions of power systems is crucial to grasp the challenges and advancements associated with modernizing them.

1.1.1 Power Generation

Power generation involves the conversion of various energy sources into electrical energy. Traditionally, fossil fuels such as coal, natural gas, and oil have been the primary sources of power generation. These fuels are burned to produce steam, which drives turbines connected to generators, thus converting mechanical energy into electrical energy. However, the detrimental environmental impact and limited availability of fossil fuels have led to the exploration of cleaner and renewable energy sources.

Renewable energy sources, such as solar, wind, hydro, and geothermal, have gained significant attention in recent years. These sources harness natural resources to generate electricity, reducing greenhouse gas emissions and dependence on finite fuel resources. The integration of renewable energy into power systems presents both opportunities and challenges, as the variability and intermittency of these sources require advanced control and management techniques.

1.1.2 Power Transmission

Once electricity is generated, it needs to be transmitted from power plants to distribution networks and ultimately to consumers. Power transmission is accomplished through an extensive network of high-voltage transmission lines. These lines transport electricity over long distances, minimizing transmission losses.

Power transmission systems face challenges related to line losses, voltage stability, and grid reliability. Advanced technologies, such as flexible AC transmission systems (FACTS) and high-voltage direct current (HVDC) transmission, have been developed to improve the efficiency and reliability of power transmission. These technologies employ intelligent control mechanisms and power electronic devices to regulate voltage, enhance transmission capacity, and enable long-distance transmission.

1.1.3 Power Distribution

Power distribution involves the final stage of delivering electricity to end consumers. Distribution systems comprise low-voltage networks, substations, transformers, and distribution lines. These systems supply electricity to residential, commercial, and industrial users, meeting their diverse energy demands.

Distribution networks face challenges related to load balancing, voltage regulation, and power quality. Load forecasting and management techniques help utilities anticipate demand patterns and optimize power distribution. Distribution automation systems, including smart meters and remote monitoring devices, enable real-time monitoring, fault detection, and efficient load management.

1.2 Power System Operation and Control

The reliable and efficient operation of power systems requires sophisticated control and monitoring mechanisms. Power system operators utilize supervisory control and data acquisition (SCADA) systems to monitor generation, transmission, and distribution processes. These systems collect real-time data, providing operators with insights into system performance and facilitating decision-making.

With the advancement of AI and IoT technologies, power system control has become more intelligent and automated. AI algorithms enable predictive maintenance, fault detection, and optimal control strategies. IoT devices and sensors provide valuable data for monitoring equipment health, grid conditions, and energy consumption patterns.

Understanding the intricacies of power systems is crucial for comprehending the challenges and opportunities associated with their modernization. Power generation, transmission, distribution, and system operation form the backbone of the power industry. As the world moves towards a cleaner and more sustainable energy future, the integration of renewable energy sources, advanced control systems, and intelligent technologies like AI and IoT will play a vital role in shaping the modernization of power systems [1].

1.3 Challenges in Conventional Power Systems

Conventional power systems, characterized by centralized generation, hierarchical control structures, and limited use of advanced technologies, face several challenges in meeting the demands of the modern energy landscape. These challenges highlight the need for the modernization and transformation of power systems into smarter and more sustainable grids. Here are some key challenges faced by conventional power systems:

1.3.1 Limited Energy Efficiency

Conventional power systems often suffer from inefficiencies in energy generation, transmission, and distribution. Power plants based on fossil fuels have lower energy conversion efficiencies, leading to significant energy losses during the generation process. Transmission and distribution networks also incur losses due to resistance and other factors. These inefficiencies contribute to higher energy consumption and increased environmental impact.

1.3.2 Lack of Flexibility and Resilience

Conventional power systems have limited flexibility to accommodate fluctuations in power generation and demand. Traditional power plants are designed for stable and predictable generation, making it challenging to integrate variable renewable energy sources. As a result, the grid may experience issues such as voltage instability, frequency deviations, and blackouts. Moreover, centralized control structures make the grid vulnerable to single points of failure, reducing its resilience to disruptions and natural disasters.

1.3.3 Aging Infrastructure

Many conventional power systems have aging infrastructure that requires significant maintenance and upgrades. Power plants, transmission lines, and distribution networks may suffer from degradation, reducing their reliability and performance.

Aging infrastructure also increases the risk of equipment failures and outages, leading to service interruptions and increased maintenance costs.

1.3.4 Limited Grid Visibility and Control

Conventional power systems often lack real-time visibility and control over grid operations. Monitoring and control systems may be limited, making it challenging to detect and respond quickly to faults, load imbalances, or grid disturbances. This limitation can result in suboptimal operation, delayed fault identification, and prolonged downtime.

1.3.5 Environmental Impact

Traditional power systems heavily rely on fossil fuels, resulting in significant greenhouse gas emissions and environmental pollution. The combustion of coal, oil, and natural gas releases carbon dioxide (CO_2), contributing to climate change. Additionally, conventional power plants may emit pollutants such as sulfur dioxide (SO_2), nitrogen oxides (NO_x), and particulate matter, leading to air pollution and negative health impacts.

1.3.6 Centralized Control and Market Structures

Conventional power systems often operate under centralized control and market structures. This centralization limits market competition and hinders the integration of distributed energy resources (DERs) and demand response programs. It can also lead to suboptimal utilization of renewable energy sources and inhibits innovation and the participation of new market players.

Addressing These Challenges Through Modernization
The modernization of power systems aims to address these challenges and create a more sustainable, efficient, and resilient energy infrastructure. Key strategies for modernization include:

- **Integration of Renewable Energy:** Incorporating renewable energy sources into the power mix reduces dependence on fossil fuels and lowers greenhouse gas emissions. This integration requires the development of advanced control and management systems to handle the variability and intermittency of renewables.
- **Advanced Grid Control and Automation:** Implementing advanced control and automation technologies, such as AI, IoT, and data analytics, enhances grid visibility, fault detection, and real-time control. These technologies enable better grid monitoring, load balancing, and optimization of power flow.

- **Decentralization and Distributed Energy Resources (DERs):** Promoting the deployment of DERs, such as solar panels, wind turbines, and energy storage systems, allows for distributed power generation and enables local resilience. Decentralized control structures and market mechanisms facilitate the integration of DERs and enable prosumers (consumers and producers) to participate actively in the energy market.
- **Smart Grid Infrastructure:** Upgrading the grid infrastructure with smart grid technologies, including advanced metering infrastructure (AMI), demand response systems, and grid sensors, improves system reliability, enhances load management, and enables efficient use of resources.
- **Electrification and Demand-Side Management:** Encouraging energy-efficient technologies, electrification of transportation, and demand-side management programs help optimize energy consumption and reduce peak demand. This approach can contribute to load balancing, grid stability, and overall energy conservation.
- **Policy and Regulatory Framework:** Developing supportive policies and regulations that incentivize renewable energy integration, grid modernization, and market competition is crucial. Governments and regulatory bodies play a significant role in encouraging innovation, ensuring fair market access, and setting standards for grid modernization.

Conventional power systems face several challenges related to energy efficiency, flexibility, aging infrastructure, limited control, environmental impact, and centralized market structures. The modernization of power systems is essential to address these challenges and create a more sustainable and resilient energy infrastructure. By embracing advanced technologies, integrating renewable energy, decentralizing control, and implementing smart grid solutions, we can pave the way for a more efficient, reliable, and environmentally friendly power system for the future [2].

1.4 Introduction to Smart Power Systems

In the twenty-first century, the modernization of power systems is becoming increasingly important to meet the growing demand for electricity, improve energy efficiency, and integrate renewable energy sources. Smart power systems, also known as intelligent or advanced power systems, are at the forefront of this transformation. These systems leverage advanced technologies, such as Artificial Intelligence (AI), the Internet of Things (IoT), and data analytics, to enable more efficient, flexible, and sustainable power generation, distribution, and consumption.

Smart power systems go beyond the traditional approach of centralized power generation and hierarchical control structures. They embrace a decentralized and intelligent paradigm that empowers consumers, optimizes grid operations, and facilitates the integration of diverse energy resources. By leveraging real-time data, automation, and advanced algorithms, smart power systems enable more informed decision-making, better grid management, and enhanced overall system performance.

1.4.1 Key Components of Smart Power Systems

Advanced Monitoring and Sensing: Smart power systems utilize a network of sensors, smart meters, and monitoring devices to collect real-time data on power generation, consumption, and grid conditions. These sensors provide granular information, enabling operators and consumers to have a comprehensive view of the power system's performance.

Data Analytics and Predictive Algorithms: Advanced data analytics techniques, including machine learning and AI algorithms, are applied to the vast amounts of data collected from smart meters and sensors. These algorithms provide insights into consumption patterns, grid stability, and fault detection, facilitating proactive decision-making and maintenance planning.

Distributed Energy Resources (DERs): Smart power systems actively integrate and manage distributed energy resources, such as solar panels, wind turbines, energy storage systems, and electric vehicle charging stations. These DERs enable localized generation, demand response, and load management, contributing to grid stability and reducing dependence on traditional power plants.

Advanced Control and Automation: Smart power systems employ advanced control and automation technologies to enable efficient power flow, load balancing, and optimal resource utilization. Real-time monitoring and control systems facilitate grid stability, voltage regulation, and quick response to changing conditions, ensuring reliable and secure power supply.

Demand-Side Management and Consumer Empowerment: Smart power systems empower consumers through demand-side management programs and technologies. Consumers can actively monitor and manage their energy consumption, participate in demand response initiatives, and make informed choices about energy usage based on real-time pricing and consumption data.

1.4.2 Benefits and Advantages of Smart Power Systems

Smart power systems, driven by advanced technologies and intelligent approaches, offer numerous benefits and advantages compared to conventional power systems. These benefits range from improved energy efficiency and grid reliability to enhanced integration of renewable energy sources and empowered consumers. Here are some key advantages of smart power systems:

Enhanced Energy Efficiency: Smart power systems leverage real-time data analytics, AI algorithms, and advanced control mechanisms to optimize energy generation, transmission, and consumption. This leads to improved energy efficiency, reducing energy waste and minimizing the environmental impact of power systems.

By optimizing power flow, load management, and demand response, smart power systems enable more efficient utilization of resources and lower energy costs.

Integration of Renewable Energy Sources: Smart power systems facilitate the seamless integration of renewable energy sources, such as solar, wind, and hydro, into the grid. These systems address the challenges associated with the variability and intermittency of renewable energy by leveraging advanced forecasting, grid management, and energy storage technologies. The integration of renewables not only reduces greenhouse gas emissions but also diversifies the energy mix, increasing energy security and reducing dependence on fossil fuels.

Grid Resilience and Reliability: Smart power systems improve grid resilience and reliability by employing advanced monitoring, fault detection, and predictive maintenance techniques. Real-time data analytics and intelligent algorithms enable early detection of grid anomalies, proactive maintenance, and quick response to system disturbances. This enhances the overall reliability of the power system, reduces downtime, and minimizes the impact of power outages on consumers.

Decentralized Control and Grid Flexibility: Unlike traditional power systems with centralized control structures, smart power systems embrace decentralized control mechanisms. Distributed energy resources (DERs) such as solar panels, wind turbines, and energy storage systems are integrated into the grid, enabling localized generation, demand response, and grid flexibility. This decentralization enhances grid stability, reduces transmission losses, and provides greater control over energy resources.

Empowered Consumers: Smart power systems empower consumers by providing them with real-time information about their energy consumption and enabling them to actively participate in demand-side management programs. Through smart meters, energy monitoring systems, and mobile applications, consumers can monitor and manage their energy usage, optimize energy efficiency, and make informed decisions based on dynamic pricing and consumption patterns. This leads to cost savings, increased energy awareness, and greater control over energy consumption.

Advanced Grid Monitoring and Management: Smart power systems leverage advanced monitoring and sensor technologies to provide real-time data on grid conditions, load profiles, and equipment health. This enhanced grid visibility enables operators to identify and respond to issues promptly, optimize power flow, and ensure grid stability. Intelligent algorithms analyze the collected data, providing valuable insights for grid planning, load forecasting, and system optimization.

Sustainability and Environmental Benefits: The integration of renewable energy sources, improved energy efficiency, and reduced reliance on fossil fuels contribute to sustainability and environmental preservation. Smart power systems enable a significant reduction in greenhouse gas emissions, air pollution, and dependence on finite fuel resources. They promote a cleaner and more sustainable energy future.

Smart power systems offer numerous benefits and advantages over conventional power systems. From improved energy efficiency and integration of renewable energy sources to enhanced grid reliability and empowered consumers, these systems pave the way for a more sustainable, flexible, and intelligent energy infrastructure. By leveraging advanced technologies, data analytics, and intelligent control mechanisms, smart power systems enable optimized resource utilization, reduced environmental impact, and a resilient energy future [3].

2 Integration of AI and IoT in Power Distribution

2.1 Intelligent Power Grid Management

Intelligent power grid management is a key component of smart power systems that leverages the integration of Artificial Intelligence (AI) and the Internet of Things (IoT) technologies. It involves the application of advanced algorithms and real-time data analytics to optimize power distribution, enhance grid efficiency, and ensure reliable operation [4]. Here are some aspects of intelligent power grid management:

(a) Grid Optimization: AI algorithms analyze data from sensors and smart meters to optimize power flow, reduce transmission losses, and balance load across the grid. By considering factors such as generation capacity, network constraints, and consumer demand, intelligent grid management algorithms make decisions to minimize energy losses and ensure efficient utilization of resources.
(b) Voltage and Frequency Control: AI-based control systems monitor voltage and frequency levels across the grid and make adjustments to maintain them within acceptable limits. This ensures stable power delivery, minimizes the risk of equipment damage, and improves the overall quality of power supply.
(c) Grid Stability and Resilience: Intelligent power grid management systems continuously monitor grid conditions, detecting anomalies or disturbances that may affect stability. AI algorithms analyze real-time data to identify potential grid disruptions and take preventive or corrective actions to ensure grid resilience. This includes automatically reconfiguring the grid, rerouting power, or initiating load shedding measures to maintain stability during contingencies.

2.2 Real-Time Monitoring and Control

Real-time monitoring and control systems play a vital role in smart power systems by leveraging IoT devices and AI algorithms. These systems enable operators to have a comprehensive view of the grid's status, monitor equipment health, and respond promptly to changing conditions [5]. Here are key aspects of real-time monitoring and control:

(a) Sensor Deployment: IoT sensors are strategically placed throughout the power distribution network to collect real-time data on voltage levels, power flow, equipment performance, and environmental conditions. These sensors provide a continuous stream of data that enables operators to monitor grid operations in real-time.
(b) Data Analytics: AI algorithms analyze the vast amounts of data collected from sensors to identify patterns, detect anomalies, and generate actionable insights. Real-time data analytics enable operators to make informed decisions, detect potential issues, and optimize grid performance.
(c) Control Systems: Real-time data, combined with AI algorithms, facilitate automated control systems that can respond to changing conditions dynamically. These systems allow for quick adjustments in power flow, load balancing, and fault detection, enhancing grid stability and reliability.

2.3 Demand Response and Load Management

Demand response and load management strategies are crucial for maintaining grid stability, managing peak loads, and optimizing energy consumption. AI and IoT technologies play a significant role in enabling effective demand response and load management solutions [6]. Here are key aspects of demand response and load management in smart power systems:

(a) Demand Monitoring and Forecasting: IoT devices and smart meters provide real-time data on energy consumption patterns. AI algorithms analyze this data to forecast demand, enabling utilities to anticipate peak demand periods and optimize load distribution accordingly.
(b) Load Balancing: AI algorithms analyze real-time demand data and grid conditions to balance loads across the distribution network. By shifting loads and redistributing power during peak demand periods, load balancing strategies help prevent grid overload and ensure reliable power supply.
(c) Demand Response Programs: IoT devices and smart meters enable demand response programs, allowing utilities to incentivize consumers to adjust their energy consumption during times of high demand. AI algorithms help automate the coordination of demand response events, ensuring efficient and reliable execution.

2.4 Predictive Maintenance and Fault Detection

Predictive maintenance and fault detection are critical aspects of power distribution systems to prevent equipment failures, minimize downtime, and optimize maintenance schedules. AI and IoT technologies enable advanced predictive maintenance and fault detection capabilities in smart power systems [7]. Here are key aspects:

(a) Condition Monitoring: IoT sensors and monitoring devices collect data on the health and performance of equipment such as transformers, switchgear, and cables. Real-time monitoring allows for the continuous assessment of equipment condition, helping identify potential issues or faults.
(b) Fault Detection: AI algorithms analyze data from sensors to detect anomalies, patterns, and potential faults in the distribution network. By detecting early warning signs of equipment degradation or failures, fault detection systems enable proactive maintenance actions, reducing the risk of system disruptions.
(c) Predictive Maintenance: AI-based predictive maintenance models leverage historical and real-time data to predict the remaining useful life of critical equipment. By identifying maintenance needs in advance, utilities can schedule maintenance activities more efficiently, reducing unplanned downtime and optimizing resource allocation.

The integration of AI and IoT technologies in power distribution brings significant advancements to smart power systems. Intelligent power grid management, real-time monitoring and control, demand response, load management, predictive maintenance, and fault detection all contribute to enhancing the efficiency, reliability, and sustainability of power distribution networks. These advancements enable utilities to optimize energy flow, balance loads, respond to demand fluctuations, and ensure reliable operation, ultimately resulting in improved grid performance and customer satisfaction.

3 Power Quality Control Using Smart Intelligence

Power quality refers to the characteristics of electrical power, including voltage, frequency, and waveform, and their deviation from the ideal sinusoidal waveform. Maintaining high power quality is crucial for the reliable operation of electrical equipment and the overall stability of the power system. In smart power systems, the integration of Artificial Intelligence (AI) and the Internet of Things (IoT) technologies enables advanced power quality control strategies. Here is a detailed explanation of power quality control using smart intelligence.

3.1 Importance of Power Quality

Power quality issues can have significant impacts on electrical equipment, affecting their performance, efficiency, and lifespan. Poor power quality can result in voltage sags, surges, harmonics, voltage fluctuations, and other disturbances. These issues can cause equipment malfunction, data loss, increased energy consumption, and even equipment damage. Power quality control is essential to mitigate these effects and ensure reliable and efficient operation of the power system [8].

3.2 AI and IoT for Power Quality Monitoring

AI and IoT technologies provide advanced tools for real-time power quality monitoring and analysis. These technologies enable the collection of data from smart sensors, smart meters, and monitoring devices throughout the power distribution network [9]. Here are key aspects of AI and IoT for power quality monitoring:

(a) Sensor Deployment: Smart sensors are deployed at critical points in the power system to measure voltage, current, and other relevant parameters. These sensors provide real-time data on power quality indicators such as voltage fluctuations, harmonic content, and power factor.
(b) Data Acquisition and Analytics: IoT devices collect and transmit the data from sensors to a centralized system. AI algorithms analyze this data to identify power quality issues, detect anomalies, and generate actionable insights. Real-time analytics enable prompt identification of power quality deviations, allowing operators to take timely corrective actions.
(c) Remote Monitoring: IoT-enabled devices facilitate remote monitoring of power quality parameters. Operators can access the data and analytics from a centralized platform, enabling real-time monitoring and analysis from any location. Remote monitoring capabilities enhance operational efficiency and enable quick response to power quality events.

3.3 Smart Sensors and Data Analytics

Smart sensors play a crucial role in power quality control by providing accurate and detailed measurements of electrical parameters. These sensors, combined with AI-based data analytics, enable advanced power quality monitoring and analysis [10]. Here are key aspects of smart sensors and data analytics:

(a) High-Resolution Data: Smart sensors capture high-resolution data, allowing for detailed analysis of power quality parameters. This includes voltage variations, harmonics, flicker, and other power quality indicators. High-resolution data enables a comprehensive understanding of power quality deviations and their impacts on electrical systems.
(b) Pattern Recognition: AI algorithms analyze the collected data to identify patterns and trends associated with power quality issues. Machine learning algorithms can recognize known patterns of power quality disturbances and classify them for further analysis and action.
(c) Anomaly Detection: AI-based anomaly detection techniques enable the identification of abnormal power quality events. By comparing the real-time data with historical patterns, the system can detect deviations and generate alerts for potential power quality issues. Early detection helps operators take corrective actions to prevent equipment damage or system disruptions.

3.4 Intelligent Control Strategies for Power Quality Improvement

Intelligent control strategies leverage AI and IoT technologies to improve power quality and mitigate power quality issues. These strategies involve real-time control and coordination of power system components to address power quality deviations [11]. Here are key aspects of intelligent control strategies for power quality improvement:

(a) Voltage Regulation: AI algorithms can adjust voltage levels in real-time to maintain them within desired limits. By monitoring power quality parameters and analyzing the data, the system can dynamically control voltage regulators, tap changers, and reactive power compensation devices to regulate voltage and minimize deviations.
(b) Harmonic Filtering: Power electronic devices, such as active filters, can be controlled intelligently to mitigate harmonics and improve power quality. AI algorithms analyze the harmonic content in real-time and determine the optimal control settings for the filters to attenuate harmonics effectively.
(c) Intelligent Switching Operations: AI algorithms optimize switching operations of network components, such as circuit breakers and switches, to minimize voltage sags, interruptions, and transient disturbances. By analyzing real-time data and predicting power quality issues, the system can intelligently perform switching operations to maintain power quality within acceptable limits.

Power quality control using smart intelligence is crucial for maintaining reliable and efficient power systems. The integration of AI and IoT technologies enables real-time monitoring, advanced analytics, and intelligent control strategies for power quality improvement. By leveraging smart sensors, data analytics, and intelligent control algorithms, power system operators can detect power quality deviations, take timely corrective actions, and enhance the overall performance of the power system. Smart power quality control contributes to the reliable operation of electrical equipment, reduces downtime, and ensures a stable and efficient power supply.

4 Smart Energy Management with AI and IoT

Smart energy management is a critical aspect of smart power systems, enabled by the integration of Artificial Intelligence (AI) and the Internet of Things (IoT) technologies. It encompasses various techniques and strategies aimed at optimizing energy generation, consumption, and distribution. By leveraging real-time data, advanced analytics, and intelligent control, smart energy management enables efficient utilization of energy resources, demand response, and seamless integration of renewable energy sources. Here is a detailed explanation of smart energy management with AI and IoT:

4.1 Energy Management Systems

Energy management systems (EMS) form the foundation of smart energy management. These systems integrate AI and IoT technologies to monitor, control, and optimize energy generation, distribution, and consumption [12]. Key components of energy management systems include:

(a) Data Acquisition: IoT devices, such as smart meters and sensors, collect real-time data on energy consumption, generation, and grid conditions. This data is transmitted to the energy management system for analysis and decision-making.
(b) Data Analytics: AI algorithms analyze the collected data to derive insights and patterns related to energy consumption, generation, and demand. These insights inform decision-making processes and optimization strategies.
(c) Energy Forecasting: AI-based forecasting models utilize historical and real-time data to predict future energy demand, consumption patterns, and renewable energy generation. Energy forecasting helps utilities and consumers plan and optimize energy usage efficiently.
(d) Control and Optimization: Based on the data analysis and energy forecasting, AI algorithms optimize energy generation, distribution, and consumption to minimize costs, enhance efficiency, and ensure grid stability.

4.2 Optimization Techniques in Energy Management

Optimization techniques play a crucial role in smart energy management, allowing for efficient allocation and utilization of energy resources [13]. AI and IoT technologies enable advanced optimization algorithms to achieve the following objectives:

(a) Demand Response Optimization: AI algorithms analyze real-time energy demand data and grid conditions to optimize demand response programs. By coordinating energy consumption patterns and load shifting, demand response optimization minimizes peak demand, balances energy supply and demand, and reduces overall energy costs.
(b) Renewable Energy Integration: Optimizing the integration of renewable energy sources into the grid is essential for balancing variable generation with demand. AI algorithms consider weather patterns, energy storage availability, and demand forecasts to optimize the utilization of renewable energy sources, minimize curtailment, and ensure grid stability.

(c) Generation and Dispatch Optimization: AI algorithms optimize the generation and dispatch of energy from various sources, including conventional power plants, renewable sources, and energy storage systems. These algorithms consider factors such as fuel costs, environmental constraints, transmission capacity, and demand profiles to minimize generation costs, reduce emissions, and maintain grid stability.
(d) Load Balancing and Energy Efficiency: AI algorithms analyze real-time energy consumption data and optimize load balancing strategies to minimize peak demand and balance energy distribution across the grid. Load balancing techniques enhance energy efficiency, reduce energy waste, and improve the overall performance of the power system.

4.3 Intelligent Energy Consumption Monitoring

Intelligent energy consumption monitoring allows consumers to track and optimize their energy usage in real-time. AI and IoT technologies provide consumers with the tools to monitor, analyze, and manage their energy consumption effectively [14]. Here are key aspects of intelligent energy consumption monitoring:

(a) Smart Meters and Sensors: IoT-enabled smart meters and sensors collect data on energy consumption at a granular level. This data is transmitted to energy management systems, enabling real-time monitoring and analysis of energy usage patterns.
(b) Energy Visualization and Feedback: AI algorithms analyze the collected energy consumption data and provide consumers with real-time feedback on their energy usage. This feedback can be visualized through user-friendly interfaces, mobile applications, or energy dashboards, allowing consumers to make informed decisions about their energy consumption.
(c) Energy Efficiency Recommendations: AI algorithms can provide personalized recommendations to consumers on how to optimize their energy usage. These recommendations may include suggestions for energy-saving practices, peak-demand reduction strategies, or equipment upgrades to improve energy efficiency.
(d) Behavioral Analysis: AI algorithms analyze historical and real-time energy consumption data to identify patterns and trends in consumer behavior. This analysis helps utilities and consumers understand energy consumption patterns, identify areas for improvement, and develop targeted energy efficiency programs.

4.4 Demand-Side Management and Renewable Integration

Demand-side management and renewable energy integration are key aspects of smart energy management, enabled by AI and IoT technologies. These strategies aim to optimize energy consumption, balance energy supply and demand, and integrate renewable energy sources seamlessly [15]. Here are key aspects of demand-side management and renewable integration:

(a) Demand Response Programs: AI and IoT technologies facilitate the implementation of demand response programs, where consumers adjust their energy consumption based on real-time pricing or grid conditions. AI algorithms analyze demand response data and optimize the coordination and execution of demand response events to ensure grid stability and minimize energy costs.
(b) Distributed Energy Resources (DERs) Integration: DERs, such as solar panels, wind turbines, and energy storage systems, are integrated into the grid through smart energy management systems. AI algorithms optimize the utilization of DERs based on energy generation, demand, and grid conditions, ensuring efficient and reliable integration of these decentralized energy resources.
(c) Microgrid Management: Microgrids, which are localized energy systems, can be optimized using AI and IoT technologies. Intelligent algorithms monitor and control microgrid components, such as generation sources, energy storage, and loads, to optimize energy generation, consumption, and storage within the microgrid. This ensures efficient energy management and enhances the resilience of the microgrid.
(d) Energy Market Integration: AI algorithms can enable efficient participation of consumers and DERs in energy markets. By analyzing energy pricing, demand, and generation data, these algorithms optimize energy trading and participation in energy markets, promoting efficient utilization of resources and enabling cost savings for consumers.

Smart energy management with AI and IoT technologies is essential for optimizing energy generation, consumption, and distribution. By leveraging real-time data, advanced analytics, and intelligent control strategies, smart energy management enables efficient resource allocation, demand response optimization, seamless integration of renewable energy, and effective energy consumption monitoring. These advancements lead to improved energy efficiency, reduced costs, enhanced grid stability, and increased utilization of renewable energy sources, ultimately contributing to a more sustainable and resilient energy future.

5 Blockchain Technology in Smart Power Systems

5.1 Introduction to Blockchain Technology

Blockchain technology, originally developed as the underlying technology for cryptocurrencies like Bitcoin, has emerged as a promising tool for various industries. In the context of smart power systems, blockchain technology offers unique advantages and possibilities for enhancing the efficiency, transparency, and security of energy transactions and management. By leveraging the decentralized and immutable nature of blockchain, smart power systems can revolutionize energy trading, grid management, and peer-to-peer energy transactions. This introduction provides an overview of blockchain technology in smart power systems [16].

5.1.1 Blockchain Basics

Blockchain is a distributed ledger technology that enables secure, transparent, and tamper-proof recording of transactions across a network of computers. Unlike traditional centralized databases, blockchain operates on a decentralized network where every participant has a copy of the ledger. Transactions are grouped into blocks, cryptographically linked to previous blocks, and added to the chain, forming an immutable record.

5.1.2 Key Features of Blockchain in Smart Power Systems

Decentralization: Blockchain eliminates the need for a central authority or intermediary in energy transactions. The decentralized nature of blockchain enables peer-to-peer energy trading, eliminating the reliance on traditional utility companies as intermediaries. This decentralized approach empowers energy producers and consumers to transact directly, leading to increased efficiency and cost savings.

Transparency and Trust: Blockchain provides transparency by allowing all participants in the network to have a copy of the ledger. Every transaction is recorded and stored in a tamper-proof manner, ensuring the integrity and immutability of the data. This transparency builds trust among participants and eliminates the need for third-party verification.

Smart Contracts: Smart contracts are self-executing contracts stored on the blockchain that automatically execute predefined conditions once they are met. In the context of smart power systems, smart contracts can enable automated and secure energy transactions, eliminating the need for manual verification and reducing transaction costs.

Energy Traceability: Blockchain technology allows for transparent tracking of the origin and movement of energy. By recording energy generation and consumption data on the blockchain, it becomes possible to track the source of renewable energy, verify its authenticity, and ensure compliance with regulatory requirements.

Grid Management and Flexibility: Blockchain can facilitate more efficient grid management by enabling real-time monitoring and control of distributed energy resources (DERs). By using blockchain to manage and coordinate DERs, grid operators can balance energy supply and demand, optimize load management, and ensure grid stability.

5.1.3 Benefits and Potential Applications

The integration of blockchain technology in smart power systems brings several benefits and opens up a range of potential applications:

Peer-to-Peer Energy Trading: Blockchain enables direct peer-to-peer energy trading, allowing consumers with solar panels or other DERs to sell excess energy to nearby consumers. This fosters a decentralized energy marketplace, promotes renewable energy integration, and enables more efficient utilization of energy resources.

Microgrid Management: Blockchain technology can facilitate the management and coordination of microgrids. Smart contracts and decentralized control mechanisms can enable secure and efficient energy sharing within microgrid communities, improving energy efficiency and enhancing grid resilience.

Renewable Energy Certificates: Blockchain can be used to track and trade renewable energy certificates (RECs), providing transparent and verifiable proof of renewable energy generation. This enables consumers and businesses to track and support renewable energy sources accurately.

Energy Data Management: Blockchain can enhance the security and privacy of energy data by enabling secure storage and sharing of data. It allows consumers to have control over their energy consumption data, while still enabling authorized parties to access the necessary information for grid management and energy planning.

Grid Billing and Settlement: Blockchain-based systems can automate and streamline energy billing and settlement processes. Smart contracts can enable automatic invoicing and payment settlements based on predefined conditions, reducing administrative costs and ensuring accurate and transparent transactions.

Blockchain technology holds immense potential for transforming smart power systems by enabling decentralized and secure energy transactions, enhancing grid management, and promoting the integration of renewable energy sources. With its features of decentralization, transparency, and smart contracts, blockchain can bring efficiency, trust, and flexibility to energy systems. As the technology continues to evolve, blockchain is likely to play a significant role in shaping the future of smart power systems and advancing the transition towards a sustainable and decentralized energy landscape.

5.2 *Decentralization and Security*

Decentralization and security are two crucial aspects of smart power systems that play a significant role in enhancing the efficiency, reliability, and resilience of the power infrastructure. With the integration of advanced technologies such as the Internet of Things (IoT), Artificial Intelligence (AI), and blockchain, smart power systems can achieve greater decentralization and improved security measures [17].

This section explores the importance of decentralization and security in smart power systems and how these concepts are implemented.

5.2.1 Decentralization in Smart Power Systems

Decentralization refers to the distribution of decision-making and control functions across the power system, reducing reliance on centralized entities and enabling a more distributed and flexible approach. Here are the key aspects of decentralization in smart power systems:

Distributed Energy Resources (DERs): Smart power systems embrace the integration of DERs, such as solar panels, wind turbines, and energy storage systems, at various points within the power grid. DERs enable localized energy generation, reducing dependency on centralized power plants and promoting energy self-sufficiency.

Peer-to-Peer Energy Trading: Decentralized energy markets powered by blockchain technology allow consumers with DERs to directly trade excess energy with nearby consumers. This peer-to-peer energy trading eliminates the need for intermediaries, empowers consumers, and promotes a decentralized marketplace.

Microgrids: Microgrids are localized energy systems that can operate autonomously or in conjunction with the main power grid. They are capable of generating, storing, and distributing energy within a specific area, providing grid resilience during disruptions and promoting local energy generation and consumption.

Decentralized Control and Decision-making: In smart power systems, decision-making and control functions are distributed across the network. Advanced algorithms and AI technologies enable autonomous decision-making at the device level, allowing for real-time optimization of energy generation, distribution, and consumption.

5.2.2 Benefits of Decentralization in Smart Power Systems

(a) Grid Resilience: Decentralization enhances the resilience of the power system by reducing the vulnerability to single points of failure. Localized energy generation and distribution capabilities of DERs and microgrids ensure continued power supply during grid disruptions or natural disasters.
(b) Efficient Resource Utilization: Decentralized power generation and local consumption reduce transmission losses, optimize resource utilization, and enable better integration of renewable energy sources.
(c) Consumer Empowerment: Decentralization allows consumers to actively participate in energy markets, make informed decisions about energy usage, and benefit from the economic advantages of peer-to-peer energy trading.

5.2.3 Security in Smart Power Systems

Security is of utmost importance in smart power systems to protect critical infrastructure, ensure data privacy, and prevent cyber threats. As the power grid becomes more interconnected and reliant on digital technologies, robust security measures are necessary. Here are key aspects of security in smart power systems:

Cybersecurity Measures: Strong cybersecurity protocols and measures are implemented to protect the power system against cyber threats, such as unauthorized access, data breaches, and attacks on control systems. This includes firewalls, encryption, intrusion detection systems, and regular security audits.

Data Privacy: Smart power systems handle large amounts of sensitive data, including energy consumption patterns, customer information, and grid operations. Robust data privacy measures are implemented to safeguard this information, including data encryption, access controls, and compliance with privacy regulations.

Blockchain Technology: The integration of blockchain technology provides enhanced security in smart power systems. The decentralized and immutable nature of blockchain ensures data integrity, prevents tampering of records, and eliminates the need for centralized authorities or intermediaries.

Secure Communication: Secure communication protocols, such as cryptographic algorithms and secure data transmission protocols, are implemented to ensure the confidentiality and integrity of data transmitted between devices, sensors, and control systems within the smart power system.

5.2.4 Benefits of Security in Smart Power Systems

(a) Protection against Cyber Threats: Robust security measures protect smart power systems from cyberattacks, preventing unauthorized access, data breaches, and potential disruptions to critical infrastructure.
(b) Data Integrity and Privacy: Security measures ensure the integrity of data and protect the privacy of consumers and grid operators. This instills trust and confidence in the system and encourages the adoption of smart power technologies.
(c) Resilience and Reliability: Enhanced security measures contribute to the overall resilience and reliability of the smart power system, ensuring the continuous operation of critical infrastructure and minimizing the risk of system disruptions.

Decentralization and security are integral components of smart power systems. Decentralization through the integration of DERs, microgrids, and peer-to-peer energy trading promotes grid resilience, efficient resource utilization, and consumer empowerment. Security measures, including robust cybersecurity protocols and blockchain technology, protect the power system from cyber threats, ensure data privacy, and enhance the reliability and trustworthiness of the system. By prioritizing decentralization and security, smart power systems can unlock the full potential of advanced technologies and pave the way for a more sustainable, efficient, and secure energy future.

5.3 Blockchain in Power System Transactions

Blockchain technology has the potential to revolutionize power system transactions by introducing transparency, security, and efficiency. By leveraging its decentralized and immutable nature, blockchain can enhance the reliability and integrity of power system transactions, ranging from energy trading and billing to grid management and asset ownership [18]. Here are key aspects of blockchain in power system transactions:

Energy Trading and Peer-to-Peer Transactions: Blockchain enables direct peer-to-peer energy trading without the need for intermediaries. Smart contracts, self-executing agreements stored on the blockchain, facilitate automated and transparent energy transactions between producers and consumers. Blockchain ensures secure and tamper-proof recording of energy transactions, eliminating the need for centralized market operators.

Decentralized Energy Marketplaces: Blockchain technology enables the creation of decentralized energy marketplaces where consumers and producers can interact directly. These marketplaces utilize blockchain's transparent and auditable ledger to facilitate the buying and selling of energy in a transparent and efficient manner. Producers can sell their excess energy to consumers at negotiated prices, promoting renewable energy integration and fostering a more sustainable energy ecosystem.

Billing and Settlement: Blockchain-based systems streamline billing and settlement processes in power systems. Smart contracts automatically generate invoices based on predefined conditions, such as energy consumption or agreed-upon rates. The immutability of the blockchain ensures transparency and accuracy in billing, reducing disputes and improving the efficiency of payment settlements.

Grid Management and Asset Ownership: Blockchain technology can be utilized for managing and tracking assets within the power system. The blockchain ledger can securely record and track the ownership, operation, and maintenance of power system assets, such as solar panels, wind turbines, and energy storage systems. This enables efficient asset management, provenance verification, and transparent asset trading.

Data Management and Privacy: Blockchain allows for secure and transparent data management in power system transactions. Energy consumption data, generation data, and other relevant information can be stored on the blockchain, ensuring data integrity, traceability, and protection against unauthorized tampering. Blockchain also provides consumers with more control over their energy data, allowing them to share it securely and selectively.

Grid Balancing and Flexibility: Blockchain technology can facilitate grid balancing by incentivizing flexible energy consumption and demand response. Smart contracts can reward consumers for adjusting their energy consumption patterns during peak demand periods or grid imbalances. This enables the optimization of energy supply and demand, reduces the need for costly infrastructure upgrades, and improves grid stability.

Renewable Energy Certificates (RECs): Blockchain enables transparent tracking and trading of renewable energy certificates (RECs). RECs represent proof of renewable energy generation and can be securely recorded on the blockchain. This ensures the integrity of renewable energy claims, provides transparency to consumers, and encourages the uptake of renewable energy sources.

5.3.1 Benefits of Blockchain in Power System Transactions

(a) Transparency and Trust: Blockchain provides a transparent and auditable ledger of power system transactions, enhancing trust among participants. All transactions recorded on the blockchain are verifiable and cannot be altered, ensuring transparency and mitigating the risk of fraud.
(b) Security and Data Integrity: The decentralized and immutable nature of blockchain enhances the security and integrity of power system transactions. The cryptographic algorithms and consensus mechanisms used in blockchain ensure the authenticity and immutability of transaction records, reducing the risk of data tampering and unauthorized access.
(c) Efficiency and Cost Savings: Blockchain streamlines and automates power system transactions, reducing the need for intermediaries and manual processes. This improves transaction efficiency, reduces administrative costs, and eliminates the potential for errors or disputes.
(d) Peer-to-Peer Interaction: By enabling direct peer-to-peer transactions, blockchain empowers energy producers and consumers, fostering a more decentralized and participatory energy ecosystem. It eliminates the need for centralized intermediaries, allowing for more efficient and cost-effective energy trading.
(e) Renewable Energy Integration: Blockchain technology promotes the integration of renewable energy sources by providing a transparent and auditable platform for renewable energy transactions. It incentivizes and facilitates the trading of renewable energy, encouraging the adoption of sustainable energy practices.

Blockchain technology holds immense potential for transforming power system transactions by introducing transparency, security, and efficiency. Through decentralized energy trading, streamlined billing and settlement processes, transparent asset management, and secure data handling, blockchain can revolutionize the way energy is traded, managed, and consumed. By leveraging the benefits of blockchain, power systems can become more resilient, sustainable, and consumer-centric, fostering a future where clean energy is seamlessly integrated into the global energy landscape.

5.4 Smart Contracts and Energy Trading

Smart contracts play a vital role in enabling efficient and transparent energy trading in the context of smart power systems. Smart contracts are self-executing agreements stored on a blockchain that automatically execute predefined terms and conditions once the specified conditions are met. When applied to energy trading, smart contracts provide several advantages, including automation, transparency, and security [19]. Here is an explanation of smart contracts and their role in energy trading:

Automation of Energy Transactions: Smart contracts automate the execution and settlement of energy transactions. They eliminate the need for intermediaries, such as energy market operators or brokers, by enabling direct peer-to-peer energy trading. Once the predefined conditions in the contract are met (e.g., energy price, quantity, time), the smart contract automatically triggers the transaction, ensuring timely and accurate execution.

Transparent and Immutable Transactions: Smart contracts leverage blockchain technology, providing transparency and immutability to energy transactions. All transaction details, including pricing, quantity, and participants, are recorded on the blockchain and can be accessed by all relevant parties. This transparency enhances trust among participants and mitigates the risk of fraud or manipulation.

Disintermediation and Cost Reduction: By eliminating intermediaries, smart contracts reduce transaction costs associated with energy trading. Participants can directly interact and negotiate terms, without the need for brokers or market operators, resulting in cost savings for both energy producers and consumers.

Real-Time Settlement: Smart contracts enable real-time settlement of energy transactions. Once the agreed-upon conditions are met, the smart contract triggers immediate payment, ensuring prompt settlement and reducing the need for extended billing cycles or delayed payments.

Integration of Renewable Energy Sources: Smart contracts facilitate the integration of renewable energy sources into the energy market. Producers of renewable energy, such as solar or wind power, can utilize smart contracts to sell excess energy to consumers, enabling a more decentralized and sustainable energy ecosystem.

Demand Response and Load Balancing: Smart contracts can incentivize demand response and load balancing strategies. Consumers can adjust their energy consumption patterns based on predefined conditions (e.g., energy price, grid demand) specified in the smart contract. By aligning energy usage with grid conditions, smart contracts optimize load balancing, reduce peak demand, and enhance grid stability.

Traceability and Provenance Verification: Smart contracts and blockchain technology enable the traceability and verification of energy sources. Energy producers can record the source and characteristics of their energy generation on the blockchain, allowing consumers to verify the origin and environmental attributes of the purchased energy, such as renewable energy certificates (RECs).

5.4.1 Benefits of Smart Contracts in Energy Trading

(a) Efficiency and Automation: Smart contracts streamline and automate energy trading processes, reducing the need for manual interventions and administrative tasks. This improves transaction speed, accuracy, and overall efficiency.
(b) Transparency and Trust: Smart contracts, recorded on a blockchain, provide transparency and immutability to energy transactions. All participants have access to the same set of data, ensuring trust and accountability among parties involved.
(c) Cost Reduction: By eliminating intermediaries and automating processes, smart contracts reduce transaction costs associated with energy trading. This benefits both energy producers and consumers, leading to more competitive pricing and cost savings.
(d) Flexibility and Customization: Smart contracts are programmable and can be customized to meet specific requirements of energy trading. Participants can define conditions and parameters according to their preferences and negotiate terms directly, allowing for greater flexibility in energy transactions.
(e) Decentralized Peer-to-Peer Trading: Smart contracts enable direct peer-to-peer energy trading, empowering energy producers and consumers to transact without the need for centralized market operators. This fosters a more decentralized and inclusive energy market.

Smart contracts are a powerful tool in facilitating efficient, transparent, and secure energy trading in smart power systems. By automating transactions, ensuring transparency, and eliminating intermediaries, smart contracts streamline energy trading processes, reduce costs, and promote renewable energy integration. As blockchain technology continues to advance, smart contracts are likely to play an increasingly prominent role in transforming the energy sector and fostering a more decentralized and sustainable energy ecosystem.

5.5 Grid Data Management and Privacy Preservation

Grid data management and privacy preservation are critical considerations in smart power systems to ensure the secure handling of sensitive data and protect the privacy of consumers and grid operators. As power systems become more interconnected and reliant on data-driven technologies, robust data management practices and privacy measures are essential [20]. Here is an explanation of grid data management and privacy preservation in smart power systems:

Data Collection and Storage: Smart power systems generate vast amounts of data from various sources, including smart meters, sensors, and grid infrastructure. Effective data management involves collecting, storing, and organizing this data in a secure and reliable manner. Data is typically stored in centralized or distributed databases, ensuring accessibility and integrity while adhering to privacy regulations.

Data Security: Robust security measures are implemented to protect grid data from unauthorized access, data breaches, and cyber threats. This includes encryption techniques, access controls, authentication mechanisms, and intrusion detection systems. Data security measures are crucial to safeguard sensitive information, prevent data tampering, and maintain the trust of consumers and stakeholders.

Data Privacy and Anonymization: Grid data often contains personally identifiable information (PII) and sensitive consumption patterns. Privacy preservation techniques, such as data anonymization and aggregation, are applied to protect the privacy of individuals. By removing or encrypting identifying information, privacy risks are minimized while allowing for meaningful analysis and grid management.

Access Controls and Consent: Access controls are implemented to ensure that only authorized personnel can access specific data based on their roles and responsibilities. Data access should be granted on a need-to-know basis, and strict controls should be in place to prevent unauthorized data access. Additionally, obtaining explicit consent from consumers for data collection and usage is essential to respect their privacy rights.

Data Sharing and Collaboration: Data sharing and collaboration among different entities in the power system ecosystem can drive innovation and improve grid operations. However, proper data sharing protocols and agreements need to be established to protect the privacy of individuals and ensure compliance with privacy regulations. Techniques such as secure data sharing platforms, data anonymization, and secure data exchange protocols can be employed to facilitate safe and privacy-preserving data sharing.

Privacy by Design: Privacy considerations should be integrated into the design and implementation of smart power systems. Privacy by design principles ensure that privacy measures are embedded throughout the system architecture, data handling processes, and algorithms. This approach promotes privacy as an inherent feature of the system, rather than an afterthought.

Regulatory Compliance: Compliance with privacy regulations, such as the General Data Protection Regulation (GDPR) and local data protection laws, is crucial for grid data management. Power system operators must adhere to legal requirements regarding data collection, storage, usage, and consent to protect consumer privacy rights.

5.5.1 Benefits of Grid Data Management and Privacy Preservation

(a) Consumer Trust and Confidence: Robust data management practices and privacy preservation measures build trust among consumers by assuring them that their data is handled securely and their privacy is protected. This fosters greater acceptance and adoption of smart power technologies.
(b) Enhanced Data Security: Effective data management practices, including encryption, access controls, and authentication mechanisms, protect grid data

from unauthorized access, data breaches, and cyber threats. This ensures the integrity and confidentiality of sensitive information.

(c) Compliance with Privacy Regulations: Adhering to privacy regulations and implementing privacy-preserving practices helps power system operators comply with legal requirements, avoiding penalties and reputational damage.

(d) Responsible Data Usage: Proper data management and privacy preservation promote responsible data usage, ensuring that data is collected and utilized for legitimate purposes and with the explicit consent of individuals.

(e) Facilitation of Data Sharing and Collaboration: By implementing secure data sharing protocols and privacy-preserving techniques, grid data can be shared among relevant stakeholders, fostering collaboration and innovation without compromising privacy.

Grid data management and privacy preservation are essential considerations in smart power systems to protect sensitive information and ensure the privacy rights of individuals. By implementing robust security measures, privacy-preserving techniques, and adhering to privacy regulations, power system operators can enhance data security, build consumer trust, and promote responsible data usage. These practices enable the efficient management and utilization of grid data while safeguarding privacy, facilitating data sharing, and driving the advancement of smart power systems.

6 Case Studies and Real-World Examples

6.1 Blockchain-Based Peer-to-Peer Energy Trading Platforms

Blockchain-based peer-to-peer energy trading platforms have emerged as a promising solution to enable direct interactions between energy producers and consumers, facilitating decentralized energy markets. These platforms leverage blockchain technology to provide transparent, secure, and efficient energy trading experiences. This section presents case studies and real-world examples that demonstrate the practical implementation and benefits of blockchain-based peer-to-peer energy trading platforms [21].

Power Ledger: Power Ledger is a prominent example of a blockchain-based peer-to-peer energy trading platform. It utilizes blockchain technology to enable consumers with rooftop solar panels to sell excess energy directly to nearby consumers. The platform allows participants to create and manage their own energy trading communities, facilitating transparent and automated energy transactions. Power Ledger has been successfully implemented in various locations, including Australia, Japan, and the United States, enabling consumers to maximize the value of their renewable energy generation and promoting the adoption of clean energy sources.

WePower: WePower is another blockchain-based platform that facilitates peer-to-peer energy trading. It enables renewable energy producers to tokenize their energy production and sell these tokens directly to consumers. WePower leverages smart contracts and blockchain technology to automate energy transactions and provide transparency and traceability to participants. The platform has been deployed in several countries, including Estonia and Spain, empowering consumers to choose clean energy sources and promoting the growth of renewable energy projects.

LO3 Energy: LO3 Energy, based in Brooklyn, New York, has developed a blockchain-based platform called Exergy. The platform allows consumers to buy and sell locally generated energy using smart contracts and blockchain technology. It enables homeowners with solar panels or other forms of distributed energy resources to directly trade their surplus energy with nearby consumers. The Exergy platform provides participants with real-time data on energy prices, availability, and consumption, enabling informed decision-making and fostering a more sustainable energy ecosystem.

Verv: Verv is an innovative energy trading platform that combines blockchain technology, artificial intelligence, and IoT devices. The platform enables consumers to track and manage their energy consumption and sell excess energy back to the grid. Verv utilizes blockchain for transparent and secure energy transactions while leveraging AI algorithms to optimize energy usage and provide personalized insights to consumers. The platform has been deployed in the UK, facilitating peer-to-peer energy trading and empowering consumers to actively participate in the energy market.

6.1.1 Benefits of Blockchain-Based Peer-to-Peer Energy Trading Platforms

(a) Increased Efficiency: By eliminating intermediaries, blockchain-based platforms streamline energy transactions, reducing costs and administrative burdens. Participants can trade energy directly, bypassing traditional energy market operators and benefiting from more efficient pricing and settlement processes.

(b) Transparency and Trust: Blockchain technology provides a transparent and immutable ledger of energy transactions, ensuring trust and accountability among participants. Consumers can verify the source and characteristics of the purchased energy, fostering transparency in the energy market.

(c) Empowerment of Energy Producers and Consumers: Peer-to-peer energy trading platforms empower energy producers, such as homeowners with solar panels, to monetize their surplus energy. Consumers have the opportunity to choose clean energy sources and support local energy producers, promoting a decentralized and sustainable energy ecosystem.

(d) Integration of Renewable Energy Sources: Blockchain-based platforms encourage the integration of renewable energy sources into the energy market. Producers of renewable energy can directly sell their excess energy to consumers, incentivizing the adoption of clean energy sources and accelerating the transition to a greener energy system.

Blockchain-based peer-to-peer energy trading platforms have demonstrated significant potential in transforming the energy market by enabling direct interactions between energy producers and consumers. These platforms leverage blockchain technology to provide transparency, efficiency, and traceability to energy transactions. Through case studies and real-world examples, we observe the successful implementation of such platforms, promoting decentralized energy markets, empowering consumers and energy producers, and facilitating the integration of renewable energy sources. These blockchain-based solutions play a crucial role in the modernization of power systems, driving the transition towards a more sustainable and consumer-centric energy future.

6.2 Decentralized Energy Management Systems

Decentralized energy management systems leverage blockchain technology to enable efficient coordination and optimization of energy resources in a decentralized manner. These systems empower individual energy producers and consumers to actively participate in managing their energy generation, consumption, and sharing [22]. This section presents case studies and real-world examples that highlight the implementation and benefits of decentralized energy management systems.

Brooklyn Microgrid: The Brooklyn Microgrid project, powered by LO3 Energy, is a notable example of a decentralized energy management system. It utilizes blockchain technology to enable peer-to-peer energy transactions within a localized community in Brooklyn, New York. The system allows participants with rooftop solar panels to sell excess energy to nearby consumers, fostering a self-sustaining energy ecosystem. The blockchain-based platform records and verifies energy transactions, ensuring transparency and facilitating secure and efficient energy trading within the microgrid.

Transactive Grid: The Transactive Grid project, developed by ConsenSys and LO3 Energy, is a decentralized energy management system implemented in the Brooklyn Navy Yard, New York. It leverages blockchain technology to enable energy trading and optimization within a microgrid. The system allows participants to buy and sell locally generated renewable energy using smart contracts. By utilizing blockchain, the Transactive Grid ensures transparency, automates energy transactions, and optimizes energy distribution within the microgrid, leading to improved grid stability and increased utilization of renewable energy resources.

FlexiDAO: FlexiDAO is a blockchain-based platform that enables decentralized energy management and optimization. It allows energy producers and consumers to interact directly and manage their energy resources efficiently. The platform leverages blockchain technology to record and settle energy transactions, provide real-time energy data, and enable automated demand response. FlexiDAO enables consumers to optimize their energy consumption patterns based on energy prices and availability, contributing to grid stability and promoting the integration of renewable energy sources.

Powerpeers: Powerpeers, a peer-to-peer energy trading platform in the Netherlands, utilizes blockchain technology for decentralized energy management. The platform enables consumers to choose their energy suppliers and trade renewable energy directly with other participants. Powerpeers leverages blockchain to record and verify energy transactions, ensuring transparency and trust among participants. This decentralized approach empowers consumers to make informed decisions about their energy sources, promotes renewable energy adoption, and fosters community engagement in energy management.

6.2.1 Benefits of Decentralized Energy Management Systems

(a) Increased Efficiency: Decentralized energy management systems eliminate the need for intermediaries, allowing for more direct interactions between energy producers and consumers. This results in streamlined energy transactions, reduced costs, and increased energy efficiency.
(b) Grid Stability and Resilience: By optimizing energy generation, consumption, and sharing within localized communities, decentralized energy management systems enhance grid stability and resilience. The systems enable efficient load balancing, demand response, and integration of renewable energy sources, reducing grid disruptions and improving overall reliability.
(c) Empowerment of Energy Producers and Consumers: Decentralized energy management systems empower individuals to actively participate in energy management decisions. Energy producers can monetize their excess energy, while consumers have the flexibility to choose their energy sources and optimize their consumption patterns based on preferences and market conditions.
(d) Promotion of Renewable Energy: By facilitating direct peer-to-peer energy trading and promoting the integration of renewable energy sources, decentralized energy management systems contribute to the adoption of clean and sustainable energy practices. They enable energy producers to monetize renewable energy generation and incentivize consumers to choose renewable energy sources.

Decentralized energy management systems, powered by blockchain technology, have emerged as effective solutions to optimize energy resources in a decentralized manner. Through case studies and real-world examples, we observe successful implementations of these systems, empowering energy producers and consumers, enhancing grid stability, and promoting the adoption of renewable energy sources. These decentralized solutions revolutionize energy management by enabling direct peer-to-peer interactions, automating energy transactions, and providing transparency and trust through blockchain. Decentralized energy management systems play a crucial role in modernizing power systems and fostering a more sustainable, efficient, and consumer-centric energy future.

6.3 Integration of Renewable Energy Sources with Blockchain

The integration of renewable energy sources with blockchain technology has gained significant attention as a means to promote transparency, traceability, and efficiency in the renewable energy sector [23]. This section presents case studies and real-world examples that showcase successful implementations of blockchain for the integration of renewable energy sources.

REGrid: REGrid, developed by Electra Commercial Vehicles, is an innovative project that combines blockchain technology with renewable energy generation and electric vehicle (EV) charging infrastructure. The project aims to enable seamless integration of renewable energy sources, such as solar and wind, with EV charging stations. The blockchain-based platform records and verifies energy generation, consumption, and charging transactions, ensuring transparency and facilitating secure and automated payments. REGrid promotes the use of renewable energy for EV charging, reducing carbon emissions and fostering a sustainable transportation ecosystem.

Energy Web Foundation (EWF): EWF is a nonprofit organization focused on accelerating the deployment of blockchain technology in the energy sector, including the integration of renewable energy sources. EWF has developed an open-source blockchain platform specifically designed for the energy sector called the Energy Web Chain. This platform enables the tracking and certification of renewable energy generation through the use of digital tokens. It allows renewable energy producers to issue Renewable Energy Certificates (RECs) and facilitates their trading in a transparent and secure manner.

RENewal: RENewal is a blockchain-based project that integrates renewable energy sources, smart contracts, and IoT devices to enable a decentralized and transparent energy marketplace. The platform facilitates the buying and selling of renewable energy among producers and consumers, while smart contracts automate the energy transactions. RENewal leverages blockchain technology to record and validate energy generation, consumption, and trading data, providing participants with an auditable and trustworthy system for renewable energy integration.

Pylon Network: Pylon Network is a blockchain-based platform that focuses on integrating renewable energy sources and empowering energy communities. The platform enables peer-to-peer energy trading among renewable energy producers and consumers. Pylon Network utilizes blockchain technology to record and validate energy transactions, ensuring transparency and eliminating the need for intermediaries. The platform also provides real-time energy data and analytics to participants, fostering informed decision-making and promoting the use of renewable energy sources.

6.3.1 Benefits of Integrating Renewable Energy Sources with Blockchain

(a) Transparency and Traceability: Blockchain technology enables transparent and traceable recording of renewable energy generation, consumption, and trading. Participants can verify the origin and characteristics of renewable energy, ensuring trust and promoting transparency in the energy market.
(b) Efficiency and Automation: Blockchain-based systems streamline energy transactions and automate processes such as billing, settlement, and certification. This reduces administrative burdens, eliminates intermediaries, and increases overall efficiency in the integration of renewable energy sources.
(c) Decentralization and Peer-to-Peer Trading: Blockchain allows for the direct peer-to-peer trading of renewable energy, bypassing traditional energy market intermediaries. This empowers energy producers and consumers, promotes decentralized energy markets, and encourages the adoption of renewable energy sources.
(d) Renewable Energy Certificates (RECs): Blockchain enables the issuance, tracking, and trading of RECs, providing a transparent and auditable system for verifying renewable energy generation. This facilitates the recognition and monetization of renewable energy production, supporting the growth of renewable energy projects.

The integration of renewable energy sources with blockchain technology offers significant advantages in terms of transparency, efficiency, and decentralization in the renewable energy sector. Through case studies and real-world examples, we observe successful implementations of blockchain for the integration of renewable energy sources, fostering transparency in energy transactions, enabling peer-to-peer trading, and supporting the growth of renewable energy projects. These blockchain-based solutions play a crucial role in modernizing power systems, promoting sustainability, and accelerating the transition to a greener energy future.

6.4 Smart Grids and Blockchain for Power Grid Security

Smart grids combined with blockchain technology have the potential to significantly enhance power grid security by ensuring data integrity, reducing vulnerabilities, and enabling secure and efficient grid operations [24]. This section explores case studies and real-world examples that showcase the application of smart grids and blockchain for power grid security.

Energy Web Foundation (EWF): The Energy Web Foundation is an organization that focuses on accelerating the adoption of blockchain technology in the energy sector. EWF has developed the Energy Web Chain, an open-source blockchain platform specifically designed for the energy industry. The platform enables the secure and decentralized management of energy data and transactions. By utilizing blockchain, the Energy Web Chain enhances power grid security by ensuring the integrity and immutability of data, protecting against unauthorized access or tampering, and facilitating secure communication between grid participants.

Singapore Power: Singapore Power, the national utility company in Singapore, has conducted a successful pilot project that combines smart grids and blockchain for power grid security. The project aimed to improve the security and efficiency of the electricity grid by utilizing blockchain technology to secure data exchanges and authenticate devices. By implementing a decentralized system based on blockchain, Singapore Power enhanced the resilience and reliability of the power grid, reducing the risk of cyberattacks and unauthorized access to critical infrastructure.

LO3 Energy and ConsenSys: LO3 Energy, in collaboration with ConsenSys, developed a project called Transactive Grid, which demonstrates the integration of blockchain technology and smart grids for power grid security. The Transactive Grid project focuses on creating a peer-to-peer energy marketplace within microgrids. Blockchain technology is utilized to secure and verify energy transactions, ensuring the integrity and transparency of data exchanged between participants. This approach enhances power grid security by reducing the reliance on centralized systems, protecting against data manipulation, and facilitating secure and automated transactions.

Australian National Electricity Market (NEM): The Australian National Electricity Market (NEM) has explored the use of blockchain technology to enhance power grid security. By utilizing a private blockchain, the NEM aims to improve the security and reliability of data exchanges among market participants. The blockchain-based system enables real-time monitoring of energy transactions, enhances data integrity and reduces the risk of unauthorized access or tampering. This approach provides increased transparency, traceability, and security within the power grid.

6.4.1 Benefits of Smart Grids and Blockchain for Power Grid Security

(a) Data Integrity and Immutability: Blockchain technology ensures the integrity and immutability of data exchanged within the power grid. By securely recording and verifying transactions on a decentralized ledger, blockchain protects against data manipulation, unauthorized modifications, and fraud.
(b) Cybersecurity: Smart grids integrated with blockchain technology enhance power grid cybersecurity by reducing vulnerabilities and protecting against cyber threats. Blockchain's decentralized nature and cryptographic algorithms provide robust protection against unauthorized access, tampering, and data breaches.
(c) Resilience and Reliability: By decentralizing control and enabling secure peer-to-peer transactions, smart grids and blockchain enhance the resilience and reliability of the power grid. This reduces the dependency on single points of failure and improves grid stability, ensuring continuous and secure energy supply.
(d) Efficient Grid Operations: The combination of smart grids and blockchain streamlines grid operations by automating processes, reducing administrative

burdens, and facilitating secure and efficient data exchanges. This enhances operational efficiency and reduces costs associated with grid management.

The integration of smart grids and blockchain technology presents significant advantages for enhancing power grid security. Through case studies and real-world examples, we observe successful implementations of blockchain in combination with smart grids, demonstrating improved data integrity, reduced vulnerabilities, and secure grid operations. These applications contribute to the resilience, reliability, and efficiency of power grids, ensuring secure data exchanges, protecting against cyber threats, and enabling the transition to a more secure and sustainable energy future.

7 Challenges and Future Perspectives

7.1 Scalability and Integration Challenges

While blockchain technology offers numerous benefits for smart power systems, there are inherent challenges related to scalability and integration that need to be addressed for its widespread adoption [25]. This section explores the scalability and integration challenges associated with blockchain technology in the context of smart power systems.

7.1.1 Scalability Challenges

Transaction Throughput: Blockchain networks, especially public ones, face limitations in terms of transaction throughput. Traditional blockchain protocols, like Bitcoin and Ethereum, have scalability limitations due to their consensus mechanisms. As the number of transactions increases, the network may experience congestion, resulting in slower transaction processing times and increased costs. Scaling solutions such as layer-two protocols (e.g., Lightning Network) and sharding aim to address these scalability concerns by offloading transactions from the main blockchain or partitioning the network into smaller pieces.

Storage Requirements: The storage requirements of blockchain networks grow as the number of transactions and participants increase. As more data is added to the blockchain, the size of the blockchain ledger becomes larger, requiring substantial storage resources. This can pose challenges for nodes participating in the network, especially those with limited storage capacity. Efficient storage mechanisms, data pruning techniques, and the use of off-chain storage solutions can help mitigate these scalability challenges.

7.1.2 Integration Challenges

Legacy Systems: Integrating blockchain technology with existing legacy systems can be challenging. Many power systems rely on established infrastructure, centralized databases, and legacy protocols. Adapting these systems to work seamlessly with blockchain technology may require significant modifications and coordination between different stakeholders. The integration process needs to consider backward compatibility, data migration, and the interoperability of blockchain networks with existing infrastructure.

Interoperability: Interoperability refers to the ability of different blockchain networks to communicate and exchange data with each other. Achieving interoperability among various blockchain platforms and protocols is crucial for seamless data transfer and collaboration in smart power systems. Standardization efforts and the development of interoperability protocols, such as cross-chain communication protocols and interoperability frameworks, are essential to overcome these integration challenges.

Regulatory and Legal Considerations: Blockchain technology introduces new regulatory and legal considerations. Smart power systems incorporating blockchain need to comply with existing energy regulations, data privacy laws, and financial regulations. Developing a regulatory framework that addresses the unique aspects of blockchain technology, such as smart contracts and decentralized transactions, is necessary to ensure compliance and provide a supportive environment for blockchain integration.

Adoption and Industry Collaboration: Widespread adoption of blockchain technology in smart power systems requires collaboration among industry stakeholders, including power grid operators, utilities, regulators, and technology providers. Collaboration is crucial for developing industry standards, establishing best practices, and fostering the necessary infrastructure for seamless integration. Overcoming potential resistance to change and fostering a collaborative ecosystem are essential for successful blockchain integration.

Addressing scalability and integration challenges is vital to unlock the full potential of blockchain technology in smart power systems. Continued research, technological advancements, and collaborative efforts among stakeholders are necessary to develop scalable blockchain solutions and overcome integration barriers. By addressing these challenges, blockchain technology can revolutionize the energy sector by enabling secure, transparent, and efficient power systems.

7.2 Regulatory and Policy Considerations

Regulatory and policy considerations play a crucial role in the adoption and implementation of blockchain technology in smart power systems. As blockchain introduces novel concepts such as decentralized transactions, smart contracts, and data privacy considerations, it is essential to establish a regulatory framework that addresses these unique aspects [26]. This section explores the regulatory and policy considerations associated with blockchain technology in smart power systems.

Data Privacy and Security: Blockchain technology raises data privacy and security concerns, as transactions recorded on the blockchain are typically immutable and visible to all participants. Regulatory frameworks need to address data privacy requirements, such as the General Data Protection Regulation (GDPR), to ensure that personal and sensitive information is appropriately handled and protected. Compliance with data privacy regulations while leveraging the benefits of blockchain technology is crucial for building trust and protecting the privacy rights of individuals.

Compliance with Energy Regulations: Energy regulations govern various aspects of power systems, including energy generation, distribution, pricing, and grid operations. When integrating blockchain into smart power systems, compliance with existing energy regulations becomes paramount. Regulatory frameworks should be adapted to account for the unique characteristics of blockchain technology, such as the use of smart contracts and decentralized transactions. Clear guidelines and standards that facilitate compliance while promoting innovation are necessary for the successful implementation of blockchain in the energy sector.

Smart Contract Legality and Enforceability: Smart contracts, which are self-executing agreements recorded on the blockchain, introduce legal considerations. Regulatory frameworks should clarify the legal status, enforceability, and liability associated with smart contracts. Establishing legal frameworks that recognize and validate the use of smart contracts, while addressing potential legal challenges and disputes, is crucial for building trust and confidence in blockchain-based transactions.

Interoperability and Standards: Interoperability among different blockchain platforms and protocols is vital for seamless data exchange and collaboration in smart power systems. Regulatory and policy frameworks can play a role in promoting interoperability by encouraging the development and adoption of industry standards. Standards for data formats, communication protocols, and smart contract interfaces can facilitate the integration and interoperability of different blockchain networks, promoting efficiency and compatibility across the industry.

Market Structure and Competition: Blockchain technology has the potential to disrupt existing market structures by enabling direct peer-to-peer transactions and eliminating intermediaries. Regulatory frameworks should consider the potential impact on market dynamics, competition, and consumer protection. Balancing innovation and competition with fair market practices and consumer rights is crucial to create a level playing field and ensure a healthy and competitive energy market.

Regulatory Sandboxes and Pilots: Regulatory sandboxes and pilot programs provide a controlled environment for testing and validating blockchain applications in the energy sector. These initiatives allow regulatory authorities to gain insights into the benefits, risks, and challenges associated with blockchain technology. By engaging in dialogue with industry stakeholders and providing a supportive regulatory environment, regulators can foster innovation while ensuring consumer protection and system integrity.

Regulatory and policy considerations play a vital role in shaping the adoption and implementation of blockchain technology in smart power systems. Regulatory frameworks need to address data privacy, energy regulations, smart contract legality, interoperability, market structure, and competition. By providing clear guidelines, supporting innovation, and fostering collaboration among stakeholders, regulators can create an enabling environment that encourages the responsible and secure use of blockchain technology in the energy sector. A balanced approach that promotes both innovation and compliance is essential to unlock the full potential of blockchain in modernizing power systems.

7.3 Collaborative Efforts and Standardization

Collaborative efforts and standardization are crucial for the successful integration and widespread adoption of blockchain technology in smart power systems. Collaboration among industry stakeholders, regulatory bodies, and standardization organizations helps establish common practices, guidelines, and interoperability standards [27]. This section explores the importance of collaborative efforts and standardization in the context of blockchain technology in smart power systems.

Industry Collaboration: Collaboration among industry stakeholders, including power grid operators, utilities, technology providers, and research institutions, is essential for driving the adoption of blockchain in smart power systems. By sharing knowledge, best practices, and experiences, stakeholders can collectively address challenges and identify opportunities for the application of blockchain technology. Collaborative efforts promote a better understanding of the technology's potential, foster innovation, and facilitate the development of practical solutions that meet the needs of the power industry.

Regulatory Alignment: Collaborative efforts between regulatory bodies and industry stakeholders are crucial to align regulatory frameworks with the unique characteristics of blockchain technology. Engaging in dialogue and collaboration allows regulators to stay informed about technological advancements, understand the benefits and risks of blockchain, and develop appropriate regulatory guidelines. Close collaboration between regulators and industry participants ensures that regulations support innovation while addressing concerns related to security, privacy, and consumer protection.

Standardization Organizations: Standardization organizations, such as the International Electrotechnical Commission (IEC), the Institute of Electrical and Electronics Engineers (IEEE), and the International Organization for Standardization (ISO), play a vital role in establishing industry-wide standards for blockchain technology in smart power systems. These organizations facilitate the development of interoperability standards, data formats, communication protocols, and security frameworks. Standardization efforts promote compatibility, enable seamless integration of blockchain solutions, and provide guidance for stakeholders in the industry.

Interoperability and Cross-Platform Collaboration: Interoperability among different blockchain platforms and protocols is crucial for the seamless exchange of data and transactions in smart power systems. Collaborative efforts aim to develop interoperability protocols, cross-platform communication standards, and common frameworks for data exchange. These initiatives enable different blockchain networks to communicate with each other, facilitating data sharing, transaction settlement, and collaboration between various stakeholders.

Knowledge Sharing and Research Collaboration: Collaboration in the form of knowledge sharing, research collaboration, and consortiums allows for a deeper understanding of blockchain technology and its application in smart power systems. Research institutions, universities, and technology providers can work together to conduct studies, pilot projects, and experiments that explore the potential benefits and challenges of blockchain integration. Sharing research findings and lessons learned fosters a collaborative learning environment, enabling stakeholders to make informed decisions and accelerate the adoption of blockchain technology.

Collaborative efforts and standardization are essential for the successful integration of blockchain technology in smart power systems. By fostering collaboration among industry stakeholders, regulatory bodies, and standardization organizations, the power industry can develop common practices, guidelines, and interoperability standards. Collaborative efforts promote innovation, ensure regulatory compliance, facilitate interoperability among different blockchain networks, and enhance the overall efficiency, security, and reliability of smart power systems. Through collaborative endeavors and standardization, the power industry can harness the transformative potential of blockchain technology to create a more sustainable, secure, and efficient energy future.

7.4 Future Trends in Smart Power Systems

Future trends in smart power systems are driven by advancements in technology, evolving consumer demands, and the need for sustainable and efficient energy solutions [28]. This section explores some of the key trends that are expected to shape the future of smart power systems.

Increased Integration of Renewable Energy: The integration of renewable energy sources, such as solar and wind, will continue to be a dominant trend in smart power systems. As renewable energy costs decline and environmental concerns grow, there will be a greater focus on harnessing clean energy sources. Smart power systems will leverage advanced technologies, such as blockchain and IoT, to efficiently manage the integration, storage, and distribution of renewable energy, promoting a more sustainable energy ecosystem.

Energy Storage Solutions: Energy storage technologies, such as batteries and advanced grid-scale storage systems, will play a crucial role in enabling a reliable and flexible power supply. As intermittent renewable energy sources become more prevalent, energy storage solutions will bridge the gap between energy generation and consumption, ensuring a continuous and stable power supply. Smart power systems will utilize innovative energy storage technologies and intelligent control mechanisms to optimize energy usage and enhance grid stability.

Demand Response and Flexibility: Demand response programs, which incentivize consumers to adjust their energy usage based on grid conditions, will become more widespread in smart power systems. By utilizing real-time data and advanced analytics, smart power systems will enable dynamic demand response, allowing consumers to participate actively in balancing energy supply and demand. This flexibility in energy consumption patterns will enhance grid stability, reduce peak loads, and optimize resource utilization.

Advanced Grid Management and Control: Smart power systems will adopt advanced grid management and control mechanisms to optimize power flow, manage grid congestion, and enhance overall system reliability. Technologies such as advanced metering infrastructure, real-time monitoring, and intelligent control algorithms will enable more efficient grid operations. By leveraging artificial intelligence (AI) and machine learning (ML) algorithms, smart power systems will autonomously analyze data, predict grid conditions, and make real-time adjustments to ensure efficient and reliable power delivery.

Internet of Things (IoT) Integration: The integration of IoT devices will be a key trend in smart power systems, enabling real-time monitoring, control, and optimization of energy resources. IoT devices, such as smart meters, sensors, and connected appliances, will collect and transmit data, enabling precise energy management and enhancing consumer engagement. The seamless integration of IoT devices with smart power systems will facilitate energy conservation, improve system efficiency, and enable personalized energy management.

Blockchain for Secure and Transparent Transactions: Blockchain technology will continue to be a transformative force in the energy sector, enabling secure, transparent, and decentralized energy transactions. Smart power systems will leverage blockchain for secure peer-to-peer energy trading, grid data management, and secure authentication of devices. Blockchain will enhance trust, enable efficient settlement of transactions, and ensure data integrity, fostering a more secure and efficient energy ecosystem.

Electric Vehicle Integration: The growing adoption of electric vehicles (EVs) will have a significant impact on smart power systems. EVs will serve as mobile energy storage units, capable of both consuming and providing electricity to the grid. Smart power systems will integrate EV charging infrastructure, vehicle-to-grid (V2G) technology, and intelligent charging algorithms to manage the charging and discharging of EVs, optimizing energy usage and supporting grid stability.

Future trends in smart power systems revolve around the integration of renewable energy, energy storage solutions, demand response programs, advanced grid management, IoT integration, blockchain technology, and electric vehicle integration. These trends aim to create sustainable, efficient, and secure energy systems that meet the evolving needs of consumers and address environmental concerns. By embracing these trends, the power industry can pave the way for a more resilient, decentralized, and intelligent energy future.

8 Conclusion

The integration of blockchain technology with smart power systems is transforming the way we generate, distribute, and consume electricity. By leveraging the power of AI, IoT, and blockchain, power systems are becoming more efficient, secure, and environmentally friendly. This article has explored various aspects of smart power systems, including power distribution, power quality control, energy management, and the significant role played by blockchain technology in creating transparent and decentralized power systems. As the demand for electricity continues to grow, the adoption of smart power systems and blockchain technology will play a crucial role in ensuring a sustainable and resilient energy future. It is essential for students, researchers, and professionals in the electrical engineering and power systems fields to stay updated with the latest advancements and understand the potential of these technologies to shape the future of the power sector. By embracing innovation and collaboration, we can accelerate the modernization of power systems and pave the way for a cleaner and smarter energy landscape.

References

1. Fadaeenejad, M., Saberian, A., Fadaee, M., Radzi, M., Hizam, H., & AbKadir, M. (2013). The present and future of smart power grid in developing countries. *Renewable and Sustainable Energy Reviews, 29*, 828–834. https://doi.org/10.1016/j.rser.2013.08.072
2. Haes Alhelou, H., Hamedani Golshan, M. E., Njenda, T. C., & Hatziargyriou, N. D. (2020). An overview of UFLS in conventional, modern, and future smart power systems: Challenges and opportunities. *Electric Power Systems Research, 179*, 106054. https://doi.org/10.1016/j.epsr.2019.106054
3. Alhelou, H. H., Hamedani-Golshan, M.-E., Zamani, R., Heydarian-Forushani, E., & Siano, P. (2018). Challenges and opportunities of load frequency control in conventional, modern and future smart power systems: A comprehensive review. *Energies, 11*(10), 2497. https://doi.org/10.3390/en11102497
4. Ramachandran, B., Srivastava, S. K., & Cartes, D. A. (2013). Intelligent power management in micro grids with EV penetration. *Expert Systems with Applications, 40*(16), 6631–6640. https://doi.org/10.1016/j.eswa.2013.06.007
5. Yin, S., Rodriguez-Andina, J. J., & Jiang, Y. (2019). Real-time monitoring and control of industrial Cyberphysical systems: With integrated plant-wide monitoring and control framework. *IEEE Industrial Electronics Magazine, 13*(4), 38–47. https://doi.org/10.1109/MIE.2019.2938025
6. Y. W. Law, T. Alpcan, V. C. S. Lee, A. Lo, S. Marusic and M. Palaniswami, *Demand Response Architectures and Load Management Algorithms for Energy-Efficient Power Grids: A Survey*. 2012 Seventh International Conference on Knowledge, Information and Creativity Support Systems, Melbourne, VIC, Australia, 2012, pp. 134–141, doi: https://doi.org/10.1109/KICSS.2012.45.
7. Divya, D., Marath, B., & Santosh Kumar, M. B. (2023). Review of fault detection techniques for predictive maintenance. *Journal of Quality in Maintenance Engineering, 29*(2), 420–441.
8. S. Bhattacharyya, J. M. A. Myrzik and W. L. Kling, *Consequences of poor power quality - An overview*. 2007 42nd International Universities Power Engineering Conference, Brighton, UK, 2007, pp. 651–656, doi: https://doi.org/10.1109/UPEC.2007.4469025.

9. Beniwal, R. K., Saini, M. K., Nayyar, A., Qureshi, B., & Aggarwal, A. (2021). A critical analysis of methodologies for detection and classification of power quality events in smart grid. *IEEE Access, 9*, 83507–83534. https://doi.org/10.1109/ACCESS.2021.3087016
10. Jaradat, M., Jarrah, M., Bousselham, A., Jararweh, Y., & Al-Ayyoub, M. (2014). The internet of energy: Smart sensor networks and big data Management for Smart Grid. *Procedia Computer Science, 56*, 592–597. https://doi.org/10.1016/j.procs.2015.07.250
11. Mahendravarman, I., Elankurisil, S. A., Venkateshkumar, M., et al. (2020). Artificial intelligent controller-based power quality improvement for microgrid integration of photovoltaic system using new cascade multilevel inverter. *Soft Computing, 24*, 18909–18926. https://doi.org/10.1007/s00500-020-05120-2
12. Zia, M. F., Elbouchikhi, E., & Benbouzid, M. (2018). Microgrids energy management systems: A critical review on methods, solutions, and prospects. *Applied Energy, 222*, 1033–1055. https://doi.org/10.1016/j.apenergy.2018.04.103
13. Thirunavukkarasu, G. S., Seyedmahmoudian, M., Jamei, E., Horan, B., Mekhilef, S., & Stojcevski, A. (2022). Role of optimization techniques in microgrid energy management systems—A review. *Energy Strategy Reviews, 43*, 100899.
14. Hussain, T., Min Ullah, F. U., Muhammad, K., et al. (2021). Smart and intelligent energy monitoring systems: A comprehensive literature survey and future research guidelines. *International Journal of Energy Research, 45*, 3590–3614. https://doi.org/10.1002/er.6093
15. Mesarić, P., & Krajcar, S. (2015). Home demand side management integrated with electric vehicles and renewable energy sources. *Energy and Buildings, 108*, 1–9. https://doi.org/10.1016/j.enbuild.2015.09.001
16. Gayathri, A., Saravanan, S., Pandiyan, P., & Rukkumani, V. (2023). Blockchain technologies for Smart Power Systems. In S. Padmanaban, S. Palanisamy, S. Chenniappan, & J. B. Holm-Nielsen (Eds.), *Artificial Intelligence-based Smart Power Systems*. https://doi.org/10.1002/9781119893998.ch17
17. Qian, Y., Jiang, Y., Chen, J., Zhang, Y., Song, J., Zhou, M., & Pustišek, M. (2018). Towards decentralized IoT security enhancement: A blockchain approach. *Computers & Electrical Engineering, 72*, 266–273. https://doi.org/10.1016/j.compeleceng.2018.08.021
18. Di Silvestre, M. L., Gallo, P., Guerrero, J. M., Musca, R., Riva Sanseverino, E., Sciumè, G., Vásquez, J. C., & Zizzo, G. (2020). Blockchain for power systems: Current trends and future applications. *Renewable and Sustainable Energy Reviews, 119*, 109585. https://doi.org/10.1016/j.rser.2019.109585
19. Aloqaily, M., Boukerche, A., Bouachir, O., Khalid, F., & Jangsher, S. (2020). An energy trade framework using smart contracts: Overview and challenges. *IEEE Network, 34*(4), 119–125. https://doi.org/10.1109/MNET.011.1900573
20. Badra, M., & Zeadally, S. (2017). Lightweight and efficient privacy-preserving data aggregation approach for the Smart Grid. *Ad Hoc Networks, 64*, 32–40. https://doi.org/10.1016/j.adhoc.2017.05.011
21. AlSkaif, T., Crespo-Vazquez, J. L., Sekuloski, M., van Leeuwen, G., & Catalão, J. P. S. (2022). Blockchain-based fully peer-to-peer energy trading strategies for residential energy systems. *IEEE Transactions on Industrial Informatics, 18*(1), 231–241. https://doi.org/10.1109/TII.2021.3077008
22. Elkazaz, M., Sumner, M., & Thomas, D. (2021). A hierarchical and decentralized energy management system for peer-to-peer energy trading. *Applied Energy, 291*, 116766. https://doi.org/10.1016/j.apenergy.2021.116766
23. Yildizbasi, A. (2021). Blockchain and renewable energy: Integration challenges in circular economy era. *Renewable Energy, 176*, 183–197. https://doi.org/10.1016/j.renene.2021.05.053
24. Liu, C., Zhang, X., Chai, K. K., Loo, J., & Chen, Y. (2021). A survey on blockchain-enabled smart grids: Advances, applications and challenges. *IET Smart Cities, 3*, 56–78. https://doi.org/10.1049/smc2.12010

25. Ma, S., Zhang, H., & Xing, X. (2018). Scalability for smart infrastructure system in smart grid: A survey. *Wireless Personal Communications, 99*, 161–184. https://doi.org/10.1007/s11277-017-5045-y
26. Watson, J., Sauter, R., Bahaj, B., James, P., Myers, L., & Wing, R. (2008). Domestic micro-generation: Economic, regulatory and policy issues for the UK. *Energy Policy, 36*(8), 3095–3106. https://doi.org/10.1016/j.enpol.2008.04.028
27. Coll-Mayor, D., Paget, M., & Lightner, E. (2007). Future intelligent power grids: Analysis of the vision in the European Union and the United States. *Energy Policy, 35*(4), 2453–2465. https://doi.org/10.1016/j.enpol.2006.09.001
28. O'Dwyer, E., Pan, I., Acha, S., & Shah, N. (2019). Smart energy systems for sustainable smart cities: Current developments, trends and future directions. *Applied Energy, 237*, 581–597. https://doi.org/10.1016/j.apenergy.2019.01.024

Effective Models for Computing Optimized Storage Systems for Energy

S. Vijayalakshmi and Vandana R. Bais

1 Introduction

The growing need for energy-efficient and high-performance storage systems has driven us to investigate innovative methods for optimizing storage solutions. Recently, interest in this area has been increasing in developing efficient models for storage systems designed to meet computational requirements. These models aim to manage data effectively while reducing energy consumption. These models aim to balance performance, energy efficiency, and cost-effectiveness, which are crucial factors in today's data-driven environment. By leveraging advanced algorithms, machine learning techniques, and adaptive strategies, we can design storage systems that not only meet the demands of large-scale data processing but also reduce the overall environmental impact. By utilizing computational tools and algorithms, we can simulate various scenarios, analyze different parameters, and ultimately identify the most effective storage system configurations. This chapter explores and evaluates different models and approaches for computing optimized storage systems that significantly improve energy efficiency and performance, contributing to a more sustainable and cost-effective storage infrastructure.

S. Vijayalakshmi (✉)
Department of Data Science, Christ (Deemed to be University) University, Pune, Lavasa, India

V. R. Bais
Department of Computer Science, Christ College, Pune, India

S. Vijayalakshmi et al. (eds.), *Smart Power Systems*, Power Systems,
https://doi.org/10.1007/978-3-031-89459-6_15

2 Literature Review

Understanding the complexities of modeling energy storage optimization requires a grasp of its fundamental concepts. Several studies have comprehensively evaluated energy storage devices examples include batteries, pumped hydro storage, and thermal energy storage, among others. These assessments lay the groundwork for future research in this field [26, 62, 63]. The optimization of storage systems faces several challenges, including capacity planning, performance optimization, and cost-effectiveness. However, recent research has emphasized the significance of tackling these challenges to fully exploit the energy storage capabilities in aiding the incorporation of renewable energy and guaranteeing grid stability [28, 45]. Various modeling approaches have been employed to optimize storage systems for energy. Mathematical optimization techniques offer rigorous methodologies for optimizing storage system operation, considering factors such as energy demand, storage capacity, and system constraints [2, 4, 18]. Machine learning (ML) approaches have also been adopted by energy storage systems to demonstrate the effectiveness of ML algorithms in forecasting energy demand, optimizing charging and discharging schedules, and improving overall system efficiency [38]. Additionally, simulation techniques offer valuable perspectives on the dynamic actions of various technologies for storing energy under different operating conditions, enabling researchers to evaluate system performance and identify optimization strategies [34].

2.1 Energy Storage Systems

Energy storage systems are growing in importance within renewable power generation, primarily because of the variable and unpredictable characteristics of energy sources like wind, solar, and to a lesser extent, hydro. With a notable decline in renewable power generation technologies costs and the EU's more ambitious de-carbonization policy, energy storage systems are becoming a crucial element of future power systems [7]. They enhance network reliability and contribute to the de-carbonization of various sectors such as industry, buildings, and transportation (Fig. 1).

Energy storage comes in different forms:

(a) **Pumped Hydro Storage:** The process of pumping hydro storage entails holding potential energy above a turbine in a reservoir. By providing balancing services in addition to baseload electricity, these plants increase system reliability [31]. However, because of complex planning processes, large capital costs, and mounting public opposition, only a few new facilities are being created.

(b) **Hydrogen fuel cells (HFCs):** When hydrogen joins with oxygen, chemical energy is released and transformed into electrical energy, which is used by hydrogen fuel cells (HFCs). They are devices that convert molecular bond-stored chemical potential energy into electrical energy. Heat, electricity, and

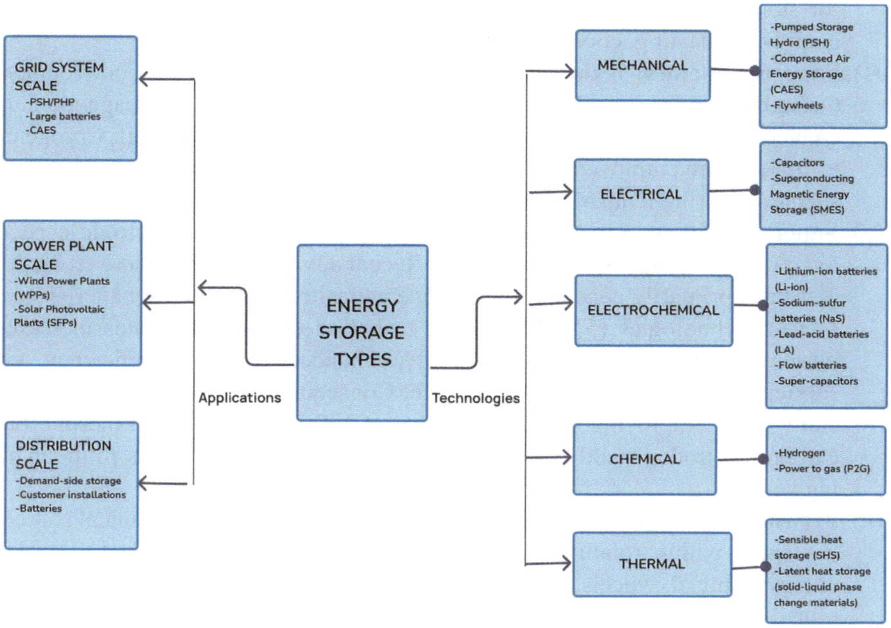

Fig. 1 Energy storage technologies and applications in a power system

water are produced by the cell's reaction. Since the atmosphere contains plenty of oxygen, the fuel cell needs hydrogen, which may be produced by an electrolysis process [10, 11]. The Secretariat plans to explore the possibilities of hydrogen-based economies in greater depth.

(c) **Compressed Air Energy Storage (CAES):** It involves storing compressed air at high pressure, typically in underground caverns. This technology, derived from gas turbine principles, utilizes electricity to compress air for storage. During the power supply phase, the compressed air that has been stored is discharged to propel a turbine, thus producing electricity [54]. Storage reservoirs can encompass underground salt caverns, mines, depleted wells, or gas chambers.

(d) **Battery banks:** Battery banks encompass various types of batteries, encompassing traditional lead-acid, sodium sulfide, lithium-ion, flow, solid-state, and various electrode batteries. They typically serve two main applications. The first is residential solutions, often combined with photovoltaic (PV) systems. The second involves systems at an industrial scale, which guarantee electricity provision for substantial consumers, provide reserve control or ancillary services, or accomplish a combination of these tasks.

(e) **Thermal (including molten salt):**-It includes systems that use molten salt, involves storing energy through temperature manipulation. This technology retains thermal energy by raising or lowering the temperature of a medium used for storage, facilitating the utilization of stored energy to be utilized subsequently for heating, cooling, or power generation purposes [3].

(f) **Superconductors:** It involves storing energy in a super-cooled environment through energy storage based on magnetic fields. It utilizes the magnetic field created by passing electric current through a superconducting coil. Cryogenic techniques are employed to cool this coil to temperatures lower than its critical superconducting temperature [51].

(g) **Supercapacitors:** Supercapacitors, utilizing porous carbon-electrode capacitors provide storage with high density. Recent advancements in these technologies have primarily concentrated on supercapacitors, lithium-ion batteries, and related technologies [51]. In contrast to batteries, it demonstrates increased power density, prolonged cycle stability, enhanced Coulombic efficiency, and shorter cycles for full charge-discharge. Consequently, supercapacitors, especially those employing carbon nanotubes (CNTs), graphene, and mesoporous carbon electrodes could become essential energy storage devices in the coming years.

(h) **Flywheel storage:** It involves storing energy derived from mechanical sources contained within rotating discs placed in a vacuum. In systems for storing energy using flywheels, electric engines rotate the flywheel at high speeds, transforming electrical energy into mechanical energy for storage purposes. When required, the flywheels subsequently operate generators to generate power [37].

2.2 *Advantages of Energy Storage Systems*

(a) **Peak Demand Reduction**: Energy storage systems provide a notable benefit by mitigating peak demand on the power grid. This is particularly evident during times of heightened electricity consumption, as if hot summer days or evenings when people return home from work, energy storage systems can supply extra power. This alleviates strain on the grid and reduces the necessity for expensive infrastructure enhancements [56].

(b) **Enhancing Grid Reliability**: Energy storage systems bolster grid reliability by furnishing backup power in instances of blackouts or grid malfunctions. In situations like natural disasters or equipment breakdowns causing disruptions to primary power sources, energy storage systems can swiftly and dependably supply power, guaranteeing the continuity of vital services [8].

(c) **Integrating Sources of Renewable Energy:** Intermittent sustainable energy sources such as solar and wind depend on variables such as weather patterns and daylight availability. Energy storage systems are essential in incorporating these sustainable energy sources into storing surplus energy when generation exceeds demand and releasing it during shortages helps support the grid [12]. This feature ensures a reliable power supply and enhances the incorporation of renewable energy into the energy mix.

(d) **Alleviating Power Interruptions:** Energy storage systems can also help alleviate power interruptions by providing localized backup power. In areas prone to frequent outages or remote regions with limited grid connectivity, these systems can act as dependable and autonomous power providers, enhancing quality of life and bolstering essential services [27].
(e) **Diverse Applications:** Energy storage systems serve various purposes, including balancing electrical grids, incorporating sustainable energy, enabling time adjustment and microgrids, offering backup power, facilitating electric vehicle charging, and optimizing energy consumption in industrial and commercial environments [58].

2.3 Challenges in Developing Computing-Optimized Storage Systems for Energy

The major challenges in developing computing-optimized storage systems for energy include:

(a) **Energy Efficiency**: Balancing the energy consumption of storage systems with their computational performance is crucial. Energy-efficient algorithms and hardware designs are needed to minimize power usage while maintaining high performance [19, 20, 59].
(b) **Scalability:** Ensuring that storage systems can scale to meet the increasing demand for energy storage and computational tasks is essential. This includes both horizontal scalability (handling larger volumes of data) and vertical scalability (supporting more complex computational tasks) [20, 61].
(c) **Data Management and Processing:** Efficient data management and processing techniques are necessary for handling and processing substantial volumes of data in real-time scenarios. This includes optimizing data storage, retrieval, and processing algorithms to minimize latency and maximize throughput [32, 59].
(d) **Cost-Effectiveness**: Developing cost-effective solutions that balance the upfront investment in hardware and infrastructure with long-term operational costs is a significant challenge. This includes optimizing hardware design, software algorithms, and deployment strategies to minimize total cost of ownership [61].
(e) **Environmental Impact:** Minimizing the environmental footprint of computing-optimized storage systems is increasingly important. This includes reducing energy consumption, minimizing waste generation, and using environmentally friendly materials in hardware design [59, 61].

3 Models for Computing Optimized Storage Systems for Energy

Optimized storage systems for energy can be crucial for efficient energy management and sustainability. Some models and algorithms commonly used for computing optimized storage systems include:

1. Energy Management Algorithms
2. Optimization Models
3. Machine Learning for Energy Optimization
4. Monte Carlo Simulation
5. Dynamic Programming Approach
6. Queuing Theory for Energy Storage\

3.1 Energy Management Algorithms

Energy management algorithms play an important role in improving the effectiveness of storage of energy by taking into account factors such as demand, supply, and storage capacity. They play a critical role in computing optimized storage systems, efficiently storing and distributing energy based on these factors [62]. Below are some typical energy management algorithms utilized in this process.

3.1.1 Peak Shaving

Peak shaving in energy management algorithms refers to a strategy used to reduce or shave off the peak energy demand during periods of high electricity consumption. This approach seeks to reduce electricity expenses and ease pressure on the grid by storing excess energy during periods of low demand and releasing it during times of peak demand [57, 60]. Here is how peak shaving typically works within an energy management algorithm: Data Collection: The algorithm gathers information on energy usage patterns, which includes identifying peak demand periods and off-peak hours.

Prediction: Using historical data and real-time information, the algorithm forecasts the timing of peak demand periods.Energy Storage During times of low demand when electricity prices are reduced, surplus energy is stored in energy storage systems like batteries or pumped hydro storage.

Discharge during Peaks: When peak demand periods are predicted, the energy accumulated in the stored energy from the storage system is utilized to fulfill a portion of the demand when discharged, diminishing the necessity to procure electricity from the grid.

Optimization: The algorithm fine-tunes the functioning of the energy storage system to minimize expenses while guaranteeing reliable fulfillment of energy demand.

3.1.2 Shifting of Loads

It entails moving energy usage from times of highest demand to periods of lower demand. This optimizes energy usage, cuts peak demand, and lowers electricity costs. The process entails arranging flexible loads, such as HVAC systems, water heaters, or electric vehicles, to function during off-peak hours when electricity prices are reduced. During peak demand periods, these loads are reduced to minimize energy consumption [48]. This strategy is supported by peak demand predictions based on historical and real-time data. Load shifting helps to reduce electricity costs, stabilize the grid by easing stress during peak times, and improve overall energy efficiency by optimizing device usage.

3.1.3 State of Charge (SOC) Control

It involves managing the charge level within energy storage systems to optimize performance and efficiency. SOC refers to the stored energy relative to total capacity [50]. Here is how SOC control works:

Monitoring SOC: The algorithm continuously monitors SOC using sensors or monitoring devices.

Setpoint Determination: Based on energy demand, grid conditions, and objectives, the algorithm determines a desired SOC setpoint.

Charge/Discharge Control: The algorithm adjusts charging and discharging to maintain SOC near the setpoint. When below, the system charges from the grid or renewables; when above, it may discharge energy.

Optimization: The algorithm enhances the efficiency of energy storage by taking into account variables like electricity costs and grid stability, ensuring equilibrium between supply and demand.

Feedback Control: Charging and discharging rates are modified according to system feedback and operational conditions.

SOC control maximizes battery life, optimizes energy usage, and can provide grid services.

3.1.4 Predictive Control

Predictive control optimizes energy consumption, generation, or storage by using predictive models to anticipate future energy demand, prices, or other variables [49]. Here's how it works:

Prediction: The algorithm forecasts future energy demand, prices, or renewable energy generation using historical data and real-time measurements. Optimization: Based on predictions, the algorithm optimizes energy usage to meet specific objectives like cost minimization or grid stability.

Feedback: It continuously updates predictions and adjusts control actions based on real-time measurements and system feedback.

Decision-Making: The algorithm decides when to charge/discharge energy storage, adjust HVAC set points, or control other loads to achieve desired outcomes.

3.1.5 Fuzzy Logic Control

Fuzzy Logic Control (FLC) utilizes fuzzy logic for managing energy systems. It deals with imprecise data by allowing for degrees of truth between 0 and 1 [22]. Here is how FLC works:

Input Variables: FLC takes input variables like energy demand, renewable energy generation, battery SOC, and electricity prices.

Fuzzy Inference System (FIS): It uses linguistic rules to model decision-making based on input variables.

Fuzzification: Input values are converted into fuzzy sets using membership functions. Rule Evaluation: FLC evaluates rules based on fuzzy set membership values.

Inference: It combines rules to determine output actions using fuzzy logic operators. Defuzzification: Fuzzy output is converted into a crisp value for control decisions.

Advantages of FLC include robustness, flexibility, and adaptability in energy management.

3.2 Optimization Models

Mathematical models can be utilized to optimize the storage of energy considering factors like cost, efficiency, and reliability. Creating optimized storage systems for energy requires a multifaceted approach, incorporating various optimization models to address different aspects of the system. Here are some effective models commonly used:

3.2.1 Linear Programming (LP)

LP is widely used for optimizing storage system operation by formulating objectives and constraints as linear equations. For energy storage systems, LP can be used to maximize revenue or minimize costs while considering constraints like storage capacity, demand, and charging/discharging rates [52].

Objective: Maximize revenue by buying low and selling high, or minimize costs. Decision Variables:-Rates of charging and discharging, State of Charge (SoC). Constraints:

- Capacity limits on SoC.
- Energy balance (supply equals demand plus losses). Operational limits on charging/discharging rates.

By solving this LP model, you can find the optimal charging and discharging schedule to achieve your objective while satisfying all constraints.

3.2.2 Integer Programming (IP)

Integer Programming builds upon Linear Programming (LP) by permitting decision variables to take integer values, which is appropriate for problems requiring decisions in discrete units. In energy storage optimization, IP can be utilized to identify the optimal sizing and placement of storage units, considering specific elements such as battery capacities or the quantity of storage units.

3.2.3 Mixed Integer Linear Programming (MILP)

MILP integrates linear (LP) as well as integer programming (IP) methods, enabling the modeling of problems with both continuous and discrete decision variables. In energy storage optimization, MILP can be used to address complex issues involving both continuous variables (e.g., energy flows) and discrete variables (e.g., binary decisions).

3.2.4 Dynamic Programming (DP)

Dynamic Programming is effective for solving optimization problems involving sequential decision-making over time. In energy storage systems, DP can be utilized to identify the optimal operation of storage units multiple times, considering factors such as energy demand, renewable energy generation, and storage degradation [47]. It is useful for problems where decisions are made over time, such as optimizing the operation of energy storage systems.

Objective: The goal can be to maximize revenue, minimize costs, or achieve other objectives over a defined time horizon.

Decision Variables: These actions correspond to operations carried out at each time interval, such as adjusting charging or discharging rates, and reflect the system's status, like the state of charge (SoC) of energy storage.

Example Problem: For energy storage scheduling, the objective might be to minimize total costs over a day while meeting demand. Decision variables include charging and discharging rates and SoC, subject to energy balance and capacity constraints.

By solving this DP problem, you find the optimal operation schedule for the storage system over the entire time horizon, considering dynamic factors such as energy prices and demand patterns. DP can be solved using techniques like backward induction or value iteration.

3.2.5 Stochastic Programming

Stochastic programming accounts for uncertainty in optimization problems by considering probabilistic distributions of uncertain parameters. In energy storage optimization, stochastic programming enables the optimization of storage operations by considering uncertain factors like energy prices, renewable energy generation, and demand fluctuations [41].**Objective:** Maximize revenue, minimize costs, or meet other goals, factoring in uncertainty.

Decision Variables: Represent decisions like energy storage operations or investment.

Uncertainty: Modeled using probability distributions for variables like energy prices or demand forecasts.**Example Problem:** Optimize the size of an energy storage system considering uncertain future energy prices and demand.

By solving this Stochastic Programming problem, you can find the optimal size of the storage system that minimizes expected costs while considering uncertainty in energy prices and demand.

3.3 Machine Learning Approaches

Methods such as neural networks (NN), support vector machines (SVM), and decision trees (DT) can be employed to develop predictive models for energy demand, renewable energy generation, and energy prices. These can be employed for predicting energy demand patterns, optimize energy storage accordingly. Machine learning methodologies provide robust tools for enhancing storage systems for energy. They enable energy storage systems to adjust dynamically to changing conditions, optimize real-time operations, and enhance overall performance and efficiency [21, 24]. Below are some efficient models and techniques within machine learning for this purpose:

3.3.1 Predictive Modeling

Predictive modeling optimizes energy storage systems by forecasting variables like energy demand, prices, and renewable energy generation.

Load and Price Forecasting:

Energy Demand: Predict future demand to optimize charging and discharging schedules. Electricity Prices:- Forecast prices to minimize costs by buying low and selling high.

Renewable Energy Forecasting:

Solar and Wind Generation: Predict renewable energy generation to optimize storage usage and grid integration.

Battery Degradation Prediction

Battery Health: Forecast degradation to optimize charging and discharging schedules, prolonging battery life.

Predictive modeling enables proactive and effective decision-making for energy storage systems, enhancing their performance and value [55].

3.3.2 Reinforcement Learning (RL)

Techniques like Q-learning and deep reinforcement learning can be employed to improve the performance of energy storage systems in dynamic environments. RL agents learn optimal control policies through trial and error, considering factors like energy prices, renewable energy availability, and storage capacity [36].

Optimal Control: RL algorithms can learn optimal charging and discharging strategies by interacting with the environment, maximizing rewards (e.g., revenue, efficiency) over time.

Energy Arbitrage: RL agents can learn when to buy and sell energy from the grid to maximize profit.

3.3.3 Deep Learning for Control

The frameworks, such as deep neural networks (DNNs) and convolutional neural networks (CNNs), are applicable for control and optimization tasks in energy storage systems. These models can learn intricate mappings between system inputs (such as energy prices and demand forecasts) and control actions (like charging and discharging), facilitating adaptive and efficient operation [23]. It offers efficient control for optimizing energy storage systems:

State Estimation: Models estimate variables like state of charge (SoC) and health (SoH). Predictive Modeling: Predicts energy demands/generation for optimal charging/discharging.

Reinforcement Learning: Learns optimal control strategies by trial and error. Fault Detection: Identifies system faults through sensor data analysis.Integration: Deep learning integrates into energy management systems for real-time decisions.

Adaptive Control: Adapts to changing conditions and uncertainties.

Model Predictive Control: Uses dynamic models for future behavior prediction and optimization.

Fast Response: Processes data quickly for real-time control.

3.3.4 Anomaly Detection

Anomaly detection algorithms, such as autoencoders and support vector machines, can identify abnormal behavior in energy storage systems, such as equipment malfunctions or performance degradation. By detecting anomalies early, maintenance interventions can be scheduled proactively, minimizing downtime and maximizing system reliability [53].

ML models can detect anomalies in energy storage system performance, helping to identify faults or inefficiencies early.

Methods such as clustering or anomaly detection algorithms may be employed to detect uncommon patterns in data, signaling potential system issues.

3.3.5 Hybrid Models

A hybrid model for optimizing energy storage systems combines physics-based and machine-learning approaches:Data Collection and Preprocessing: Gather and clean historical data. Feature Engineering: Extract relevant features like time of day and weather patterns. Modeling: Use physics-based models to understand system behavior. Train machine-learning models for prediction and optimization. Combine outputs for hybridization.

Evaluation and Optimization: Fine-tune the hybrid model for accuracy and efficiency. Deployment: Implement the optimized model for real-time decision-making. It improved accuracy, flexibility, and efficiency though challenges include data quality and model interpretability [17].

3.3.6 Multi-Agent Systems

These systems, in which independent agents collaborate to accomplish shared objectives, can be employed to optimize decentralized energy storage systems. Each agent can represent a storage unit or a component of the system, employing machine-learning techniques to adaptively adjust its behavior based on local observations and global objectives [30, 35]. It optimizes energy storage systems by:

Distributed Control: Agents act autonomously based on local information for efficient system control.

Coordination: Agents communicate to avoid conflicts and optimize system performance. Dynamic Load Balancing: Adjusts operations to balance grid load and reduce waste.

Fault Tolerance: Redistributes tasks to maintain system operation during failures. Adaptability: Adapts to changes in the environment or system conditions.Learning and Optimization: Agents optimize strategies over time using learning algorithms.

Scalability: Easily scales for large-scale systems and complex grid environments. Resource Allocation: Allocates resources efficiently to meet system-wide objectives.MAS enhances efficiency, reliability, and sustainability in energy storage systems.

3.4 Monte Carlo Simulation

It is a valuable technique for modeling various energy storage scenarios to determine the most efficient ones. This technique is particularly useful for complex systems with uncertain variables, making it ideal for optimizing energy storage systems [5]. By using Monte Carlo simulation, stakeholders can analyze and optimize energy storage systems under uncertainty, allowing for informed decision-making and the design of robust systems capable of adapting to changing conditions effectively.

3.4.1 Uncertainty Analysis

Monte Carlo simulation enables the analysis of uncertain factors impacting energy storage systems, including energy demand, renewable energy generation, and energy prices. By sampling from probability distributions representing these uncertainties, Monte Carlo simulations can provide insights into the range of possible outcomes and their associated probabilities [16].

3.4.2 Optimization under Uncertainty

Energy storage systems often operate in dynamic environments with uncertain inputs. Monte Carlo simulation can be integrated into optimization models to account for this uncertainty and identify robust solutions [46]. By simulating multiple scenarios and their outcomes, Monte Carlo optimization techniques can help determine storage system configurations and operating strategies that perform well across a range of conditions.

3.4.3 Risk Assessment

Monte Carlo simulation can be employed to evaluate the risk linked to investments and operations in energy storage. By quantifying the probability distribution of key performance metrics, such as revenue, cost, or return on investment, Monte Carlo simulations help stakeholders understand the likelihood of different outcomes and identify strategies to mitigate risks [33]. This information is valuable for making investment decisions and developing risk management strategies.

3.4.4 Optimal Sizing and Siting

Monte Carlo simulation aids in determining the best size and location for energy storage systems by considering uncertain variables like load profiles, renewable energy generation patterns, and grid conditions. By simulating the performance of storage systems under different scenarios, Monte Carlo techniques can help identify the most cost-effective configurations and locations to maximize benefits such as peak shaving, energy arbitrage, and grid support [14]. Analysis of results allows stakeholders to identify optimal sizing and siting options while considering associated risks. This iterative process ensures informed decision-making to maximize the value and effectiveness of energy storage systems amidst uncertainty.

3.5 Dynamic Programming (DP)

It is a potent method for solving optimization problems by decomposing them into smaller sub-problems and solving them recursively. In the context of computing optimized storage systems for energy, a DP model considers all possible states and decisions over time. It defines state variables such as state of charge and energy prices, determines decisions at each state to maximize an objective function, and solves the Bellman equation recursively [39].This approach helps in making decisions on energy storage at different time intervals to achieve the optimal solution. DP ensures accuracy and effectiveness over time by incorporating storage dynamics, operational constraints, and iterative refinement, making it valuable for long-term planning in dynamic environments.

3.6 Queuing Model

Queuing models can be applied to analyze waiting times in energy storage systems and optimize them for efficiency. To optimize energy storage systems with queuing theory, define the system, model energy demand arrival and service processes, and create a queuing model. Then, optimize using algorithms, validate through simulation, and implement the optimized policies while monitoring performance. It offers a structure for examining and enhancing the efficiency of systems featuring waiting lines or queues, rendering it pertinent for computing optimized energy storage systems [34].To apply the queuing theory-computing model for optimizing an energy storage system, you would consider the following aspects:

Arrival Process: Model the arrival of energy demands into the system. This could be based on historical data or stochastic processes such as Poisson arrivals.

Service Process: Model how the energy storage system serves these demands. This involvesdefining the charging and discharging processes, including rates and efficiencies.

Queue Discipline: Decide how energy demands are prioritized in the queue, if there is one. For instance, First-In-First-Out (FIFO) or another policy.

System Capacity: Specify the capability limitations of storage system of energy, encompassing maximum rates for charging and discharging

Objective Function: Establish the optimization goal, such as minimizing energy costs, maximizing system efficiency, or ensuring a reliable energy supply.

Optimization Algorithms: Use algorithms to find the optimal operating policies. Dynamic programming, Markov decision processes, or reinforcement learning can be employed.

Validation and Simulation: Validate the model using historical data or simulations. This guarantees that the model faithfully reflects the behavior of the energy storage system.

Implementation and Monitoring: Implement optimized policies in real energy storage systems and continuously monitor their performance. Adjust the policies as needed based on changing conditions or new data.

By creating a computing model that considers these aspects, queuing theory can be effectively utilized to optimize energy storage systems, leading to improved efficiency, cost savings, and better management of energy resources.

4 Examples

4.1 Enhancing Storage and Renewable Energy Systems through REopt Optimization

REopt™, a model created at NREL, is used to ascertain the ideal size and dispatch strategy for Behind The Meter (BTM) or off-grid applications. REopt acts as energy-planning platform, aiding clients in accomplishing their goals related to cost savings, resilience, and energy performance. It is structured as a mixed-integer linear optimization framework and is employed for the techno-economic assessment of renewable and conventional generation, energy storage, dispatch able technologies, and energy efficiency [25, 29, 6] (Fig. 2).

4.2 Evaluation Tool for Battery Storage

It is a computerized model designed to simulate the operation of energy storage system for various goals. It optimizes battery control to maximize total benefits, such as energy arbitrage, balancing services, capacity value, and more. This tool evaluates 1 year of battery storage operations, considering minute-by-minute power exchange and utilizing input data like battery parameters, energy prices, and outage information [9]. The tool can simulate various locations and service combinations

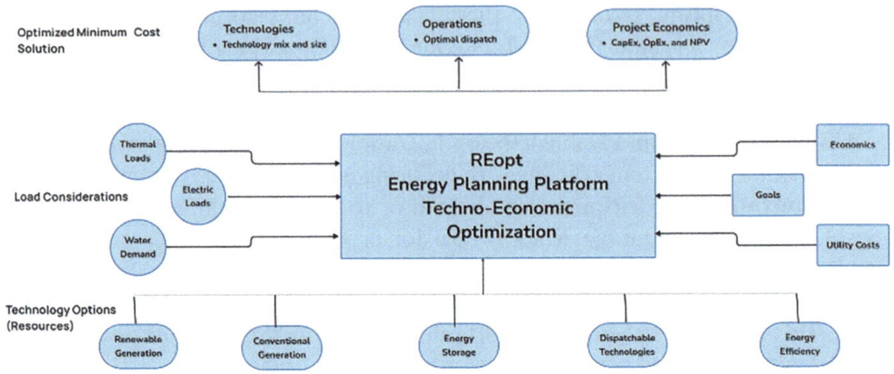

Fig. 2 REopt energy planning platform for techno-economic optimization in energy system

using 50 balancing price scenarios. It calculates the annual value of each service and determines the duration for which the storage system provides each service. These discoveries guide the optimal utilization of energy storage within current infrastructure and offer insights for future setups, ultimately reducing both time and costs while ensuring efficient operation of the power systems [15].

4.3 Battery Storage Optimal Sizing Tool for Grid Applications

It is customized for consumer-side energy storage systems. It evaluates the advantages of installing a storage system powered by batteries behind the consumer's meter located on their premises, as opposed to integrating it within the electric grid. This approach is increasingly popular for businesses looking to reduce energy and demand charges cost-effectively [1, 13].The tool addresses two main objectives:

1. Evaluating the financial benefits of the storage system.
2. Establishing optimal energy and power capacity for the system in economic terms by analyzing historical energy usage and utility tariffs.

By analyzing trade-offs between benefits and costs, the tool optimizes battery size, guiding businesses on purchasing and utilizing behind-the-meter energy storage systems. The Optimal Sizing Tool distinguishes itself with its ability to improve the power and energy capabilities of battery storage in behind-the-meter applications. Meanwhile, these tool aids utilities in building adaptable, dependable, and resilient power systems by enabling the efficient deployment of energy storage solutions. Both programs use open-source software, which is simple to set up and run [40, 42, 43, 44].

References

1. Alsaidan, I., Khodaei, A., & Gao, W. (2018). A comprehensive battery energy storage optimal sizing model for Microgrid applications. *IEEE Transactions on Power Systems, 33*(4), 3968–3980. https://doi.org/10.1109/tpwrs.2017.2769639
2. *Benefits and applications of energy storage systems.* (n.d.). Veolia UK. https://www.veolia.co.uk/benefits-and-applications-energy-storage-systems
3. Breidenbach, N., Martin, C., Jockenhöfer, H., & Bauer, T. (2016). Thermal energy storage in molten salts: Overview of novel concepts and the DLR test facility TESIS. *Energy Procedia, 99*, 120–129. https://doi.org/10.1016/j.egypro.2016.10.103
4. Byrne, R. H., Nguyen, T. A., Copp, D. A., Chalamala, B. R., & Gyuk, I. (2018). Energy management and optimization methods for grid energy storage systems. *IEEE Access, 6*, 13231–13260. https://doi.org/10.1109/access.2017.2741578
5. Cercueil, A., & Francois, O. Monte Carlo simulation and population-based optimization. *Proceedings of the 2001 Congress on Evolutionary Computation (IEEE Cat. No.01TH8546).* https://doi.org/10.1109/cec.2001.934389
6. Elgqvist, E. M., Anderson, K. H., Cutler, D. S., DiOrio, N. A., Laws, N. D., Olis, D. R., & Walker, H. A. (2017). *Optimizing storage and renewable energy systems with REopt.* https://doi.org/10.2172/1415353
7. Energy Community. https://www.energy-community.org/dam/jcr:5602c889-cee2-4d45-86ce-e64fc2f9de58/Energy_Storage_Background_052020.pdf.
8. . (2011). Energy storage in the power system itself. *Energy Storage for Power Systems*, 173–176. https://doi.org/10.1049/pbpo063e_ch12
9. *Energy storage optimization tools.* PNNL. https://www.pnnl.gov/available- technologies/energy-storage-optimization-tools
10. *Energy storage technologies and the challenges to face.* (2023, July 19). AZoNano. https://www.azonano.com/article.aspx?ArticleID=6513
11. Environmental and Energy Study Institute (EESI). *Hydrogen fuel cells.* Environmental and Energy Study Institute | Ideas. Insights. Sustainable Solutions. https://www.eesi.org/topics/hydrogen-fuel-cells/description
12. Venayagamoorthy, G. K., Ahmadi, S. C., & Amanullah, M. M. (2020). Energy storage systems for integration of renewable energy sources, 255–275. https://doi.org/10.1049/pbpo146e_ch18
13. Gholami, M., Muyeen, S., & Abokhamis Mousavi, S. (2024). Optimal sizing of battery energy storage systems and reliability analysis under diverse regulatory frameworks in microgrids. *Energy Strategy Reviews, 51*, 101255. https://doi.org/10.1016/j.esr.2023.101255
14. Gupta, R., & Sossan, F. (2023). Optimal sizing and siting of energy storage systems considering curtailable photovoltaic generation in power distribution networks. *Applied Energy, 339*, 120955. https://doi.org/10.1016/j.apenergy.2023.120955
15. Hallmann, M., Wenge, C., & Komarnicki, P. (2020). Evaluation methods for battery storage systems. *2020 12th International Conference and Exhibition on Electrical Power Quality and Utilisation-(EPQU).* https://doi.org/10.1109/epqu50182.2020.9220321
16. Heck, N., Smith, C., & Hittinger, E. (2016). A Monte Carlo approach to integrating uncertainty into the levelized cost of electricity. *The Electricity Journal, 29*(3), 21–30. https://doi.org/10.1016/j.tej.2016.04.001
17. Hemavathi, S. (2021). Modeling and energy optimization of hybrid energy storage system. *Hybrid Renewable Energy Systems*, 97–114. https://doi.org/10.1002/9781119555667.ch4
18. Hussain Rizvi, S. S., Chaturvedi, K. T., & Lal Kolhe, M. (2023). A Review on Peak Shaving Techniques for Smart Grids. *AIMS Energy, 11*(4), 723–752. https://doi.org/10.3934/energy.2023036
19. IJETS – International Journals. https://www.ijets.in/Downloads/Published/E0202401004.pdf
20. https://www.researchgate.net/publication/351453233_Machine_Learning_Based_Optimization_Model_for_Energy_Management_of_Energy_Storage_System_for_Large_Industrial_Park

21. https://www.sciencedirect.com/topics/engineering/energy-storage-modeling
22. *Low-cost fuzzy logic-controlled home energy management system.* (2023, June 15). SpringerOpen. https://jesit.springeropen.com/articles/10.1186/s43067-023-00100-6
23. *Machine learning and deep learning in energy systems: A review.* (2022, April 18). MDPI. https://www.mdpi.com/2071-1050/14/8/4832
24. *Machine learning for energy systems optimization.* (2022, June 3). MDPI. https://www.mdpi.com/1996-1073/15/11/4116
25. Mancò, G., Tesio, U., Guelpa, E., & Verda, V. (2024). A review on multi energy systems modelling and optimization. *Applied Thermal Engineering, 236*, 121871. https://doi.org/10.1016/j.applthermaleng.2023.121871
26. Mitali, J., Dhinakaran, S., & Mohamad, A. (2022). Energy storage systems: A review. *Energy Storage and Saving, 1*(3), 166–216. https://doi.org/10.1016/j.enss.2022.07.002
27. Molina, M. G. (2019). Grid Energy Storage Systems. *Power Electronics in Renewable Energy Systems and Smart Grid*, 495–583. https://doi.org/10.1002/9781119515661.ch10
28. Márquez, J. M., Manzano, F. S., & Luengo, J. R. (2023). *Energy storage systems: Fundamentals, classification and a technical comparative*. Springer Nature.
29. National Renewable Energy Laboratory (NREL). NREL. https://www.nrel.gov/docs/fy18osti/70333.pdf
30. Nazaralizadeh, S., Banerjee, P., Srivastava, A. K., & Famouri, P. (2024). Battery energy storage systems: A review of energy management systems and health metrics. *Energies, 17*(5), 1250. https://doi.org/10.3390/en17051250
31. Nikolaos, P. C., Marios, F., & Dimitris, K. (2023). A review of pumped hydro storage systems. *Energies, 16*(11), 4516. https://doi.org/10.3390/en16114516
32. *Opportunities and challenges in energy storage in India.* IndustryOutlook: Outlook of Industry Business | Industry Outlook. https://www.theindustryoutlook.com/manufacturing/panorama/opportunities-and-challenges-in-energy-storage-in-india-nwid-1536.html#google_vignette
33. Pereira, E. J., Pinho, J. T., Galhardo, M. A., & Macêdo, W. N. (2014). Methodology of risk analysis by Monte Carlo method applied to power generation with renewable energy. *Renewable Energy, 69*, 347–355. https://doi.org/10.1016/j.renene.2014.03.054
34. Raeis, M., Burchard, A., & Liebeherr, J. (2020). Analysis of a queueing model for energy storage systems with self-discharge. *ACM Transactions on Modeling and Performance Evaluation of Computing Systems, 5*(3), 1–26. https://doi.org/10.1145/3422711
35. Raju, L., Vishwa, V. R., E Vishal., & K M Vimalan. (2023). Energy management system using multi agent systems with IoT and machine learning. *2023 International Conference on Circuit Power and Computing Technologies (ICCPCT).* https://doi.org/10.1109/iccpct58313.2023.10245738
36. *Reinforcement learning for energy-storage systems in grid-connected Microgrids: An investigation of online vs. offline implementation.* (2021, September 9). MDPI. https://www.mdpi.com/1996-1073/14/18/5688
37. *Research on flywheel energy storage system for power quality.* (IEEE Xplore. https://ieeexplore.ieee.org/document/1053592
38. Shehzad, M., & Gueniat, F. (2021). Optimal operation of renewable energy microgrids considering lifetime characteristics of battery energy storage system. *2021 60th IEEE Conference on Decision and Control (CDC).* https://doi.org/10.1109/cdc45484.2021.9683478
39. Sioshansi, R., Denholm, P., Arteaga, J., Awara, S., Bhattacharjee, S., Botterud, A., Cole, W., Cortes, A., Queiroz, A. D., DeCarolis, J., Ding, Z., DiOrio, N., Dvorkin, Y., Helman, U., Johnson, J. X., Konstantelos, I., Mai, T., Pandzic, H., Sodano, D., et al. (2022). Energy-storage modeling: State-of-the-art and future research directions. *IEEE Transactions on Power Systems, 37*(2), 860–875. https://doi.org/10.1109/tpwrs.2021.3104768
40. Sioshansi, R., Madaeni, S. H., & Denholm, P. (2014). A dynamic programming approach to estimate the capacity value of energy storage. *IEEE Transactions on Power Systems, 29*(1), 395–403. https://doi.org/10.1109/tpwrs.2013.2279839

41. *Stochastic optimization method for energy storage system configuration considering self-regulation of the state of charge.* (2022, January 5). MDPI. https://www.mdpi.com/2071-1050/14/1/553
42. UKERC EDC. https://ukerc.rl.ac.uk/Landscapes/EnergyStorage.pdf
43. Vaka, S. S., & Matam, S. K. (2021). Optimal sizing and management of battery energy storage systems in Microgrids for operating cost minimization. *Electric Power Components and Systems, 49*(16–17), 1319–1332. https://doi.org/10.1080/15325008.2022.2061641
44. Vaka, S. S., & Sailaja Kumari, M. (2023). Optimal sizing of battery energy storage systems considering degradation effect for operating and electricity cost minimization in Microgrids. *Energy Technology, 12*(1). https://doi.org/10.1002/ente.202300704
45. Yao, L., Yang, B., Cui, H., Zhuang, J., Ye, J., & Xue, J. (2016). Challenges and progresses of energy storage technology and its application in power systems. *Journal of Modern Power Systems and Clean Energy, 4*(4), 519–528. https://doi.org/10.1007/s40565-016-0248-x
46. Zheng, Y., Jenkins, B. M., Kornbluth, K., & Træholt, C. (2018). Optimization under uncertainty of a biomass-integrated renewable energy microgrid with energy storage. *Renewable Energy, 123*, 204–217. https://doi.org/10.1016/j.renene.2018.01.120
47. https://www.researchgate.net/publication/261045728_A_Dynamic_Programming_Approach_to_Estimate_the_Capacity_Value_of_Energy_Storage
48. https://www.researchgate.net/publication/314491173_Impacts_of_Load_Shifting_on_Renewable_Energy_Integration
49. https://www.researchgate.net/publication/341559053_A_Model_Predictive_Control-Based_Energy_Management_Scheme_for_Hybrid_Storage_System_in_Islanded_Microgrids
50. https://www.researchgate.net/publication/356698394_State_of_Charge_SoC_Estimation_of_Battery_Energy_Storage_System_BESS_Using_Artificial_Neural_Network_ANN_Based_on_IoT-_Enabled_Embedded_System
51. https://www.researchgate.net/publication/366249386_COMPARISON_OF_SUPERCAPACITORS_AND_SUPERCONDUCTING_MAGNETS_AS_ENERGY_STORAGE_SYSTEMS
52. https://www.researchgate.net/publication/44260394_Optimal_Operation_of_Energy_Storage_Using_Linear_Programming_Technique
53. https://www.sciencedirect.com/science/article/abs/pii/S0045790623000642
54. https://www.sciencedirect.com/science/article/abs/pii/S1364032122005901
55. https://www.sciencedirect.com/science/article/pii/S2589004220311330
56. Kumagai, Y., Sumita, J., Iwamoto, S., & Murotani, K. (2008). Application of energy storage systems to peak load leveling. *IEEE Transactions on Power Delivery, 23*(2), 1007–1015.
57. Liu, Y., Watson, J. P., & Ye, Y. (2020). Energy storage systems for power grid resilience: A review. *Applied Energy, 277*, 115624.
58. Beaudin, M., Zareipour, H., Schellenberg, J., & Rosehart, W. (2015). The role of energy storage in modern power systems. *IEEE Transactions on Sustainable Energy, 6*(3), 724–731.
59. Salameh, Z., Alzahrani, S., & Al-Shehri, M. A. (2021). Energy storage systems: Challenges and applications. *Energies, 14*(10).
60. Ghazi, R., Lehtonen, M., & Syed, M. (2019). A review of energy storage technologies for peak shaving in renewable based stand-alone microgrid systems. *Energies, 13*(12).
61. https://www.frontiersin.org/articles/10.3389/fenrg.2023.1254371/full
62. A. K. Gahlawat, S. Jain, V. Kumar (2015) Optimization of energy storage system operation: A review
63. M. A. Ortega-Vazquez, M. K. Sheikh, S. Ruiz-Vega, et al. (2016) Energy storage optimization in power systems: A review

Enhancing Smart Power System through Blockchain Technology: Innovation and Implication

S. Suganyadevi and M. Nivedha

1 Introduction

Blockchain is a decentralised ledger that gives every transaction an unalterable audit record. Anyone on the network may access this shared information store. A distributed ledger called the blockchain records transactions in units called blocks, each of which is linked to the one before it. Because of this, it's very impossible to modify or remove information from the blockchain. The term "smart power system" refers to electrical grids that have been upgraded with digital technology to be more eco-friendly, cost-effective, and efficient [1]. They gather information about the grid by using sensors, actuators, and communication networks. Using this information, grid operators can better manage supply and demand, deal with power outages, and protect against cyberattacks.

Blockchain technology can be used in smart power systems to improve a variety of aspects, including:

- Security: Blockchain can be used to secure the grid against cyberattacks. Hackers will have a difficult time changing or erasing information stored on a blockchain because of its immutable nature.
- Transparency: Blockchain technology has the potential to increase grid transparency. Every transaction is recorded on the blockchain, which also makes the data accessible to all network users. This facilitates monitoring of energy flows and the detection of issues.
- Efficiency: The usage of blockchain technology may boost the grid's efficiency. With its aid, for instance, the exchange of energy between generators and consumers may be automated.

S. Suganyadevi (✉) · M. Nivedha
Assistant Professor, Department of Computer Science, Sri Krishna Arts and Science College, Coimbatore, Tamilnadu, India

S. Vijayalakshmi et al. (eds.), *Smart Power Systems*, Power Systems,
https://doi.org/10.1007/978-3-031-89459-6_16

- Reliability: The grid's dependability may be enhanced by using blockchain technology. Data regarding the state of the grid, for instance, might be stored there so that failures may be anticipated and avoided.

Despite the fact that the use of blockchain technology to smart power systems is still in its infancy, it has the potential to fundamentally alter how energy is managed and then used.

Following are some specific ways that blockchain technology is being used in smart power grids:

- P2P energy trading: Blockchain technology may be used to simplify P2P energy trade. Customers may then trade their surplus energy with the utility company or other customers.
- Demand response: Demand response programmes may make use of blockchain technology. By offering incentives during times of high demand, these programmes encourage people to cut down on their energy use.
- Distributed Energy Resources [DERs]: Solar panels and batteries are examples of Distributed Energy Resources (DERs), both of which may be managed by blockchain technology. This facilitates DERs' safer and more effective grid connection.
- Grid security: Blockchain can be used to improve the security of the grid against cyberattacks. The blockchain's immutability makes it incredibly difficult for hackers to change or delete the data.

Blockchain is a distributed ledger system that ensures transactions are private, public, and unchangeable. It's a common repository of information accessible by anybody on the network. A block is a database entry that stores a single transaction, and each block in the database is connected to the one before it. This makes it very difficult to add, edit, or delete any data on the blockchain.

The term "smart power system" refers to electrical grids that have been upgraded with digital technology to be more eco-friendly, cost-effective, and efficient. They gather information about the grid by using sensors, actuators, and communication networks. Supply and demand can be balanced, outages can be managed, and cyberattacks can be prevented with the use of this information. The current power grid is facing a number of challenges, including:

- Increasing demand for electricity: By 2050, the demand for energy is anticipated to increase by 50% globally. This growth is being driven by factors such as population growth, economic development, and the increasing use of electric vehicles.
- Ageing infrastructure: The power grid is ageing and in need of investment. Many of the components of the grid are nearing the end of their lifespan and need to be replaced.
- Climate change: Climate change is putting a strain on the power grid. The frequency and severity of severe weather events, including as heat waves, droughts, and floods, are increasing the number of outages and grid damage.

1.1 Blockchain Era: Reshaping Industries

Although it is still in its infancy, blockchain technology in smart power systems has the potential to fundamentally change how energy is managed and used. The following are a few examples of technical blockchain applications in smart power systems:

- Decentralised energy trading: Blockchain technology may be used to support decentralised energy trading, allowing consumers and generators to transact directly with one another. This has the potential to enhance productivity while reducing costs.
- Demand response: The use of demand response programmes, which may be done with the use of blockchain technology, may encourage consumers to reduce their energy consumption during peak demand hours. This might increase grid dependability while lowering the need for further investments in physical infrastructure.
- Distributed energy resources (DERs) management: DERs like solar panels and batteries can be managed using blockchain technology. This may aid in the safe and effective incorporation of DERs into the grid.
- Grid security: Blockchain can be used to improve the security of the grid against cyberattacks. Since the blockchain can't be altered, it is very difficult for hackers to remove or change data.
- Green energy trading: Green energy, such as solar and wind power, may be traded using blockchain technology. This may be useful in encouraging the use of green energy.

These are but a few examples of how blockchain technology may be used in intelligent power grids [2]. Future years should provide even more cutting-edge uses as technology advances.

One use of blockchain technology in smart power grids is shown below.

- VPP or Virtual Power Plant: A collection of DERs, such as solar panels, batteries, and electric vehicles, that cooperate to provide energy to the grid is known as a virtual power plant (VPP). Blockchain can be used to manage the VPP and ensure that it is operating efficiently and reliably.

For example, the Grid Singularity company is using blockchain technology to manage a VPP in Europe. The VPP consists of over 100 DERs, and it is used to provide backup power to the grid during outages.

Direct exchange between users and generators is made possible by the transactive energy system. Blockchain technology has the potential to increase the safety and openness of active energy transactions.

The Australian startup Power Ledger, for instance, is using blockchain technology to enable transactional energy trading. The firm provides a marketplace where energy buyers and sellers may interact.

1.2 Blockchain Architecture

The architecture of a blockchain is the plan behind how a blockchain works. As shown in Fig. 1, it describes the system's components, how they interact, and the safeguards in place to keep the system safe.

The architecture components of a blockchain system are:

1. Nodes: Computers that take part in the blockchain network are called "nodes". They act as data repositories and consensus hubs for the blockchain.
2. Transactions: Transactions are the data that is stored on the blockchain. They typically represent an exchange of value, such as a transfer of money or a smart contract execution.
3. Blocks: A block is a group of connected transactions that connect altogether. Each next block contains a hash of the one before it, making the data in the chain irreversible.
4. Consensus mechanism: In order to get an agreed-upon opinion on the blockchain's current state, a consensus method is used. It guarantees that the blockchain is same on every node in the network.
5. Cryptography: Cryptography is used to secure the blockchain system. It helps protect the data from manipulation and unauthorised access.

There are two primary components to a blockchain's structure: the application layer and the underlying infrastructure.

The apps created on top of the blockchain fall within the purview of the application layer. It specifies the protocols and data structures that enable communication with the blockchain.

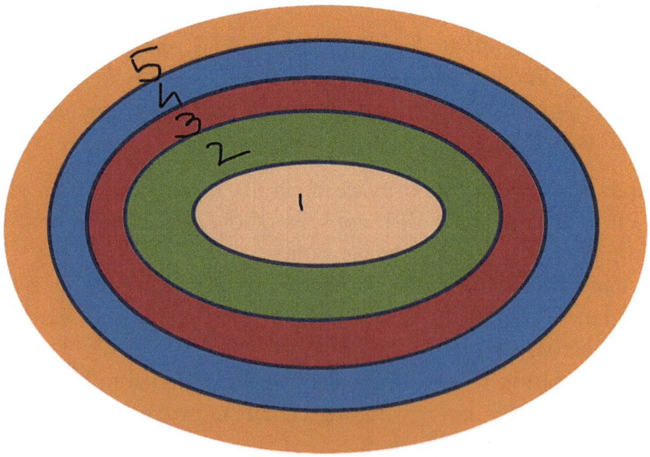

Fig. 1 Blockchain Architecture

The infrastructure layer is responsible for the underlying blockchain technology. It defines the network topology, the consensus mechanism, and the cryptography that are used to secure the system.

There are many different blockchain architectures that have been proposed. The most common architecture is the public blockchain. In a public blockchain, anybody may join in on the action. Common uses include cryptocurrency and supply chain management, which both benefit greatly from a high level of public scrutiny.

Private blockchains are another popular setup. Permissioned, or private, blockchains allow only approved users to join the network. Financial and other applications that need to keep sensitive data safe are common places to find them in use.

There are also hybrid blockchain architectures that combine the features of public and private blockchains. With these frameworks, public and private blockchains' scalability and security features are meant to be integrated.

The blockchain architecture is still evolving. We anticipate the emergence of novel and revolutionary designs as the technology develops.

Here are some of the challenges in blockchain architecture:

- Security: Blockchain systems are designed to be secure, but they are not immune to attack. There have been a number of successful attacks on blockchain systems, and there is always the potential for new attacks to be developed.
- Scalability: Blockchain systems can be slow and inefficient, especially as they grow in size. This can make them unsuitable for some applications, such as real-time trading.
- Complexity: Blockchain systems are complex and can be difficult to understand and manage. This can make them a challenge to deploy and use.
- Regulation: Blockchain systems are not yet well-regulated. This can make it difficult for businesses to adopt blockchain technology.

Blockchain architecture remains a promising technology despite these obstacles, having the potential to disrupt multiple industries.

1.3 Understanding Blockchain Networks: Securely Verifying Transactions and Beyond

A blockchain network is a group of computers working together to track and verify transactions. Using cryptography, the ledger links together an ever-expanding collection of records (blocks). Every block has its own timestamp and a reference to the one before it. The impossibility of erasing or changing information on the blockchain is a direct result of this feature.

Blockchain networks illustrated in Fig. 2 are often used to record transactions of digital assets, such as cryptocurrencies. However, they can also be used to record other types of data, such as contracts, medical records, and supply chain data.

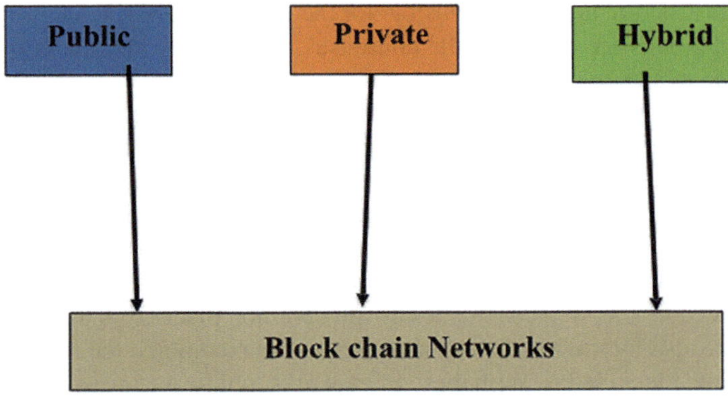

Fig. 2 Blockchain Network system

Blockchain networks may be broken down into various categories. Some of the most common types include:

- Public blockchains: Anyone may join a public blockchain and add their own transactions. Common uses include cryptocurrency and supply chain management, which both benefit greatly from a high level of public scrutiny.
- Private blockchains: Permissioned, or private, blockchains allow only approved users to join the network. They are frequently used in highly secure contexts, such as financial ones.
- Hybrid blockchains: Hybrid blockchains combine the benefits of both public and private blockchains. With these frameworks, public and private blockchains' scalability and security characteristics are meant to be integrated.

The benefits of blockchain networks include:

- Security: Blockchain networks are very secure. Information recorded on a blockchain cannot be easily modified or removed.
- Transparency: All transactions on the blockchain are public. This makes it easy to track the flow of funds and ensure that there is no fraud.
- Efficiency: Blockchain networks can be used to automate transactions. This can save time and money.
- Disintermediation: Blockchain networks can be used to eliminate the need for intermediaries. This can reduce costs and improve efficiency.

The challenges of blockchain networks include:

- Complexity: It may be challenging to comprehend and keep track of blockchain networks due to their natural complexity.
- Scalability: Blockchain networks can be slow and inefficient, especially as they grow in size.
- Regulation: Blockchain networks are not yet well-regulated. This can make it difficult for businesses to adopt blockchain technology.

Despite these challenges, blockchain networks are a promising technology with the potential to revolutionise many industries. These problems will likely be solved as blockchain technology advances, leading to wider use of blockchain networks.

Here are some of the applications of blockchain networks:

- Financial services: Blockchain networks can be used to record transactions of digital assets, such as cryptocurrencies. They can also be used to automate payments and settlements.
- Supply chain management: The supply chain of any given product or resource may be tracked via a blockchain network. There may be gains in both effectiveness and transparency as a result of this.
- Healthcare: Medical records may be stored and shared via a blockchain network. This has the potential to save costs without compromising care quality.
- Government: Blockchain networks can be used to record land titles, property deeds, and other legal documents. This can improve transparency and efficiency.
- Internet of Things (IoT): Blockchain networks may be used to control and protect Internet of Things devices. Fraud and hacks may be avoided with this measure in place.

1.4 Pros and Cons

Pros:

- Security: The blockchain system is very safe and reliable. Information on a blockchain is incredibly difficult to change or remove. Since each block in the blockchain contains a connection to the one before it, altering any information in one block would need altering all of the ones that followed.
- Transparency: All transactions on the blockchain are public. This makes it easy to track the flow of funds and ensure that there is no fraud.
- Efficiency: Blockchain can be used to automate transactions. This can save time and money.
- Disintermediation: Blockchain can be used to eliminate the need for intermediaries. This can reduce costs and improve efficiency.
- Immutability: Once data is recorded on a blockchain, it cannot be altered in any way. Because of this, blockchain may be used to securely store vital documents like property titles and medical records.
- Scalability: Blockchain technology can be easily expanded. This paves the way for its usage in processing several transactions simultaneously.

Cons:

- Complexity: The blockchain is a sophisticated and often difficult to handle technology.
- Energy consumption: Blockchain mining consumes a lot of energy. This is a major environmental concern.
- Regulation: Blockchain technology is not yet well-regulated. This can make it difficult for businesses to adopt blockchain technology.

- Hacking: Blockchain is not completely immune to hacking. There have been a number of successful attacks on blockchain systems.
- Slow transaction speeds: Blockchain transactions can be slow, especially on public blockchains. This can make it unsuitable for some applications, such as real-time trading.

Overall, blockchain technology is a promising technology with the potential to revolutionise many industries.

2 Block

Smart power systems based on blockchain technology have been around since the early 2000s, when scientists first started investigating its potential to enhance the safety and reliability of electrical grids. The Bitcoin whitepaper from 2008 exposed the world to blockchain technology and its potential uses in the realm of digital currency. An increasing number of initiatives were initiated in the years that followed to investigate the feasibility of using blockchain technology in smart power grids. One of the earliest projects to use blockchain for smart power systems was the Brooklyn Microgrid project, which was launched in 2015. This project employed blockchain technology to build a decentralised marketplace where people in Brooklyn, New York, could buy and sell energy from one another. The experiment demonstrated that blockchain technology has the potential to improve the market's transparency and effectiveness for trading in electricity.

Australia initiated the Power Ledger initiative in 2016. Renewable energy trade is made possible with this initiative by using the blockchain technology. The project has been successful in attracting a number of participants, and it has been used to trade over 100,000 megawatt hours of energy.

These are just a few examples of the numerous ongoing initiatives using blockchain technology in smart power grids. It's reasonable to assume that as the technology advances, more and more creative uses will emerge.

The following are some of the main advantages of blockchain technology for smart power systems:

- Security: Blockchain is a very secure technology. On a blockchain, data is recorded and cannot be easily changed or deleted. This makes it ideal for recording important data, such as energy transactions.
- Transparency: All transactions on the blockchain are public. This makes it simple to track the movements of energy and check for fraud.
- Efficiency: Blockchain can be used to automate energy trading. This can save time and money.
- Disintermediation: Blockchain can be used to eliminate the need for intermediaries. This can reduce costs and improve efficiency. Immutability: Once information is recorded on a blockchain, it cannot be altered in any way. This makes blockchain a perfect medium for legally enforceable agreements like energy contracts.

- Scalability: Blockchain technology can be easily expanded. This paves the way for its usage in processing several transactions concurrently.

Blockchain's use in smart power systems is in its infancy, but it might dramatically alter the way we handle and make use of energy. Power grids may be made more efficient, transparent, and secure with the use of this technology, and new markets for trading energy can be established as a result [3].

Before blockchain is extensively implemented for smart power systems, there are a number of obstacles that must be overcome.

- Complexity: The blockchain is an advanced and often difficult to handle technology.
- Energy consumption: Blockchain mining consumes a lot of energy. This is a major environmental concern.
- Regulation: Blockchain technology is not yet well-regulated. This can make it difficult for businesses to adopt blockchain technology.
- Hacking: Blockchain is not completely immune to hacking. There have been a number of successful attacks on blockchain systems.
- Slow transaction speeds: Blockchain transactions can be slow, especially on public blockchains. This can make it unsuitable for some applications, such as real-time trading.

2.1 *Transactive Energy*

The term "transactive energy" refers to a system that facilitates the bilateral exchange of energy between generators and users. This can be done through a blockchain-based platform, which would record all transactions securely and transparently.

Blockchain technology, a distributed ledger system, allows for the secure and transparent recording of transactions. Because of this, it is well-suited for transactive energy applications, since it may provide a unified understanding of the energy market among all stakeholders.

The benefits of using blockchain technology for transactive energy include:

- Security: Blockchain is a very secure technology. Information on a blockchain might be incredibly hard to change or remove. Because of this, energy transactions can be recorded with ease.
- Transparency: All transactions on the blockchain are public. This enables monitoring energy use and detecting any instances of fraud.
- Efficiency: Blockchain can be used to automate energy trading. This can save time and money.
- Disintermediation: Blockchain can be used to eliminate the need for intermediaries. This can reduce costs and improve efficiency.

- Immutability: Once data is recorded on a blockchain, it cannot be altered in any way. This makes blockchain a perfect medium for legally binding documents like energy contracts.
- Scalability: The blockchain may be expanded forever. This paves the way for its usage in processing several transactions simultaneously.

Some of the applications of transactive energy in smart power systems include:

- Peer-to-peer energy trading: Peer-to-peer energy trading platforms may be developed using blockchain technology. This would eliminate the need for a middleman in energy transactions, allowing consumers and generators to deal directly with one another.
- Demand response: Demand response programmes may make use of blockchain technology. As a result, utilities might provide discounts or other incentives to customers during times of high energy demand.
- Distributed Energy Resources (DERs) management: DERs like solar panels and batteries can be managed using blockchain technology. This would improve DERs' ability to compete in the energy market.
- Grid security: Blockchain can be used to improve the security of the grid. This would make it more difficult for hackers to disrupt the grid or steal energy.
- Green Energy Trading: Green energy, such as solar and wind power, may be traded using blockchain technology. The widespread use of renewable energy would be aided by this.

The use of blockchain technology for energy transactions, albeit it is still in its infancy, has the potential to radically transform our perspective on this resource. Power grids may benefit from the increased efficiency, transparency, and security that this technology provides.

2.1.1 Microgrid Management

Independent of the larger power grid, a microgrid provides electricity to a specific area. Microgrids tend to be installed in out-of-the-way places or regions that frequently experience power outages [4].

Blockchain technology has several applications in microgrid management. Some applications of blockchain technology include:

- Track energy transactions: Blockchain technology may be used to monitor the distribution of electricity in a microgrid. This may aid in preventing energy theft and ensuring that everyone is responsible for paying for their use.
- Automate demand response: Blockchain can be used to automate demand response programmes. This implies that users might be encouraged to use less energy during moments of high demand. Microgrid performance and reliability may both benefit from this.
- Manage distributed energy resources: Solar panels and batteries are examples of distributed energy resources (DERs), both of which may be controlled by block-

chain technology. This may aid in the safe and effective incorporation of DERs into the microgrid.
- Provide grid security: Blockchain technology may be utilised to protect grid infrastructure. Because of the immutability of the blockchain, the microgrid may be shielded from hacking attempts.

Here is an example of a company that is using blockchain for microgrid management:

- Power Ledger: In order to facilitate the exchange of renewable energy, Power Ledger is creating a blockchain-based platform. All DERs and other users of the platform may participate in the buying and selling of energy with one another. The platform is currently being used in a number of microgrids around the world.

Blockchain is an exciting new technology that might dramatically alter how power is produced, distributed, and consumed. One way that blockchain technology may be utilised to enhance the reliability, safety, and longevity of the electrical system is via its use in microgrid management.

2.2 *Cyber Security*

- Distributed ledger: Blockchain's decentralised ledger technology makes it hard for hackers to modify data in transit. It is because every block in the blockchain is interconnected with every block before it, making it almost hard to change any data without also changing every block after it.
- Encryption: Blockchain uses encryption to secure data. Data is encrypted and may only be read by approved persons.
- Immutability: Once information is recorded on a blockchain, it cannot be altered in any way. This makes it more difficult for hackers to erase or change information.
- Transparency: All transactions on the blockchain are public. This makes it easy to track and monitor activity on the network.
- Auditability: The blockchain ledger is transparent and immutable, making it easy to audit and trace transactions. This can help to identify and prevent fraud and cyberattacks.

As a case study of how blockchain technology might be utilised to strengthen the cyber security of a smart power system, consider the following:

- A smart power grid could use blockchain to track the flow of energy. This would make it more difficult for hackers to disrupt the grid by tampering with the data.
- Blockchain could also be used to store data about the grid's assets, such as transformers and generators. It would be more challenging for hackers to steal or destroy these assets if this were implemented.

- Blockchain could also be used to create a secure communication network for the grid. This would make it more difficult for hackers to intercept or disrupt communications.

When applied to smart power systems as a whole, blockchain has the potential to greatly enhance their cyber security. The blockchain, however, is not a panacea, it is crucial to remember that. Since it is still in its infancy, there are significant security issues that must be worked out before it can be widely used.

2.3 Smart Grid

2.3.1 Intelligent Power Network

The term "smart grid" refers to a kind of intelligent electrical grid that uses digital technology to improve the grid's efficacy, dependability, and safety. Data about the grid is collected and monitored by smart grids using sensors, communication networks, and data analytics. Using this information, the grid's functioning may be optimised, electricity quality enhanced, and outages minimised.

A distributed ledger called blockchain may be used to publicly and trustlessly record commercial transactions. Blockchain is not controlled by any central authority, making it a tamper-proof and immutable record of transactions.

2.3.2 How Blockchain Can be Used in Smart Grids

Many aspects of a smart grid may benefit from the usage of blockchain technology, including the grid's efficiency, dependability, and security.

- Decentralised energy trading: Using blockchain technology, it is possible to build a decentralised marketplace for the selling of energy between customers and producers. As a result, the grid's efficiency may be increased and dependence on central power plants decreased.
- Demand response: The use of demand response programmes, which may be done with the use of blockchain technology, may encourage consumers to reduce their energy consumption during peak demand hours. In turn, this may assist ease grid stress and improve dependability.
- Fraud detection: Blockchain technology may be used to monitor power use and identify fraudulent transactions. Losses may be minimised and safety enhanced in this way.
- Asset management: Blockchain technology may be used to record who owns which grid assets and where they are located. This has the potential to increase productivity and lessen the possibility of theft.

- Security: The grid can be protected against hackers with the use of blockchain technology. Blockchain's immutability makes it difficult for hackers to modify or remove data.

2.3.3 The Future of Smart Grids

The smart grid might be completely transformed by blockchain technology. There will likely be an increase in blockchain's use in smart grids as the technology develops. These applications will help to make the grid more efficient, reliable, and secure [5, 6].

Here are some of the specific use cases of blockchain in smart grids:

- Providing transparency and accountability: Energy transfers may be tracked in a transparent and open manner using blockchain technology. As a result, customers, producers, and grid operators may become more trusted.
- Improving grid security: Blockchain can be used to secure the grid against cyberattacks. Hackers find it difficult to add, delete, or change data, thanks to blockchain's immutability.
- Facilitating peer-to-peer energy trading: The use of blockchain technology may make it easier for people to deal in energy. As a result, the grid's efficiency may be increased and dependence on central power plants reduced.
- Optimising grid operations: Blockchain can be used to optimise grid operations. In addition to saving cash, this may also boost dependability.
- Managing distributed energy resources: Two examples of distributed energy resources that might be handled using blockchain technology are solar panels and wind turbines. This could allow the more effective incorporation of these resources into the grid.

These are just a few of the many ways that blockchain can be used in smart grids.

2.4 Energy Flexibility

- Traditional power systems are centralised and rely on large power plants to generate electricity. This can lead to problems such as blackouts and brownouts when there is not enough electricity to meet demand.
- Digital technologies are used by smart power systems to make the grid more efficient and reliable. Smart metres, which may collect data on energy usage, are examples of distributed energy resources (DERs), along with solar panels and batteries that can generate power locally.
- The adaptability of smart power grids may be further enhanced by using blockchain technology. A distributed ledger called blockchain is capable of accurately and transparently recording transactions. This makes it perfect for monitoring power distribution in a smart grid [7, 8].

The following are some concrete applications of blockchain technology that increase energy flexibility in smart power systems:

- Demand response: Blockchain technology may improve demand response programmes, which provide incentives for customers to reduce their energy usage during times of high demand. Brownouts and power outages may be avoided in this way.
- Virtual power plants: Blockchain can be used to create virtual power plants, which aggregate the distributed energy resources of many different consumers and businesses. This may aid in enhancing the dependability of the grid and facilitating the provision of grid balancing services.
- Green certificates: Energy's environmental characteristics, such as its source and accompanying emissions, may be tracked using blockchain technology. This can help to promote the use of renewable energy.
- Peer-to-peer energy trading: The usage of blockchain technology enables peer-to-peer energy trading, in which clients buy and sell energy directly with one another. The grid's effectiveness may be raised as a consequence, and fewer central power plants will be needed.

Blockchain technology's use in smart power grids is in its infancy, but it might significantly alter the energy industry and the lives of its consumers. By improving energy flexibility, blockchain technology can help to make the grid more resilient, efficient, and sustainable.

The following benefits of blockchain technology may also apply to smart power systems:

- Increased transparency: By creating an immutable and transparent record of energy transactions, blockchain has the potential to boost confidence in the entire system.
- Reduced costs: Grid management costs, as well as costs associated with energy trading and settlement, might all be decreased with the use of blockchain technology.
- Enhanced security: Blockchain technology may provide a reliable and verifiable means of storing energy data, shielding it from hacking and other forms of fraud.

All things considered, the development of future smart power grids may be significantly impacted by blockchain technology. By improving energy flexibility, the grid may be made more dependable, effective, sustainable, and secure with the use of blockchain technology. [9] We must solve the following challenges in order to fully realise the potential of blockchain technology in smart power systems.

- Scalability: Blockchain networks can be slow and expensive to scale, which can be a barrier to widespread adoption.
- Security: Blockchain networks are vulnerable to cyberattacks, and there is a need to develop secure and robust solutions.
- Regulation: There is a need for clear and uniform blockchain legislation, since the regulatory landscape continues to evolve.

3 Blockchain-Based Smart Grid

The term "blockchain-based smart grid" refers to a system that employs blockchain technology to make the electricity grid more reliable, secure, and transparent. A distributed ledger called blockchain enables immutable and secure transaction recording. This makes it ideal for tracking the flow of electricity in a smart grid [10].

The use of blockchain technology in a smart grid has the following advantages:

- Increased transparency: Blockchain has the potential to increase confidence between parties by providing an immutable and auditable record of energy transactions.
- Reduced costs: Energy trade and settlement costs, as well as grid management costs, may all be reduced with the use of blockchain technology.
- Enhanced security: Blockchain innovation may provide a trustworthy and verifiable means of storing energy data, shielding it from hacking and other forms of fraud.
- Improved efficiency: Peer-to-peer energy trading and optimised energy flow are two ways in which blockchain technology might assist increase grid efficiency.
- Increased reliability: By offering a secure and dependable means of storing and transmitting energy data, blockchain technology may contribute to improving the grid's resilience.

Blockchain technology has a wide range of possible smart grid applications. Some of the most common applications include [11, 12]:

- Demand response: With the use of blockchain technology, demand response programmes, which encourage customers to reduce their energy usage during times of high demand, may be made much easier. Blackouts and brownouts are less likely to occur if this is done.
- Virtual power plants: Blockchain can be used to create virtual power plants, which aggregate the distributed energy resources of many different consumers and businesses. This could help in enhancing the dependability of the grid and facilitating the provision of grid balancing services.
- Green certificates: Energy's environmental characteristics, such as its source and accompanying emissions, may be tracked using blockchain technology. This may support encouraging the usage of renewable energy sources.
- Peer-to-peer energy trading: Through the use of blockchain technology, peer-to-peer energy trading, in which users buy and sell energy directly with one another, is made possible. As a consequence, we may improve grid efficiency and reduce the demand for centralised power plants.

Even while blockchain-based smart grid applications are still in their infancy, this technology has the potential to fundamentally transform how power is generated, delivered, and then used. Blockchain technology may make the grid more robust, sustainable, and cost-effective by enhancing the grid's efficiency, security, and transparency [13].

In order to fully realise the promise of blockchain technology in smart grids, the following obstacles must be overcome:

- Scalability: Blockchain networks can be slow and expensive to scale, which can be a barrier to widespread adoption.
- Security: Blockchain networks are vulnerable to cyberattacks, and there is a need to develop secure and robust solutions.
- Regulation: In the blockchain technology sector, which is still in its infancy from a legal standpoint, there is a need for clear and consistent laws.

Blockchain technology's potential advantages in smart grids are substantial, notwithstanding these challenges. Future energy industries are projected to make further use of this technology as it continues to advance [14].

- 2016: The Brooklyn Microgrid project was launched in New York City. This project used blockchain technology to create a microgrid that could island from the main power grid during outages.
- 2017: The Transactive Energy Coalition was formed. More than 50 firms and institutions have banded together to create and launch transactive energy apps that make use of blockchain technology.
- 2018: The Energy Web Foundation was launched. This group is supporting the creation and dissemination of open-source blockchain-based energy industry solutions.
- 2019: In Australia, they started working on the Power Ledger. This project utilises blockchain technology to facilitate peer-to-peer transactions between users for the purchase and sale of solar energy.
- 2020: Europe is the birthplace of the Grid Singularity initiative. In order to provide a decentralised market for exchanging energy, this project employs blockchain technology.
- 2021: The WePower project was launched in Estonia. This project allows consumers to invest in renewable energy projects using blockchain technology.

These are but a handful of the numerous active blockchain-based smart grid initiatives. In the years to come, we may expect to see even more cutting-edge implementations of this technology.

Here are some of the current scenarios where blockchain is being used in smart grids:

- Demand response: Blockchain technology may improve demand response programmes, which provide incentives for customers to reduce their energy usage during times of high demand. Blackouts and brownouts are less likely to occur if this is done.
- Virtual power plants: Blockchain can be used to create virtual power plants, which aggregate the distributed energy resources of many different consumers and businesses. This can help to provide grid balancing services and improve the reliability of the grid.

- Green certificates: Energy's environmental impacts, such as where it comes from and any pollutants produced, may be traced using blockchain technology. This may assist encouraging the usage of renewable energy sources. Peer-to-peer energy trading: The adoption of blockchain technology makes peer-to-peer energy trading—where clients buy and sell energy directly to one another—possible. The grid's effectiveness may be raised as a consequence, and fewer central power plants will be required.
- Security: Blockchain technology may be used to secure the grid from hackers. This is due to blockchain's distributed ledger, which is incredibly secure and cannot be changed.
- Transparency: The energy company is one area where blockchain technology might increase openness. This is due to the fact that blockchain technology permits reliable and verifiable monitoring of energy transactions.

All things considered, blockchain technology may improve smart grids in terms of efficacy, reliability, safety, and transparency. The energy industry might rely more and more on this technology in the future as it continues to advance.

3.1 Challenges and Opportunities

Some of the challenges and opportunities in blockchain-based smart grids:

Challenges

- Scalability: Blockchain networks can be slow and expensive to scale, which can be a barrier to widespread adoption.
- Security: Blockchain networks are vulnerable to cyberattacks, and there is a need to develop secure and robust solutions.
- Regulation: In the blockchain technology sector, which is still in its infancy from a legal standpoint, there is a need for clear and consistent laws.
- Data privacy: Since blockchain is a distributed ledger, everyone in the system can see every transaction that takes place. Protecting sensitive data, including client information, may be difficult in situations like these.
- Interoperability: There are many different blockchain platforms, and it can be difficult to get them to interoperate with each other. This can be a challenge for creating a truly decentralised and unified smart grid.

Opportunities

- Increased transparency: By creating an irreversible and auditable record of energy transactions, blockchain has the potential to boost confidence in the entire system.
- Reduced costs: Energy trade and settlement expenses, as well as grid management costs, may all be reduced with the use of blockchain-based technologies.

- Enhanced security: Blockchain technology may provide a trustworthy and verifiable means of storing energy data, shielding it from hacking and other forms of fraud.
- Improved efficiency: Peer-to-peer energy trading and optimised energy flow are two ways in which blockchain technology might assist increase grid efficiency.
- Increased reliability: Blockchain technology's capacity to reliably store and transfer energy data may greatly improve the grid's reliability.
- Sustainability: Blockchain technology may facilitate the tracking and trading of green energy credits, which in turn can assist to increase the usage of renewable energy.

In general, there is a lot to gain from using blockchain technology to smart grids. There are, however, obstacles that must be overcome before blockchain technology can be extensively utilised.

4 Blockchain Ethics

Concerned with the moral issues raised by blockchain technology, blockchain ethics is its own subfield of ethics. A distributed ledger called blockchain may be used to publicly and trustlessly record commercial transactions. Despite the potential for widespread disruption, this presents a variety of moral issues [15].

Some of the key ethical issues associated with blockchain include:

- Using anonymisation techniques to protect privacy.
- Developing more secure blockchain protocols.
- Minimising the energy consumption of blockchain networks.
- Promoting the development of decentralised blockchains.
- Ensuring that blockchain systems are designed to be fair and transparent.

In other words, blockchain ethics is the study of the moral and ethical implications of blockchain technology. The distributed ledger known as blockchain may be used to consistently and transparently record commercial transactions. Numerous businesses might be profoundly affected by this, but it also presents some serious ethical questions.

Some of the key ethical issues associated with blockchain include:

- Privacy: Since blockchain is a distributed ledger, everyone in the system can see every transaction that takes on. This can raise concerns about privacy, especially for sensitive data such as financial transactions or medical records.
- Security: Blockchain networks are vulnerable to cyberattacks. This might be used to sabotage transactions, steal data, or even take over the network.
- Energy consumption: Blockchain networks require a lot of energy to operate. The environment could be negatively impacted by this.
- Centralization: Blockchain technology can be used to create decentralised systems. However, it is also possible to create centralised blockchains that are

controlled by a single entity. This could lead to new forms of centralization and concentration of power.

- Fairness: Blockchain technology could be used to create more fair and transparent systems. However, it is also possible to use it to create systems that are less fair and more opaque.

These moral issues must be resolved if blockchain technology is to be used in an ethical way. The field of blockchain ethics is still in its early stages, but it is an important area of research that will need to be addressed as blockchain technology continues to develop.

Here are some of the ethical principles that can be applied to blockchain technology:

- Autonomy: Individuals should have the right to control their own data and transactions.
- Transparency: Blockchain transactions should be transparent and accessible to all stakeholders. Accountability: Those who participate in blockchain networks should be held accountable for their actions.
- Fairness: Blockchain systems should be designed to be fair and equitable for all participants.
- Sustainability: Blockchain networks should be designed to minimise their environmental impact.
- Immutability: Since the blockchain itself is irreversible, data added to it cannot be changed once it has been recorded there. This can be seen as a positive, as it can help to ensure the accuracy and transparency of data. However, it can also be seen as a negative, as it can make it difficult to correct errors or remove outdated data.
- Privacy: Since blockchain is a distributed ledger, everyone in the system can see every transaction that takes place. This may be seen favourably since it increases openness and trust. However, it can also be seen as a negative, as it can expose sensitive data, such as financial information or medical records.
- Energy consumption: The mining process used to create new blocks on the blockchain network can be very energy-intensive. This has led others to worry about the ecological effects of blockchain technology.
- Centralization: Some blockchain networks are more centralised than others. This can raise concerns about the control of the network and the potential for censorship.
- Cyber security: Blockchain networks are vulnerable to cyberattacks. There is now cause for alarm regarding the safety of information kept on the blockchain.

These are only a few of the major moral concerns that have been voiced in relation to blockchain. New ethical concerns are expected to surface as the technology advances. These problems must be understood, and action must be taken to resolve them.

5 Future of Smart Power Systems with Blockchain

With blockchain, the future of intelligent power grids seems bright. Blockchain technology may significantly alter the ways in which power is produced, transmitted, and used. Blockchain technology's ability to increase the grid's efficiency, security, and transparency will go a long way towards making it more robust, sustainable, and affordable [16].

6 Conclusion

The smart electricity system might be completely transformed by blockchain technology by making it more transparent, efficient, and safe. Peer-to-peer energy trading, demand response programme facilitation, virtual power plant creation, and more are all possible with blockchain technology. Scalability, security, and legal issues still need to be resolved before blockchain can be widely used in the smart power system.

Generally speaking, the future growth of smart power grids may be significantly impacted by blockchain technology. Blockchain technology's ability to increase the grid's efficiency, security, and transparency will go a long way towards making it more robust, sustainable, and cost-effective.

References

1. Aggarwal, S., Chaudhary, R., Aujla, G. S., Kumar, N., Choo, K.-K. R., & Zomaya, A. Y. (2019). Blockchain for smart communities: Applications, challenges and opportunities. *Journal of Network and Computer Applications, 144*, 13–48.
2. Aghimien, D. O., Aigbavboa, C., Edwards, D. J., Mahamadu, A.-M., Olomolaiye, P., Nash, H., & Onyia, M. (2020). A fuzzy synthetic evaluation of the challenges of smart city development in developing countries. Smart sustain. *Built Environment, 11*, 405.
3. Ahmed, S., Shah, M. A., & Wakil, K. (2020). Blockchain as a trust builder in the smart city domain: A systematic literature review. *IEEE Access, 8*, 92977–92985.
4. Banerjee, M., Lee, J., & Choo, K.-K. R. (2018). A blockchain future for internet of things security: A position paper. *Digital Communications and Networks, 4*, 149.
5. Bernal Bernabe, J., Luis Canovas, J., Hernandez-Ramos, J. L., Torres Moreno, R., & Skarmeta, A. (2019). Privacy-preserving solutions for blockchain: Review and challenges. *IEEE Access, 7*, 164908.
6. Bhushan, B., Khamparia, A., Sagayam, K. M., Sharma, S. K., Ahad, M. A., & Debnath, N. C. (2020). Blockchain for smart cities: A review of architectures, integration trends and future research directions. *Sustainable Cities and Society, 61*, 102360.
7. Calvillo, C. F., Sánchez-Miralles, A., & Villar, J. (2016). Energy management and planning in smart cities. *Renewable and Sustainable Energy Reviews, 55*, 273–287.
8. Caragliu, A., Bo, C. D., & Nijkamp, P. (2011). Smart cities in Europe. *Journal of Urban Technology, 18*(2), 65–82.

9. Chaudhary, R., Jindal, A., Aujla, G. S., Aggarwal, S., Kumar, N., & Choo, K.-K. R. (2019). *BEST: Blockchain-based secure energy trading in SDN-enabled intelligent transportation system* (Vol. 85, p. 288). Computers & Security.
10. Chiu, W.-Y., & Meng, W. (2021). EdgeTC - A PBFT blockchain-based ETC scheme for smart cities. *Peer-to-Peer Networking and Applications, 14*(5), 2874–2886.
11. Dwivedi, A. D., Srivastava, G., Dhar, S., & Singh, R. (2019). A decentralized privacy-preserving healthcare blockchain for IoT. *Sensors (Switzerland), 19.*
12. Gong, S., Tcydenova, E., Jo, J., Lee, Y., & Park, J. H. (2019). Blockchain-based secure device management framework for an internet of things network in a smart city. *Sustainability, 11*(14), 3889.
13. Gupta, R., Kumari, A., & Tanwar, S. (2021). A taxonomy of blockchain envisioned edge-as-a-connected autonomous vehicles. *Transactions on Emerging Telecommunications Technologies, 32*(6), e4009.
14. Hammi, M. T., Hammi, B., Bellot, P., & Serhrouchni, A. (2018). Bubbles of trust: A decentralized blockchain-based authentication system for IoT. *Computers & Security, 78*, 126–142.
15. Hamed Taherdoost, *Smart contracts in blockchain technology: A critical review*. MDPI Emerging Industrial Applications: Orchestration of Machine Learning, the IoT, and Blockchain (2023). https://doi.org/10.3390/info14020117.
16. Hang, L., & Kim, D.-H. (2020). Reliable task management based on a smart contract for runtime verification of sensing and actuating tasks in IoT environments. *Sensors, 20*(4), 1207.

An Impression of Firewall Technologies along with Secure Electronic Transaction

Prashant Kumar, Chitra Kushwaha, Dinesh Kumar Yadav, and Solomon Raju Kota

1 Introduction

Firewalls play a crucial role in ensuring the security and protection of computer networks from unauthorized access and malicious threats. As the first line of defense, firewalls act as a barrier between internal networks and the external world, monitoring and controlling incoming and outgoing network traffic based on predetermined security rules. The primary objective of a firewall is to prevent unauthorized access to a network while allowing legitimate traffic to pass through. By examining packets of data, firewalls can enforce security policies and filter out potentially harmful or suspicious traffic. This proactive approach helps in mitigating various cyber threats such as unauthorized access, malware infections, distributed denial-of-service (DDoS) attacks, and data breaches. Firewalls employ different techniques to enforce security policies. Some common types of firewalls include packet-filtering firewalls, stateful inspection firewalls, application-level gateways (proxies), and next-generation firewalls (NGFW). Each type offers varying levels of security and functionality, catering to different network environments and security requirements [1]. The need for firewall arose in the late 1980s as the computer networks became more interconnected and the number of cyber threats increased. The

P. Kumar (✉)
Bennett University (The Times Group), Greater Noida, Uttar Pradesh, India

C. Kushwaha
DRDO - Institute for System Studies & Analyses, Civil Lines, Delhi, India

D. K. Yadav
Netaji Subhas University of Technology, Delhi, India

S. R. Kota
CSIR - Fourth Paradigm Institute, Bangalore, Karnataka, India

S. Vijayalakshmi et al. (eds.), *Smart Power Systems*, Power Systems,
https://doi.org/10.1007/978-3-031-89459-6_17

rapid growth of the internet and the expansion of computer networks in organizations created new challenges in terms of network security.

During that time, packet-filtering firewalls emerged which examines packets of data based on predefined rules, such as source and destination IP addresses, port numbers, and protocols, to determine whether to allow or block the traffic. Stateful inspection firewalls add an extra layer of intelligence by keeping track of the state of network connections, ensuring that only legitimate and authorized connections are established. Application-level gateways, or proxies, operate at the application layer of the network stack and act as intermediaries between internal and external networks. They provide enhanced security by inspecting application-specific data and validating its integrity before forwarding it to the intended recipient. Next-generation firewalls (NGFW) combine the capabilities of traditional firewalls with advanced features like deep packet inspection, intrusion detection and prevention systems (IDPS), and application awareness along with this, it offers enhanced visibility, control, and threat detection capabilities, allowing organizations to better protect their networks from sophisticated attacks. Firewalls are vital components of network security infrastructure, providing essential protection against cyber threats. By monitoring and controlling network traffic, firewalls help safeguard sensitive information, prevent unauthorized access, and maintain the integrity and availability of computer networks. The continuous evolution of firewall technologies aims to address emerging threats and ensure the highest level of network security in today's ever-changing digital landscape [2].

2 Evolution of Firewalls

The history of firewalls dates back to the late 1980s and early 1990s when the need for network security became increasingly evident. Let's delve into the detailed history of firewalls [3, 4]:

2.1 Early Development

In 1987, the first known firewall-like system, called the DEC SEAL, was developed by Digital Equipment Corporation (DEC). It provided basic packet filtering capabilities to protect internal networks. In 1988, the Morris Worm incident highlighted the vulnerability of computer networks, further emphasizing the need for security measures.

2.2 Commercial Firewalls

In the early 1990s, commercial firewalls began to emerge, offering more advanced features and better network protection. In 1992, Check Point Software Technologies introduced FireWall-1, one of the first commercial firewall products. In 1994, Cisco Systems introduced the PIX firewall, which became a popular choice for network security.

2.3 Evolution of Firewall Technologies

2.3.1 Stateful Inspection

In the mid-1990s, firewalls incorporating stateful inspection capabilities were introduced. These firewalls could examine the context and history of network connections, providing better security.

2.3.2 Application-Level Gateways

Also known as proxy firewalls, these firewalls operated at the application layer, allowing more granular control and inspection of network traffic.

2.3.3 Intrusion Detection Systems (IDS)

In the late 1990s, firewalls started integrating IDS capabilities to detect and respond to network intrusions.

2.3.4 Unified Threat Management (UTM)

In the 2000s, UTM firewalls emerged, combining various security features like firewalling, intrusion prevention, antivirus, and content filtering into a single appliance.

2.4 Next-Generation Firewalls (NGFW)

NGFWs, introduced in the late 2000s, offered enhanced security features such as deep packet inspection, SSL inspection, and application awareness. NGFWs focused on identifying and blocking advanced threats, including malware, botnets, and application-layer attacks.

2.5 Cloud-Based Firewalls

With the rise of cloud computing, cloud-based firewalls gained popularity. These firewalls provided scalable security solutions for cloud environments, protecting virtualized networks and applications.

2.6 Modern Firewalls

Today, firewalls continue to evolve with advancements in technology. They incorporate machine learning, AI, and threat intelligence to detect and respond to sophisticated cyber threats. Firewalls are often integrated with other security tools and technologies as part of a comprehensive security strategy [5].

Overall, firewalls have evolved from simple packet filters to complex, intelligent security solutions. They play a critical role in protecting networks, data, and systems from unauthorized access and cyber threats in the ever-changing landscape of cybersecurity.

3 Definition of Firewall

A device that is used for security of network filters all the incoming and outgoing packets. It has its own protocols on the basis of these protocols it decides which packet should be blocked and which should be allowed. Its main purpose is to secure the network from any type of security threat that could arise in the system because of outside data [6].

4 Features of Firewall

1. It provides multilevel protection to network and also protects against malware.
2. It has the capability to block the unauthorized access and ensure a secure system and an OS.
3. It provides access only to valid data packets, acts as a protection layer over a defined network, and strengthens the network security.
4. It also prevents leakage of internal information.

5 Types of Firewalls

Firewalls can be of software, hardware and combined of software and hardware. So, these firewalls are classified in three types of firewalls. Structurally, firewalls can be software, hardware, or a combination of both:

1. Software Firewall
2. Hardware Firewall.
3. Hybrid Firewall

5.1 *Software Firewall*

In this type of firewall, a software resides in the same system, it is protecting and do not require separate hardware. IT has some security rules those are predefined for each profile it can also be customized. This software is separately installed on the device in which security is required. Its advantages are that it requires less cost. Mainly it is used for personal or company laptops. It also uses some memory of the system [7].

5.1.1 Features of Software Firewall

Software firewalls offer a wide range of features that enhance network security and protect against threats. They provide application control, allowing administrators to monitor and control network traffic generated by specific applications. Port filtering enables access control to specific ports, ensuring only authorized services or applications can communicate through them. Real-time traffic monitoring provides visibility into network activities, enabling the identification of suspicious behavior. Intrusion Detection and Prevention System (IDPS) functionality helps detect and prevent network-based attacks by analyzing traffic for known attack signatures. Web filtering allows organizations to control access to websites and web content, blocking malicious or inappropriate sites. Integration with antivirus and anti-malware software provides an additional layer of protection against malware. User authentication ensures that only authorized users can access network resources, and VPN support enables secure remote access. Logging and reporting functionalities allow administrators to review network activities, security events, and policy violations, aiding in incident identification and compliance maintenance. Overall, software firewalls offer customizable and adaptable security measures to meet specific organizational needs [8].

5.1.2 Top 10 Software Firewall

I. Sophos Firewall
II. Check Point Next Generation Firewalls
III. FortiGate NGFW
IV. WatchGuard Network Security
V. pfSense
VI. SonicWall
VII. Zscaler Internet Access
VIII. Huawei Firewall
IX. Palo Alto Next-Generation Firewall
X. NG Firewall [9]

5.2 Hardware Firewall

A separate hardware unit that can be placed on a network between segments of the network. In hardware firewall, there is no need of any extra device it runs on its own machine. It is established behind the router so that it can filter all the incoming packets towards server. It acts as an antivirus solution for the server [10].

5.2.1 Features of Hardware Firewall

Hardware firewalls offer a range of features that collectively enhance network security and provide robust protection against threats. These features include packet filtering, stateful inspection, network address translation (NAT), VPN support, intrusion prevention system (IPS), content filtering, bandwidth management, and logging/reporting capabilities. By inspecting and filtering network traffic, maintaining connection states, implementing secure remote access, detecting and blocking malicious activities, controlling access to web content, managing bandwidth, and providing detailed logs and reports, hardware firewalls ensure the integrity, confidentiality, and availability of network resources. These features make hardware firewalls a vital component of a comprehensive cybersecurity strategy, enabling organizations to safeguard their networks from unauthorized access, malicious attacks, and other potential security risks [11].

5.2.2 Top 10 Hardware Firewall

XI. Bitdefender BOX: This firewall protects all internet connected devices in your digital life, at home, and on the go.
XII. Cisco ASA 5500-X.
XIII. CUJO AI Smart Internet Security Firewall.

XIV. Fortinet FortiGate 6000F Series.
XV. Netgear ProSAFE.
XVI. Palo Alto Networks PA-7000 Series.
XVII. Netgate pfSense Security Gateway Appliances.
XVIII. SonicWall Network Security Firewalls.
XIX. Sophos XG Firewall.
XX. WatchGuard Firebox [12].

5.3 Hybrid Firewalls

Hybrid firewalls represent a convergence of hardware and software firewall technologies, leveraging the respective strengths of each approach to deliver comprehensive network security. These firewalls combine the robustness of hardware firewalls, which feature dedicated security hardware and high-performance packet processing capabilities, with the flexibility and advanced features of software firewalls [13]. The structure of the hybrid firewall can be seen in Fig. 1. From a technical perspective, hybrid firewalls can be categorized into two primary types:

5.3.1 Physical Hybrid Firewalls

This type involves a hardware appliance that integrates software-based security functionalities. The hardware component of the firewall provides efficient packet filtering, stateful inspection, and optimal network performance. The software

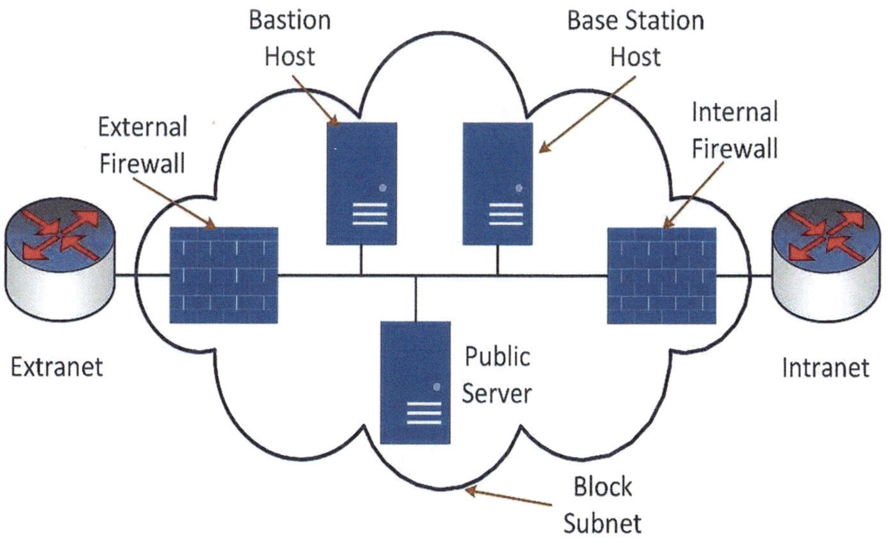

Fig. 1 Placement of hybrid firewall [14]

component adds advanced security features, including intrusion detection and prevention systems (IDPS), web filtering, and application control. By combining the hardware's strength in processing network traffic with the software's flexibility and advanced security capabilities, organizations can achieve high-performance network security while benefitting from the adaptability of software-based features.

5.3.2 Virtual Hybrid Firewalls

In virtualized or cloud environments, hybrid firewalls comprise a combination of virtual firewalls and physical firewalls. Virtual firewalls are software-based instances that run within virtual machines or cloud infrastructure, securing the virtualized network environment. Physical hardware firewalls complement these virtual counterparts by safeguarding the network perimeter and managing external network traffic. This hybrid approach capitalizes on the scalability and agility of virtual firewalls, along with the robustness and performance of physical firewalls. It provides organizations with a comprehensive security solution suitable for complex and dynamic network environments.

Hybrid firewalls offer numerous benefits by combining the strengths of hardware and software firewalls. They provide robust network protection, advanced security features, and scalability, making them well-suited for organizations that prioritize a balance between performance, flexibility, and security within their network infrastructure.

5.3.3 Top 10 Hybrid Firewalls

I. Cisco Adaptive Security Appliance (ASA)
II. Fortinet FortiGate
III. Palo Alto Networks Next-Generation Firewalls
IV. Check Point Security Gateway
V. SonicWall NSA Series
VI. Juniper Networks SRX Series
VII. Sophos XG Firewall
VIII. Barracuda Networks NextGen Firewall
IX. WatchGuard Firebox

6 Top 10 Hybrid Firewall Firewall Security Techniques

Firewalls have evolved over the years and today are quite sophisticated. Basically, there are three security techniques, which are used in firewalls [15].

6.1 Packet Filter Firewall Technique

The first firewalls called packet filter firewall was simply a router enabled with special filtering capabilities. This firewall works on the third layer of OSI model which is network layer. This type of firewalls makes processing decisions whether to allow or deny individual packet based on the network address, ports or protocols. Speed of these firewall is fast but not secure because it does not do any internal inspection of the traffic, it observes ports only. As it filters only source and destination address, source and destination port, and checks only headers of the packet. So anyone can easily bypass this by using some spoofing techniques and gives impractical results also. It means if any traffic is flowing in approved port, then it will not be blocked even it is malicious traffic.

6.2 Application Gateway

Special software was written for servers, and these types of firewalls were able to filter much more diligently. This firewall works on the application-level data and examines the data at every layer. These firewalls were greatly improved firewall capabilities in comparison of packet filter firewall. Multiple application gateways can run on a single host. Initially, the user connects to the application gateway using an application based on the TCP/IP model. Then after checking the authenticity of user, application gateway accesses the host remotely on the behalf of the user. So, in this way it hides network topology from the attacker. One application gateway technique's example is Proxy Server that is a system or router that provides a gateway between user and the internet. A proxy server has its own IP address. Whenever the computer sends any request to the internet then it passes through proxy server, it filters the messages at application layer.

6.3 Circuit Level Gateway

This setup is done with the help of two TCP connections; one is inside connection for inner host and another is outer connection for outer host (Fig. 2). After connection, establishment gateway transmits the TCP segments. This prepares a table to validate the connections and checks that which network packet is having data to send. At the time of terminating the connection, it removes those entries from the table and the virtual circuit connection is closed that was created between two nodes. This works on the session layer of the OSI model and between the application layer and transport layer.

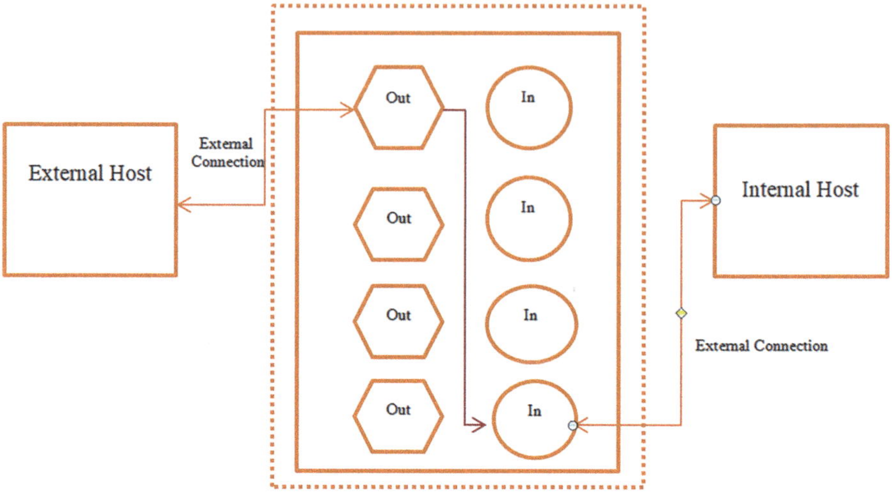

Fig. 2 Working of circuit level gateway [16]

7 Limitations of Firewall

Firewalls serve as an important line of defense in network security, but they have limitations. They cannot prevent users from accessing malicious websites, protect against virus-infected files or software transfers, prevent misuse of passwords, safeguard against misconfigurations, defend against social engineering, block attackers using modems, or secure already infected systems. It is crucial to supplement firewalls with additional defense systems to ensure comprehensive protection against evolving cyber threats.

In conclusion, firewalls play a crucial role in protecting networks from both external and internal attacks. They can be implemented as hardware or software solutions, following a specific set of rules to safeguard the network from viruses and other malicious threats. We have examined various aspects of firewalls, including their categories, components, design, implementation, and notable firewall software used in the networking industry. By deploying effective firewall systems, organizations can enhance their network security and mitigate potential risks.

Here we are going to see in detail the features and working of Sophos XG Firewall [17]:

7.1 Sophos XG

The Sophos firewall stands out as a network security solution that excels in identifying the source of infections on a network and taking immediate action to restrict the infected device's access to other resources. However, it has certain limitations in terms of its overall security capabilities. While Sophos offers a stable and comprehensive firewall solution, there may be challenges when it comes to integrating it with other products. Additionally, there is room for improvement in meeting the requirements of SD-WAN (Software-Defined Wide Area Network) implementations.

7.2 Features of Sophos XG650

7.2.1 Visibility and Protection

The Sophos XG650 firewall offers a range of powerful features for comprehensive network security. Deep Packet Inspection enables detailed analysis of network traffic, while support for encrypted traffic ensures protection even for secure communications. The firewall utilizes advanced techniques such as zero-day and machine learning protection to detect emerging threats in real time. The inclusion of a cloud sandbox allows for safe execution and detection of suspicious files. Web protection safeguards against web-based threats, while synchronized security integrates with other Sophos products for enhanced defense. Advanced threat protection mechanisms further enhance the firewall's capabilities, providing robust visibility and protection for networks.

7.2.2 Connectivity and Security

Sophos XG650 firewall is equipped with a range of networking and access features that enhance connectivity and security. The support for SD-WAN technology allows organizations to optimize network performance and effectively manage multiple network connections, resulting in improved efficiency and cost savings. The central SD-WAN orchestration capabilities simplify the management and configuration of SD-WAN deployments across multiple sites, streamlining network operations. Secure site-to-site VPN connections enable organizations to securely connect multiple locations over public networks, ensuring the confidentiality and integrity of data transmission. The implementation of Zero Trust Network Access principles in Sophos XG650 ensures secure access control and authentication, minimizing the risk of unauthorized access to network resources. Additionally, the remote access VPN functionality enables remote workers or traveling employees to securely connect to the corporate network, enabling productivity and collaboration from anywhere while maintaining a high level of security. These networking and access features provided by Sophos XG650 facilitate secure and efficient connectivity between different sites, while also implementing robust security measures to protect the network and ensure authorized access.

7.2.3 Management and Reporting

The Sophos XG650 firewall offers a range of management and reporting features to simplify network administration and provide valuable insights. The firewall can be centrally managed through Sophos Central, a cloud-based platform that provides a unified console for managing multiple Sophos security products. With central SD-WAN orchestration capabilities, network administrators can efficiently configure and manage SD-WAN deployments across multiple sites from a centralized interface. Central firewall reporting allows administrators to gather and analyze data on network traffic, security events, and user activity. The customizable dashboard provides a comprehensive overview of the network's security status and performance, while alerts notify administrators of potential threats or system issues. Additionally, the firewall includes free reporting capabilities, enabling administrators to generate detailed reports on network usage, security incidents, and other relevant metrics. These features empower administrators to effectively monitor and manage network security, detect potential threats, and ensure compliance with organizational policies.

7.3 Sophos XG650 Specifications

The Sophos XG650 firewall offers high-performance specifications, making it suitable for demanding network environments. This table provides a visual representation of the Sophos XG650 firewall specifications, showcasing the different throughput capacities and performance capabilities.

Specifications	Throughput (Mbps)
Firewall throughput	85,000
Firewall latency	5Us
IPsec VPN throughput	9000
New connections/s	220,000
Concurrent connections	3 million
Threat protection throughput	7700
NGFW throughput	4.6 Gbps
IPS throughput	20,250
Xstream SSL concurrent connections	512,000
Xstream SSL decryption + threat protection	1350
Firewall IMIX	34,500

7.4 Sophos XG650 Physical Interface

The physical interface and connectivity options of the Sophos XG650 firewall are as follows:

Specifications	Description
Display	The firewall is equipped with a multi-function LCD module that provides convenient access to system information and status
Power supply	It supports a power supply range of 100–240 VAC and operates at a frequency of 50–60 Hz, Ensuring compatibility with different power sources
I/O ports	The firewall features two USB 2.0 ports on the front panel, one USB 3.0 port on the rear panel, two management ports (eth0/eth1) on the front panel, one COM port (RJ445) on the front panel, and one VGA port on the rear panel
Flexi port slots	The Sophos XG650 has the flexibility to accommodate 4 or 6 optional interface cards, allowing for additional connectivity options based on specific networking requirements
Mounting	The firewall includes 2 U sliding rails for easy installation and mounting in a rack-mounted configuration
Weight	The unpacked weight of the firewall is 17.8 kg (39.24 lbs.), while the packed weight is 27 kg (59.53 lbs)
Dimensions	The firewall has dimensions of 438 mm (width) × 600 mm (depth) × 88 mm (height), or approximately 17.24 × 23.62× 3.46 in.
Ethernet interfaces	It is equipped with 8 GbE (gigabit Ethernet) copper ports that allow for high-speed network connectivity
Storage	The Sophos XG650 includes two integrated hot-swappable SSDs (Solid State Drives) configured in a RAID (Redundant Array of Independent Disks) setup. These drives are used for local quarantine and logs storage
Connectivity modules (optional)	The firewall supports optional connectivity modules, such as the SFP DSL module for VDSL2 (very-high-bit-rate digital subscriber line) connections and SFP/SFP+ transceivers for fiber-optic connectivity

These physical interface and connectivity features make the Sophos XG650 firewall a versatile and robust network security solution suitable for various deployment scenarios.

8 Introduction to E-payment Systems

8.1 Definition of E-Payment

E-payment refers to the electronic transfer of funds or the exchange of monetary value between buyers and sellers through digital channels. It eliminates the need for physical cash or checks and enables secure and convenient transactions using electronic devices such as computers, smartphones, or payment cards [18].

8.2 Advantages of E-payment Systems

E-payments, or electronic payments, offer a range of benefits that contribute to improved security, streamlined processes, greater flexibility, easier access to global markets, and increased visibility. Here's how e-payments provide these advantages [19]:

8.2.1 Improved Security

E-payments often incorporate advanced security measures such as encryption, multi-factor authentication, and tokenization. These technologies help protect sensitive financial information and reduce the risk of fraud, unauthorized access, and data breaches. Compared to traditional paper-based methods, e-payments provide a more secure way to transfer funds and conduct transactions.

8.2.2 Streamlined Processes

E-payments automate and digitize payment processes, reducing the need for manual intervention and paperwork. This leads to faster and more efficient payment processing, minimizing errors and delays that can occur with manual handling. Businesses can optimize their operations by automating invoice generation, payment collection, and reconciliation.

8.2.3 Greater Flexibility

E-payments offer various payment options, including credit and debit cards, digital wallets, bank transfers, and mobile payments. This flexibility accommodates diverse customer preferences and enables businesses to cater to a broader audience. Customers can choose the most convenient and secure payment method for their needs.

8.2.4 Easier Access to Global Markets

E-payments facilitate cross-border transactions by eliminating geographical barriers and reducing currency conversion complexities. Businesses can accept payments from customers around the world, expanding their customer base and tapping into new markets. This accessibility helps drive international trade and economic growth.

8.2.5 Increased Visibility

E-payments generate digital records that are easier to track, monitor, and reconcile. Businesses can gain real-time insights into their financial transactions and cash flows. This increased visibility enhances financial planning, budgeting, and decision-making. Additionally, auditing and compliance processes are simplified as electronic records are readily accessible and easier to verify.

8.2.6 Reduced Costs

E-payments can lead to cost savings by minimizing the need for physical infrastructure, such as paper-based forms, postage, and manual processing. Businesses can also reduce administrative overhead and operational expenses associated with handling and reconciling paper payments.

8.2.7 Environmental Impact

Shifting to e-payments reduces the need for paper, contributing to environmental sustainability. Fewer paper checks and invoices mean less paper waste and a smaller carbon footprint, aligning with eco-friendly practices.

E-payments offer a range of benefits (Fig. 3) that contribute to improved security, streamlined processes, greater flexibility, easier access to global markets, increased visibility, enhanced data analytics, cost savings, and environmental sustainability. As businesses and consumers continue to adopt digital payment methods, these advantages will continue to drive the evolution of the financial landscape.

8.3 Challenges of E-payment Systems

E-payment systems face various security threats (Fig. 4), including fraud, identity theft, data breaches, and malicious attacks. Ensuring robust security measures is crucial to protect sensitive financial information and maintain user trust. Effective

Fig. 3 Advantages of e-payment system [19]

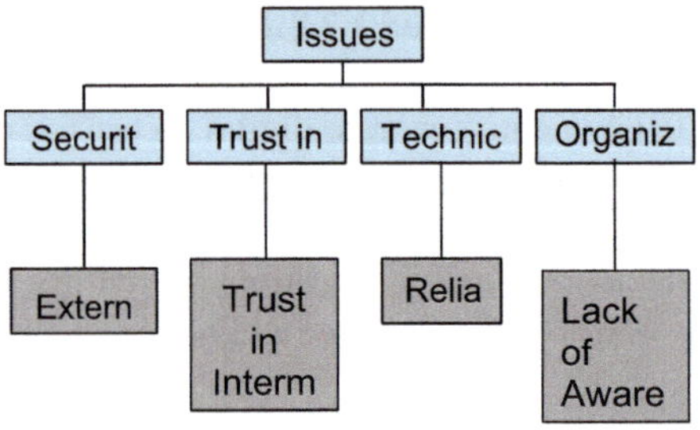

Fig. 4 Issues with e-payment system [20]

e-payment systems require a reliable technological infrastructure, including internet connectivity, secure servers, and payment gateways. Inadequate infrastructure can lead to transaction failures or delays. The success of e-payment systems relies on user adoption and trust. Overcoming user resistance to change, addressing concerns about data privacy, and building confidence in the security of e-payment systems are important challenges. It must comply with relevant financial regulations and data protection laws to ensure the security and privacy of user data. Compliance with standards such as the Payment Card Industry Data Security Standard (PCI DSS) is essential for organizations handling payment card information [21, 22].

8.4 Importance of Security in E-Payment Systems

Security is a fundamental aspect of e-payment systems to protect against fraud, unauthorized access, and data breaches. E-payment systems handle sensitive financial data, such as credit card details or personal identification information. Robust security measures are essential to prevent unauthorized access and misuse of this information. Secure e-payment systems guarantee the integrity and authenticity of transactions, minimizing the risk of tampering or unauthorized alterations. Strong security measures instill confidence in users, encouraging their adoption and continued usage of e-payment systems. Effective security protocols help mitigate the risk of financial losses due to fraud or data breaches, safeguarding the interests of both businesses and consumers [23, 24].

By understanding the advantages, challenges, and importance of security in e-payment systems, businesses and consumers can navigate the digital landscape while leveraging the benefits of convenient and secure electronic transactions.

8.5 Types of E-payment Systems

There are mainly two types of e-payment systems available. These are [25]:

8.5.1 Internet-Based Payment Systems

These payment systems utilize the internet as a platform for conducting transactions. They enable individuals and businesses to make payments electronically. Common models of internet-based payment systems include: E-cash, Credit card, Debit card, Smart Card.

8.5.2 Electronic Transaction-Based Payment Systems

These payment systems focus on secure electronic transactions and may involve specific protocols or technologies for secure processing. Such models are Secure Electronic Transaction (SET), Cyber Cash, Net Bill, and First Virtual Holdings, prominent in todays of e-commerce but have diminished in usage over time.

Smart Card

Smart cards, also known as chip cards or integrated circuit cards, are plastic cards embedded with a microchip that can store and process data. They are widely used for various payment purposes, including online transactions. Smart cards can be contact-based, requiring physical contact with a card reader, or contactless, allowing for proximity payments. The microchip on the card securely stores payment information and performs cryptographic operations to ensure the security of transactions. Examples of smart card-based payment systems include EMV (Europay, MasterCard, Visa) cards used for secure credit and debit card transactions. Overall, smart cards provide improved security, convenience, and versatility in e-payment transactions [26, 27].

Prepaid Cards

Prepaid cards in e-payments are payment cards that are loaded with a specific amount of money in advance. These cards are not linked to a bank account or a credit card. In this user can only spend the funds available on the card. Prepaid cards are a popular form of electronic payment, offering convenience and security for various transactions. To use a prepaid card for e-payments, the user must first load money onto the card. This can be done through various methods, such as direct deposit, bank transfer, cash reload at designated locations, or through online platforms. Once the cardholder exhausts the funds on the card, they cannot make additional purchases unless they reload the card with more money. These prepaid cards provide one extra layer of security as they are not directly linked to bank account. It the card is lost or stolen; the user's primary bank account remains protected from potential unauthorized transactions. In addition to physical cards, some prepaid card exists as virtual cards that are specifically designed for online purchases. These virtual cards are generated with unique card details and can only be used for online transactions, providing an extra layer of security for e-commerce purchases. These cards offer a cashless and secure payment option for users who prefer not to use traditional credit or debit cards or do not have access to a bank account. They provide the flexibility to manage expenses and make digital transactions easily, making them a popular choice for various financial needs [28, 29] . The following in Fig. 5 few advantages of prepaid cards are depicted.

Credit and Debit Cards

Credit cards for e-payments allow cardholders to borrow funds from the card issuer to complete transactions. The amount spent accumulates as a balance owed, which is recorded in billing cycles. If the full balance is not paid by the due date, interest charges are applied. Each credit card has a credit limit, and cardholders must make minimum payments to avoid penalties. While the debit cards are directly linked to the cardholder's bank account and provide immediate access to the available funds. When used for e-payments, the amount spent is deducted directly from the cardholder's bank account, with no borrowing or debt involved. Debit card transactions may require PIN verification or a signature for authorization. Real-time updates on the bank account balance are provided after each transaction. Unlike credit cards, there are no interest charges associated with using a debit card for e-payments. Debit cards offer convenient and secure access to one's own funds for instant e-payments [30, 31] .

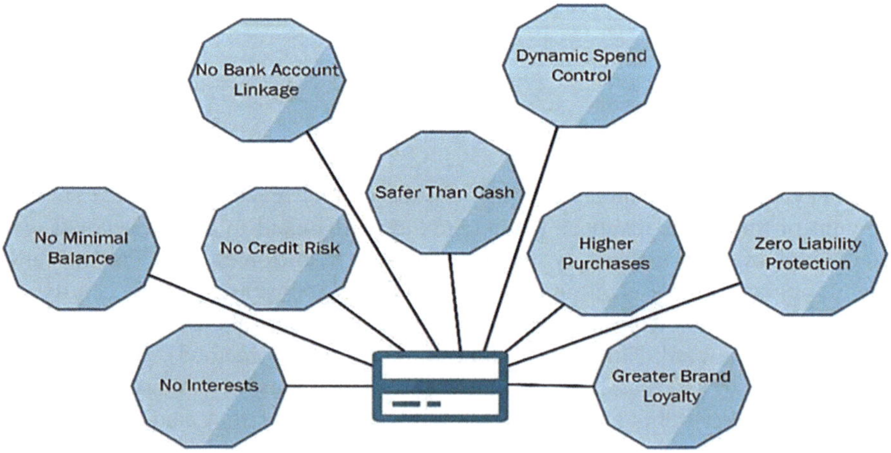

Fig. 5 Advantages of prepaid cards [28]

Mobile Payments

Mobile payment, also known as mobile money or mobile wallet, refers to the use of mobile devices, such as smartphones or tablets, to make financial transactions. It enables individuals to pay for goods and services, transfer money, and perform other financial transactions using their mobile devices instead of traditional payment methods like cash or credit cards. Mobile payment systems typically rely on wireless technologies such as Near Field Communication (NFC) or mobile apps to facilitate the transfer of funds between the payer and the recipient. These systems securely store the user's payment information, such as credit card details or bank account information, within the mobile device or a cloud-based platform. Many banks offer mobile banking apps that enable customers to transfer money, pay bills, and make purchases directly from their bank accounts using their mobile devices. P2P payment apps like Venmo, PayPal, or Zelle enable users to send and receive money directly to/from other individuals using their mobile phone numbers, email addresses, or usernames. Some mobile payment systems use Quick Response (QR) codes, which are scanned by the user's smartphone camera to initiate a payment. The QR code may contain payment details, merchant information, or a link to a payment gateway. It offers several advantages, including convenience, speed, and enhanced security. User can make payments anytime, anywhere, without the need for physical cash or cards. Additionally, these systems often employ encryption and tokenization techniques to protect sensitive payment information, reducing the risk of fraud [4, 32].

Online Banking

It is a type of electronic transaction-based system that allows customers to access their bank accounts and conduct various financial transactions over the internet. It enables users to manage their finances and perform banking activities without the need to physically visit a bank branch. In this, first users login to their online bank accounts securely using an authentication method to access their account information. After logging in, they can perform various operations like balance inquiry, fund transfers. This system uses various security measures, such as encryption, two-factor authentication, and secures login procedures to protect customers' sensitive information and ensure secure transactions. Online banking has become increasingly popular due to its convenience and accessibility. Customers can perform financial tasks from the comfort of their homes or on the go, saving time and reducing the need for physical paperwork. However, users are encouraged to take necessary precautions, such as using secure networks and keeping login credentials confidential, to maintain the security of their online banking activities [33, 34].

Digital Wallets

Digital wallets, also known as mobile wallets or e-wallets, are virtual applications or platforms that allow users to store, manage, and conduct various financial transactions securely using their mobile devices or computers. These wallets store sensitive payment information, such as credit card details, debit card information, and sometimes even digital currencies, providing a convenient and secure way to make payments and manage finances digitally. Digital wallets employ security measures like encryption, tokenization, and biometric authentication to safeguard users' payment information and protect against fraudulent activities. Digital wallets have gained widespread adoption due to their convenience, speed, and enhanced security compared to traditional payment methods. As technology continues to evolve, digital wallets are likely to play an increasingly significant role in the future of financial transactions and digital commerce [35, 36].

Bank Transfer

A bank transfer (Fig. 6) refers to the electronic transfer of funds from one bank account to another. It allows individuals or businesses to send money securely and conveniently without the need for physical cash or paper-based transactions. Bank transfers are an essential part of electronic systems, facilitating various financial transactions, such as paying bills, transferring money to friends or family, and conducting business transactions.

Bank transfers can be categorized into main types:

1. Internal bank transfers: This type of transfer involves moving funds between accounts within the same bank. For example, transferring money from a checking account to a savings account within the same banking institution.
2. External bank transfers: In an external bank transfer, funds are moved between accounts held at different banks. This type of transfer is commonly used for sending money to individuals or businesses who bank with a different financial institution.

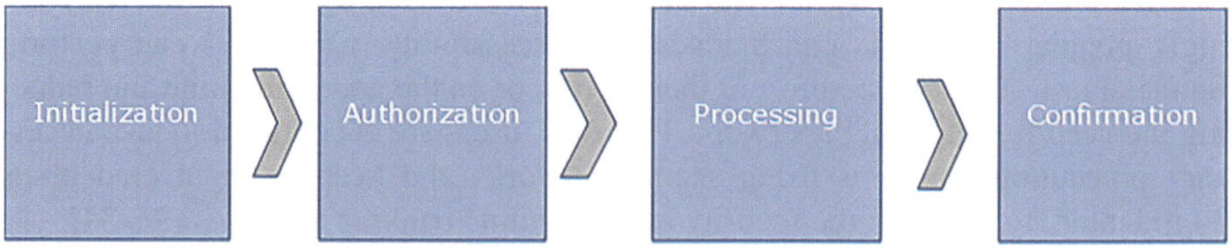

Fig. 6 Steps involved in Bank Transfer

Cryptocurrencies

These are digital or virtual currencies that use cryptography for secure and decentralized financial transactions. Unlike traditional currencies issued and controlled by central banks, cryptocurrencies operate on decentralized networks, typically based on block chain technology. This technology ensures that transactions are secure, transparent, and immutable. Cryptocurrencies operate on a decentralized network of computers, known as nodes, spread across the globe. The network collectively maintains a public ledger called the block chain. It is a chronological and immutable record of all transactions ever conducted with a specific cryptocurrency. Each transaction is bundled together in a block and linked to the previous block, forming a chain of blocks. This ensures transparency and prevents any unauthorized alterations to transaction history. It relies on cryptographic techniques to secure transactions and control the creation of new units. In this, public and private keys are used to verify and sign transactions, providing authentication and encryption. Users interact with cryptocurrencies through digital wallets, these cryptocurrency wallets are software application or hardware device that stores private keys, enabling users to access and manage their cryptocurrency holdings. These wallets can be online (web based), desktop, mobile or hardware based, each offering different levels of security and convenience. Each cryptocurrency wallet has a unique public address to make payment using cryptocurrencies; the sender initiates a transaction by specifying the recipient's public address and the amount to be sent. This transaction request is broadcast to the decentralized network [37].

9 Security Threats in E-Payment

E-payment systems offer convenience and efficiency, but they also present various security threats (Fig. 7) that users and businesses must be aware of. Some common security threats in e-payment systems include [38]:

9.1 Phishing Attacks

Phishing is a fraudulent technique where attackers attempt to trick users into revealing sensitive information, such as login credentials or payment card details, through deceptive emails, websites, or messages. It is a type of cyber-attack in e-payment that involves deceptive techniques used by malicious actors to trick individuals into revealing sensitive information, such as usernames, passwords, credit card details, or other personal data. These attackers often pretend to be trustworthy entities, such as banks, e-payment providers, or legitimate websites, to gain the trust of their victims. Phishing attacks commonly target users of online banking platforms, digital wallets, or e-commerce websites. The attackers send fraudulent emails, messages,

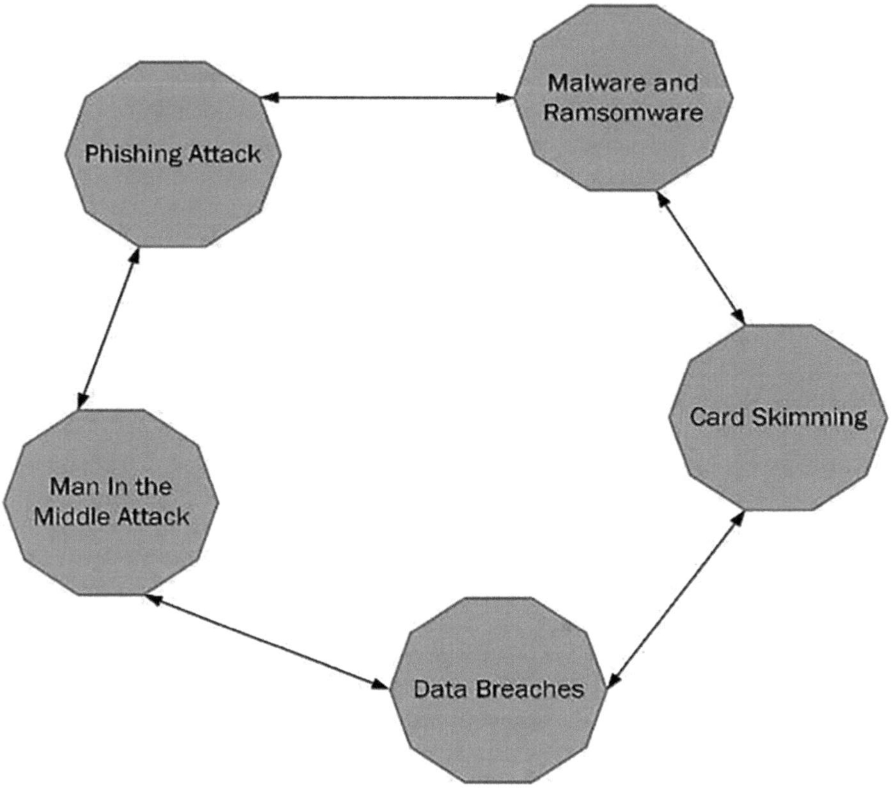

Fig. 7 Common threats in e-payment system

or create fake websites that closely resemble the legitimate ones. These phishing attempts aim to lure unsuspecting users into providing their confidential information under the guise of a genuine request or warning [39, 40].

Here's how a typical phishing attack in e-payment works:

- The attackers send phishing emails, SMS messages, or direct messages on social media, pretending to be a well-known e-payment service or financial institution.
- The phishing messages often contain urgent or enticing content to prompt immediate action from the recipient. They may claim that the user's account is compromised, a payment is due, or a special offer is available.
- Phishers create fake websites that closely resemble the legitimate e-payment platforms or financial institutions. These websites often use similar logos, layouts, and domain names to deceive users.
- When users click on the links in the phishing messages and visit the fake websites, they are prompted to enter their login credentials, payment card details, or other sensitive information.

- Once the victims submit their information, the attackers capture and store the data. They can then use this information for identity theft, unauthorized e-payments, or sell it on the dark web.

To protect against phishing attacks in e-payment, users should:

- Be cautious of unsolicited emails or messages, especially those that request personal or financial information.
- Verify the legitimacy of the sender's email address or domain before clicking on any links or responding to messages.
- Use official apps or websites of e-payment providers and avoid clicking on links from unknown sources.
- Check for the secure "https://" connection and a padlock icon in the browser's address bar when visiting e-payment websites.
- Enable multi-factor authentication (MFA) whenever possible to add an extra layer of security to their accounts.

Additionally, e-payment providers and businesses should educate their users about phishing threats, implement robust security measures, and continuously monitor for suspicious activities to protect their customers from falling victim to phishing attacks.

9.2 Malware and Ransomware

Malware, short for malicious software, encompasses various harmful programs designed to disrupt, damage, or gain unauthorized access to computer systems and networks. This includes viruses, which attach to legitimate files and replicate to infect other systems, causing data corruption and deletion. Malicious software, such as viruses, Trojans, and ransomware, can infect devices and steal sensitive data, including e-payment information, or encrypt files and demand a ransom for decryption. Worms are self-propagating programs that exploit vulnerabilities to spread across networks. Trojans disguise themselves as legitimate software, enabling unauthorized access or data theft. Spyware covertly monitors user activities and gathers sensitive information for third parties without consent. Adware displays unwanted and intrusive advertisements on users' devices. Malware can infiltrate systems through malicious email attachments, infected downloads, compromised websites, or removable media, emphasizing the need for robust cybersecurity measures and user awareness to prevent infection and protect valuable data [41]. Ransomware is a specific type of malware designed to encrypt the victim's data, rendering it inaccessible until a reason is paid to the attackers. Once the victim's data is encrypted, the ransomware displays a ransom note demanding payment in exchange for a decryption key to unlock the data. Ransomware attacks can be devastating for individuals and organizations, causing data loss, disruption of operations, and financial losses. Some high-profile ransomware attacks have targeted large corporations, government agencies, and healthcare institutions, resulting in significant impacts [42].

Malware and ransomware attacks involve a strategic sequence of steps aimed at infiltrating systems, establishing control, and often demanding a ransom. These actions, orchestrated by cybercriminals, exploit vulnerabilities and human behavior to devastating effect [43].

1. Attackers commonly use phishing emails as a vector. These deceitful messages appear genuine, luring users to click on malicious attachments or links. Malware can also be delivered via malicious websites that trigger drive-by downloads, exploiting browser or software vulnerabilities.
2. Interacting with the malicious payload initiates execution. The malware exploits system vulnerabilities, granting unauthorized access. Trojans, spyware, or ransomware might be executed, each serving distinct malicious purposes.
3. The malware establishes a connection with the attacker's Command and Control communication server, enabling remote control and data exfiltration. Personal data, credentials, or sensitive information might be stolen.
4. Some malware self-replicates, spreading to connected systems. Persistence is achieved through system modifications, registry entries, or backdoor installations, ensuring malware remains active.
5. In ransomware attacks, the victim's files are encrypted using strong algorithms. This encryption renders files inaccessible without a decryption key, which attackers retain.
6. Victims receive a ransom note detailing the encryption and demanding payment for the decryption key. Payment instructions often involve cryptocurrencies for anonymity.
7. Victims face a choice: pay the ransom for a decryption key, though there's no guarantee of compliance, or endure data loss.
8. Paying the ransom may not prevent data leaks. After payment or refusal, cleanup involves removing malware, restoring systems from backups, and addressing vulnerabilities.

Malware and ransomware are types of malicious software designed to harm or exploit computer systems, networks, and users. Both pose significant cybersecurity threats and can cause severe damage to individuals and organizations. To protect against malware and ransomware, individuals and organizations should implement strong cybersecurity practices, including:

- Regularly updating software and operating systems to patch known vulnerabilities.
- Using reputable antivirus and anti-malware software to detect and block malicious activities.
- Backing up critical data regularly to an external and secure location to mitigate the impact of ransomware attacks.
- Being cautious of suspicious emails, links, and attachments, and not clicking on or downloading content from untrusted sources.

- Educating employees and users about the risks of malware and ransomware and promoting a cybersecurity-aware culture.

9.3 Man in the Middle (MITM) Attacks

It is a type of cyber-attack where a malicious actor secretly intercepts and relays communications between two parties who believe they are directly communicating with each other. The attacker positions themselves between the sender and the receiver, making it appear as if they are the legitimate endpoint for both parties. It can be executed on various communication channels, including web browsers, email clients, instant messaging apps, and even phone calls. Here's how a Man-in-the-Middle attack works [44]:

The attacker secretly gains access to the communication channel between the sender and the receiver. This can be achieved through various methods, such as eavesdropping on unencrypted Wi-Fi networks, compromising routers, or exploiting vulnerabilities in communication protocols. Once the attacker is in the middle of the communication flow, they can intercept, relay, and even modify the messages exchanged between the sender and the receiver without either party knowing. In some cases, the attacker may impersonate the sender to the receiver and vice versa, pretending to be each party to manipulate the communication. The attacker captures sensitive information, such as login credentials, financial data, or other confidential information, which can be later used for malicious purposes, such as identity theft or financial fraud.

A Man-in-the-Middle (MitM) attack in the context of e-payment via spoofing is a type of cyber-attack where an attacker intercepts and manipulates communication between two parties engaged in an electronic payment transaction. The attacker positions themselves between the victim (the payer) and the legitimate payment gateway or service provider (the recipient), giving them the ability to eavesdrop on, modify, or even divert the transaction.

Here's how an MitM attack in e-payment via spoofing typically works [45]:

1. The attacker sets up a fake wireless network or access point (AP) that appears legitimate, often with a name similar to a legitimate network. The attacker may use techniques like ARP (Address Resolution Protocol) spoofing to redirect traffic from the victim's device to their own system.
2. The victim unknowingly connects to the attacker's fake network, thinking it's a legitimate public Wi-Fi hotspot, or the attacker compromises an existing network that the victim is using.
3. The attacker intercepts the communication between the victim's device and the payment gateway or service provider.

The attacker can capture sensitive data, such as payment card details, login credentials, and transaction information.

4. The attacker can modify the intercepted data before forwarding it to the legitimate payment gateway, making it appear as if the transaction was legitimate. The attacker might alter the recipient's account details, transaction amount, or even redirect the payment to their own account.
5. The victim's device receives a confirmation message that the payment was successful, further convincing them that the transaction went as intended.
6. The attacker may allow the legitimate transaction to proceed, ensuring that the victim doesn't immediately become suspicious.
7. After successfully executing the attack, the attacker may terminate the MitM setup and disappear with the stolen payment card details or other sensitive information. Alternatively, the attacker could use the stolen data for fraudulent transactions or sell it on the black market.

To defend against Man-in-the-Middle attacks, users and organizations should:

- Use encrypted communication channels such as HTTPS, SSL/TLS, and VPNs.
- Be cautious when connecting to public Wi-Fi networks, especially for sensitive transactions.
- Regularly update software and applications to patch known vulnerabilities. Verify the authenticity of websites and digital certificates before entering sensitive information.
- Implement strong authentication methods, such as two-factor authentication (2FA).
- Employ intrusion detection and prevention systems to identify and block suspicious network activities. By adopting these measures, users can reduce the risk of falling victim to MITM attacks and ensure the privacy and security of their communication and sensitive data.

9.4 Data Breaches

E-payment systems can be targeted by cybercriminals seeking to breach databases and steal valuable customer information, including payment card data, usernames, and passwords. A data breach is an incident in which sensitive, confidential, or protected information is accessed, disclosed, or stolen by unauthorized individuals or entities. Data breaches can occur in various contexts, such as cyber-attacks on organizations, accidental exposure of data, or intentional insider actions. When a data breach occurs, personal or sensitive information, such as names, addresses, social security numbers, financial data, login credentials, or health records, may be compromised. The severity of a data breach depends on the type and amount of data exposed, as well as the potential harm it can cause to individuals or organizations. Once a data breach is detected or discovered, affected individuals and relevant authorities should be notified promptly to take appropriate actions, such as mitigating potential harm, notifying customers or users, and implementing measures to prevent future breaches. Data breaches can have severe consequences for individuals and organizations, including financial losses, damage to reputation, legal liabilities, and compliance violations. Consequently, organizations must implement robust cybersecurity measures, conduct regular security audits, and educate employees about data security best practices to minimize the risk of data breaches and protect sensitive information from falling into the wrong hands. Additionally, individuals should practice strong password management, enable two-factor authentication when available, and be cautious about sharing personal information online to protect themselves from potential data breaches. Preventing data breaches requires a comprehensive and proactive approach to cybersecurity [46].

Here are some essential steps to help safeguard our data and reduce the risk of breaches that we should [47]:

- Regularly assess for organization's security risks and vulnerabilities. We must identify potential weaknesses in systems and processes and prioritize addressing them. Enforce the use of complex passwords and regular password changes and consider implementing multi-factor authentication (MFA) to add an extra layer of security to user accounts.
- Regularly update all software and operating systems to patch known vulnerabilities. Unpatched software is a common entry point for hackers.
- Protect network with firewalls, intrusion detection systems (IDS), and intrusion prevention systems (IPS) and regularly monitor network traffic for suspicious activities.
- Grant access to data and systems on a need-to-know basis. Restrict administrative privileges to only those who require them for their roles. Perform regular data backups and store them securely offline. In case of a breach, having a backup ensures you can recover essential information without paying a ransom.
- Implement security measures on mobile devices, including passcodes, biometric authentication, and remote wiping capabilities. Develop a detailed incident response plan that outlines how your organization will respond to a data breach or security incident. Test and update the plan regularly.
- Continuously monitor our systems, networks, and logs for any signs of unauthorized access or unusual activity and regularly conduct security audits to identify potential weaknesses.
- Assess the security practices of your vendors and third-party partners, especially those who have access to your data. Ensure they meet your security standards. Stay informed about relevant data protection regulations and ensure your organization complies with them. This includes regulations like GDPR, HIPAA, or CCPA, depending on your location and industry.
- Promote a culture of cybersecurity awareness within your organization. Encourage employees to report suspicious activities and reward proactive security practices.

Preventing data breaches is an ongoing effort that requires vigilance and commitment from everyone in the organization. By implementing these best practices, we can significantly reduce the risk of data breaches and better protect your sensitive information.

9.5 *Card Skimming*

Card skimming, also known as credit card skimming, is a type of fraud in which criminals use devices to capture and steal data from the magnetic stripe of credit cards, debit cards, or other payment cards. The stolen data can then be used to create counterfeit cards or make unauthorized transactions, potentially leading to financial losses for the cardholders. Card skimming devices are often covertly placed on legitimate card readers, such as those found on ATMs, gas pumps, point-of-sale (POS) terminals, or even handheld card readers used by merchants. The skimming devices are designed to look like part of the legitimate card reader, making them difficult for users to detect [48].

9.5.1 Here's How Card Skimming Typically Works

Criminals physically attach skimming devices over or inside the actual card reader. These devices can be small and inconspicuous, making them hard to notice. When a cardholder inserts their payment card into the compromised card reader, the skimming device reads the data from the magnetic stripe. This includes the card number, expiration date, and cardholder's name. In some cases, the skimming device may transmit the stolen data wirelessly to the criminals in real time. In other cases, the criminals return later to retrieve the skimming device and access the stolen data. With the stolen card data, criminals can create counterfeit cards or use the stolen information for online purchases or other unauthorized transactions. To protect yourself from card skimming [49]:

- Inspect card readers before use. Look for any signs of tampering or loose parts that could indicate the presence of a skimming device.
- Whenever possible, use ATMs and payment terminals from reputable sources, like banks and established retailers.
- When entering your PIN at an ATM or POS terminal, use your other hand or body to shield the keypad from prying eyes or hidden cameras.
- Review your account statements regularly for any unauthorized transactions and report any suspicious activity to your bank or card issuer immediately.
- Contactless payment methods like EMV chip cards or mobile payments (e.g., Apple Pay, Google Pay) are more secure than magnetic stripe transactions and can help protect against skimming. By being vigilant and adopting safe payment practices, you can reduce the risk of falling victim to card skimming and other forms of payment card fraud.

To mitigate these security threats, users and businesses should implement best practices such as using strong passwords, enabling multi-factor authentication, regularly updating software and devices, being cautious of suspicious emails or messages, and using reputable and secure e-payment platforms. Additionally, businesses should prioritize cybersecurity measures, including encryption, regular security audits, and employee training to protect both their systems and their customers' sensitive data.

9.6 Real-World Security Breaches in E-Payment System

There have been several real-world security breaches in e-payment systems. Here are some notable incidents [50]:

1. SolarWinds (2020): This cybersecurity incident affected various organizations, including several US government agencies and major companies. The attackers inserted malicious code into the SolarWinds software, compromising the software supply chain and leading to data breaches which affected more than 30,000 public and private organization [51].
2. Capital One (2019): A hacker gained unauthorized access to Capital One's cloud infrastructure and stole personal information of over 100 million customers in the United States and Canada. The breach exposed names, addresses, credit scores, and social security numbers [52].
3. Wawa (2019): The convenience store chain Wawa experienced a data breach that affected its payment processing systems. The breach exposed credit and debit card information of thousands of customers who used their cards at Wawa locations [53].
4. Ticketmaster (2018): In this breach, attacker compromised a third party Chatbot used by Ticketmaster's website. The attackers accessed customer data, including payment details, of around 40,000 users [54].
5. British Airways (2018): Cybercriminals targeted British airways' 'website and mobile app, stealing payment card information over 380,000 customers. The breach involved the interception of card details during the payment process [55].
6. Saks Fifth Avenue and Lord & Taylor (2018): Hackers breached the parent company of Saks Fifth Avenue and Lord & Taylor stores, gaining access to their payment processing systems and compromising around 5 million payment cards [56].
7. OnePlus (2018): Chinese smartphone manufacturer OnePlus experienced a breach in its online store's payment system, affecting over 40,000 customers. Credit card details were stolen during the incident [57].
8. Dunkin' Donuts (2018): Dunkin' Donuts' mobile app was targeted by hackers, who accessed customer accounts using stolen login credentials and made unauthorized transactions, compromising users' payment information [58].
9. Chipotle (2017): Chipotle faced a data breach affecting payment card information used in its restaurants. The breach impacted customers across various locations in the United States [59].
10. Intercontinental Hotels Group (2017): A malware attack on the payment card processing systems of several Intercontinental Hotels Group properties resulted in the theft of guests' credit card data [60].

11. Equifax (2017): Equifax, one of the major credit reporting agencies, suffered a massive data breach that exposed sensitive information of approximately 147 million Americans. The breach included names, social security numbers, birth dates, addresses, and in some cases credit card information [61].

10 Smart Authentication Methods

Smart authentication methods in e-payment systems typically refer to advanced and secure techniques used to verify the identity of users during electronic payment transactions. These methods aim to enhance security and prevent fraudulent activities in online payment processes. It requires users to provide two different types of authentication factors before completing a transaction. These factors usually fall into three categories: something you know (e.g., a password or PIN), something you have (e.g., a mobile device or smart card), and something you are (e.g., biometric data like fingerprints or facial recognition). By combining two of these factors, the system adds an extra layer of security and also called as two-factor authentication [62, 63].

10.1 Dynamic CVV or CVC

Dynamic CVV (Card Verification Value) or CVC (Card Verification Code) is an advanced security feature used in some payment cards to enhance the security of card-not-present transactions, such as online or over-the-phone purchases. Unlike the static CVV/CVC found on the back of most payment cards, which remains the same throughout the card's validity period, the dynamic CVV/CVC changes periodically [64].

10.1.1 How It Works

The card issuer or payment network generates a dynamic CVV/CVC for the card and associates it with the cardholder's account. The dynamic CVV/CVC changes at regular intervals, such as every few hours or after specific triggering events (e.g., after each successful transaction). It is different from the printed static CVV/CVC on the back of the card. During an online transaction, the cardholder is prompted to enter the current dynamic CVV/CVC along with other card details. The merchant verifies the validity of the dynamic CVV/CVC with the card issuer before completing the transaction (Fig. 8).

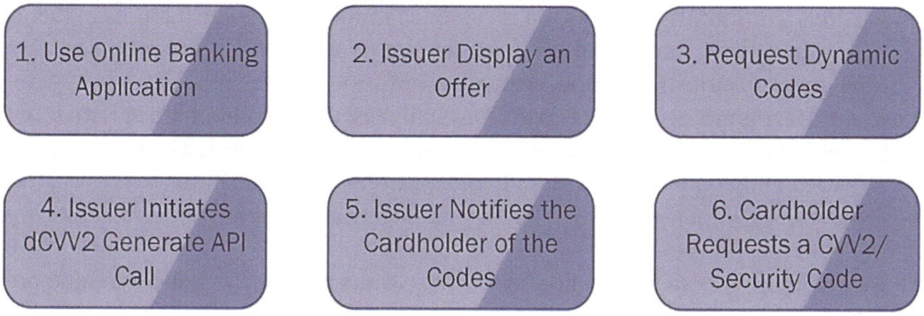

Fig. 8 Working of CVV2 Code [65]

10.1.2 Benefits

Since the dynamic CVV/CVC changes frequently, it provides an additional layer of security against card-not-present fraud. Even if a fraudster obtains the static CVV/CVC from a compromised source, it becomes useless after a short period. In case of a data breach or card information compromise, the dynamic CVV/CVC limits the usefulness of the stolen data for fraudulent transactions. By requiring the current dynamic CVV/CVC for each transaction, the risk of unauthorized or fraudulent transactions is significantly reduced. It's essential for cardholders to keep their cards secure and protect access to the dynamic CVV/CVC, just like any other sensitive card information. Dynamic CVV/CVC is not yet widely adopted on all payment cards but is gradually gaining popularity as a security enhancement measure in the payment industry.

10.2 Biometric Authentication

Biometric authentication in e-payment systems involves using unique biological characteristics or behavioral traits to verify the identity of users during electronic payment transactions. Biometrics provides a high level of security and accuracy as they are difficult to forge or replicate. Common biometric features used for authentication in e-payment systems include [66]:

- Fingerprint Recognition: This method analyzes the unique patterns on an individual's fingertip. Users need to place their finger on a fingerprint sensor, and the system matches the scanned fingerprint against a pre-registered template to verify identity.
- Facial Recognition: Facial recognition uses algorithms to analyze facial features, such as the distance between the eyes or the shape of the face. Users' faces are captured by a camera, and the system compares the captured image with the stored biometric data for authentication.

- Iris Recognition: Iris recognition involves scanning the unique patterns in the iris of the eye. This method offers high accuracy and is used in some advanced e-payment systems for secure authentication.
- Voice Recognition: Voice recognition analyzes the unique characteristics of a person's voice, such as pitch, tone, and other vocal patterns. Users' voices are recorded and compared to a stored voiceprint for authentication.
- Behavioral Biometrics: This category includes various behavioral traits like typing patterns, mouse.
- Movements, gait analysis, and more. These traits are used to build a unique profile of the user.
- Verify their identity during e-payment transactions.

The biometric data collected during the authentication process is usually converted into a digital template and stored securely on the server. It is important to note that reputable e-payment systems implement strong encryption and data protection measures to safeguard biometric information. Biometric authentication in e-payment systems offers several advantages:

10.2.1 Strong Security

Biometrics provides a high level of security since each individual's biometric traits are unique and difficult to replicate.

10.2.2 Convenience

Biometric authentication eliminates the need for users to remember passwords or carry physical tokens, making the authentication process more convenient.

10.2.3 Speed and Efficiency

Biometric authentication is generally quick and seamless, allowing for faster payment processing and reducing checkout times.

10.2.4 Reduced Fraud

The use of biometrics can significantly reduce fraudulent transactions, as forged biometric data is challenging to produce.

However, it's essential to consider privacy concerns when implementing biometric authentication. Users must trust that their biometric data is handled securely and used only for authentication purposes. As such, e-payment systems must adhere to strict data protection regulations and implement robust security measures to ensure the privacy and confidentiality of biometric information.

10.3 Tokenization

Tokenization in an e-payment system refers to the process of substituting sensitive payment card information, such as credit card numbers, with unique identification symbols called "tokens". These tokens are randomly generated strings of characters that have no intrinsic value and are used as a surrogate for the actual payment card data. The purpose of tokenization is to enhance security and protect the sensitive payment information during transactions and storage. When a customer initiates a payment using their credit card through an e-payment system, the system captures the card details (e.g., card number, expiration date, CVV) and passes them to a secure tokenization service. This service generates a unique token corresponding to that specific payment card. The token is then sent back to the merchant's system, which stores it securely in its databases instead of storing the actual payment card data. By doing so, the merchant reduces the risk associated with handling sensitive information since the token itself holds no value and cannot be used for fraudulent transactions. When a subsequent payment is needed (e.g., for a subscription renewal or a recurring payment), the token is used as a reference to the customer's card on file. The e-payment system can then use the token to perform the transaction without the need to access the actual card information, making the process more secure and reducing the chances of exposing sensitive data to potential attackers. In summary, tokenization in e-payment systems provides an additional layer of security by substituting sensitive payment card data with unique tokens, helping to protect customers' financial information from unauthorized access and fraud [67, 68].

10.3.1 How Does It Work

Tokenization in e-payment systems involves a series of steps to secure and process payment transactions? Here's a more detailed explanation of how it works: The process begins when a customer initiates a payment on a website, mobile app, or any other digital platform. They provide their payment card information, which typically includes the card number, expiration date, and CVV (Card Verification Value). The payment application or merchant's system sends the payment card details to a tokenization service or payment processor. This service is responsible for handling the sensitive information and generating tokens. The tokenization service receives the card information and uses encryption and other security measures to protect the data. It then generates a random and unique token that corresponds to the provided card details. This token is a string of characters that has no relation to the actual card number and holds no intrinsic value. The tokenization service sends the generated token back to the merchant's system. The merchant's system stores this token securely in its database along with other necessary transaction information. Crucially, the actual payment card data is not stored, minimizing the risk of data breaches. When a subsequent payment is required, such as for a subscription renewal or a recurring transaction, the customer's card on file is referenced using the token. The token is sent to the tokenization service or payment processor. The tokenization service decrypts the token and retrieves the original payment card details associated with that token. It then uses the decrypted information to process the payment, just like a regular card transaction. The payment processor returns the transaction response to the merchant's system, indicating whether the payment was successful or not. The merchant can then complete the transaction process accordingly, and the payment is settled between the customer's bank and the merchant's bank [69].

Key Benefits of Tokenization in E-Payment Systems:

- Enhanced Security: Tokenization helps protect sensitive card data from being exposed or intercepted during transactions and while stored in databases.
- Reduced PCI DSS Scope: By storing tokens instead of actual card data, merchants can reduce the scope of PCI DSS (Payment Card Industry Data Security Standard) compliance audits.
- Seamless Customer Experience: Tokenization allows for smooth and frictionless payments, especially in scenarios where recurring transactions are involved, as customers don't need to re-enter their payment details every time.

Fraud Prevention: Since tokens have no intrinsic value and cannot be used outside the specific merchant's system, they are of little to no use for potential attackers attempting to gain unauthorized access to customer card data. Overall, tokenization plays a crucial role in ensuring secure and convenient e-payment transactions, benefiting both merchants and customers by safeguarding sensitive information and mitigating the risks associated with handling payment data.

10.4 Device Fingerprinting

In the context of e-payment, device fingerprinting refers to the process of uniquely identifying electronic devices, such as smartphones, tablets, or computers, based on their characteristics and attributes. E-payment providers and platforms use device fingerprinting as an additional security measure to help detect and prevent fraudulent transactions and ensure a secure payment environment. Here's how device fingerprinting works in e-payment [70, 71]:

I. Data Collection: When a user attempts to make an e-payment transaction, the e-payment system gathers various information about the user's device. This information can include hardware details (e.g., device model, manufacturer, IMEI, MAC address), software information (e.g., operating system, software versions, installed applications), browser data (e.g., browser type, version, plugins), time zone, language settings, IP address, and other relevant attributes.

II. Fingerprint Generation: The collected data is processed and analyzed to create a unique identifier, known as the device fingerprint. This fingerprint represents a combination of the device's attributes and serves as a distinct profile for that particular device.
III. Matching and Verification: The device fingerprint is compared to a database of known device fingerprints. This database may contain information about previously identified fraudulent devices or patterns associated with suspicious activities. If the device fingerprint matches any known fraudulent patterns, the system may trigger additional authentication measures or block the transaction to prevent potential fraud.
IV. Security and Fraud Prevention: Device fingerprinting adds an extra layer of security to e-payment systems. It helps identify and flag suspicious activities or devices that may be involved in fraudulent transactions. By tracking and recognizing devices associated with suspicious behavior, e-payment providers can take appropriate actions to protect users and prevent financial losses.
V. User Experience and Personalization: Device fingerprinting can also be used to enhance the user experience and personalize services. Recognizing a user's device can allow for a more seamless and frictionless payment process, as the system can remember preferences and settings associated with that particular device.

It's important to note that device fingerprinting, like any technology involving user data, raises privacy concerns. While device fingerprints themselves do not contain personal information, they can still be used for tracking user behavior and profiling. To address these concerns, e-payment providers must follow strict data protection regulations, maintain transparency about data collection and usage, and implement robust security measures to safeguard the device fingerprinting data [72].

10.5 Behavioral Analysis

Behavioral authentication methods in e-payments involve using patterns of user behavior to verify the identity of individuals during electronic payment transactions. These methods analyze how users interact with devices and systems, looking at their unique behavioral traits to distinguish authorized users from potential fraudsters. The behavioral data collected during the authentication process creates a behavioral profile for each user, which can be used to detect anomalies and ensure secure transactions. Here are some common behavioral authentication methods in e-payments [73, 74]:

I. Keystroke Dynamics: This method analyzes the unique typing patterns of users, including factors such as typing speed, key press duration, and time intervals between keystrokes. By monitoring these patterns over time, the system can recognize the legitimate user based on their specific typing behavior.

II. Mouse Movement Patterns: Mouse movement dynamics, such as cursor speed, acceleration, and click patterns, can be used to create a unique behavioral profile. These patterns are difficult for fraudsters to mimic accurately, making this method effective for authentication.
III. Swipe and Touch Gestures: For touchscreen devices, behavioral authentication can involve analyzing the way users' swipe, tap, or perform specific touch gestures on the screen. Each individual tends to have distinct patterns for these actions. Navigation and Interaction Behavior: This method assesses how users navigate through an application or website, including the sequence of actions they perform and the pages they access. Deviations from the established behavioral profile can raise security alerts.
IV. User Location and Device Usage: Behavioral authentication can take into account the typical geolocation patterns and device usage patterns of users. If a transaction originates from an unexpected location or a new device, additional verification may be required.
V. Time-Based Behavior: Analyzing the time of day or week when users typically perform transactions can help establish their behavioral profile. Unusual transaction timings may trigger additional authentication steps.
VI. User Response Patterns: Behavioral authentication may involve analyzing how users respond to specific challenges or security prompts during the payment process. For instance, the time taken to respond to a one-time password (OTP) request can be a part of the behavioral profile.

The strength of behavioral authentication lies in its ability to adapt and learn from user behavior over time. By continually updating the behavioral profile based on legitimate user interactions, the system becomes more accurate in identifying anomalies and detecting potential fraud. Behavioral authentication complements other authentication methods, such as biometrics and two-factor authentication, to create a multi-layered security approach. Implementing behavioral authentication in e-payments enhances security, minimizes the risk of fraud, and provides a seamless user experience by reducing the reliance on traditional passwords and security tokens. However, like any authentication method, it requires robust data protection measures to ensure user privacy and maintain the confidentiality of behavioral data.

11 Secure Communication Protocols

Protocols are essential in e-payment systems to safeguard sensitive data during transactions and authentication processes. These protocols ensure that the data exchanged between various components in the e-payment ecosystem remains confidential, secure, and protected from potential threats. Here are some common secure communication protocols used in smart authentication methods for e-payment systems [75]:

11.1 Transport Layer Security (TLS)/Secure Sockets Layer (SSL)

TLS (Transport Layer Security) and SSL (Secure Sockets Layer) are cryptographic protocols used to provide secure communication over a computer network, such as the internet. They are commonly employed to ensure the security and privacy of sensitive data during online transactions, including e-payments. SSL was the original protocol developed to provide secure communication over the internet (Fig. 9). It establishes an encrypted link between a client (such as a web browser) and a server (such as an e-commerce website). This encryption ensures that any data transmitted between the client and server remains confidential and cannot be easily intercepted by malicious actors. TLS is an updated and more secure version of SSL. It was developed as a successor to SSL due to security vulnerabilities discovered in earlier versions of SSL. TLS operates in a similar manner to SSL, creating a secure communication channel between a client and server. It uses cryptographic algorithms to authenticate the parties involved, negotiate encryption parameters, and ensure the integrity and privacy of data [77, 78]. When it comes to e-payments, TLS and SSL play a critical role in securing the exchange of sensitive information, such as credit card details, personal identification numbers (PINs), and other payment-related data. Here's how they work [79]:

11.1.1 Step 1 Handshake Initiation

The client (e.g., your web browser) initiates a connection to the e-payment server by requesting a secure connection. The server responds with its digital certificate, which contains its public key and some additional information, including the certificate's issuer (Certificate Authority).

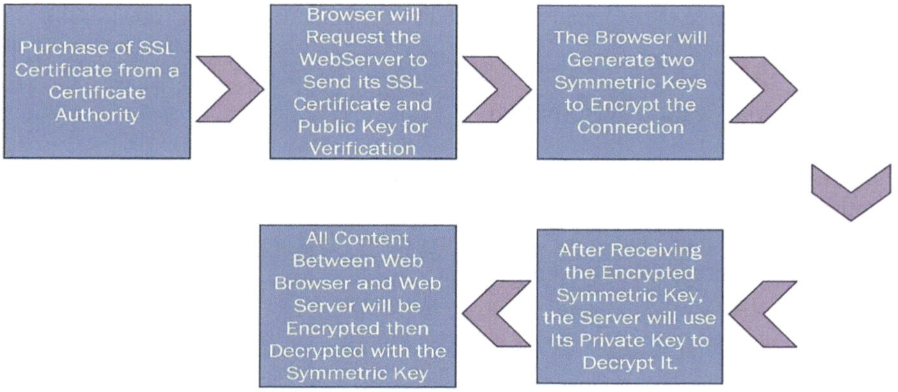

Fig. 9 Use of SSL Certificate by the browser [76]

11.1.2 Step 2 Certificate Verification and Key Exchange

The client verifies the authenticity of the server's certificate. It checks the certificate's validity, expiration date, and whether it was issued by a trusted Certificate Authority (CA). The client generates a random symmetric encryption key (session key) to be used for the rest of the session.

11.1.3 Step 3 Key Exchange and Encryption Parameters

The client encrypts the session key with the server's public key (from the server's certificate) and sends it back to the server. This process ensures that only the server, with its corresponding private key, can decrypt the session key. The server decrypts the session key using its private key.

11.1.4 Step 4 Secure Communication

With the session key established, both the client and server switch to using symmetric encryption. This type of encryption is faster and more efficient than asymmetric encryption used for the initial key exchange. All data transmitted between the client and server is encrypted and decrypted using the shared session key. This ensures the confidentiality and integrity of the data.

11.1.5 Step 5 Data Transmission

The client and server can now exchange e-payment information, such as credit card details, transaction amounts, and other sensitive data, over the secure connection. The encrypted data is sent in packets, each containing both the data and a message authentication code (MAC) that helps detect any tampering during transmission.

11.1.6 Step 6 Session Termination

Once the e-payment transaction is complete, either the client or the server can initiate the termination of the secure session. The shared session key is discarded, making it difficult for an attacker to decrypt any intercepted data, even if they managed to capture the encrypted communication.

Encryption

TLS/SSL protocols encrypt the data transmitted between the user's device (like a computer or smartphone) and the e-payment server. This encryption ensures that even if intercepted, the intercepted data would be unreadable and useless to attackers.

Authentication

TLS/SSL protocols help establish the authenticity of the e-payment server and, in some cases, the client's device. This prevents man-in-the-middle attacks, where a malicious actor intercepts the communication between the client and the server and poses as one of the parties.

Integrity

TLS/SSL also ensures the integrity of the data being transmitted. It detects any tampering or alteration of the data during transit.

Trust

E-payment services often use digital certificates issued by trusted Certificate Authorities (CAs) to verify the identity of the server. This adds another layer of trust to the communication.

In summary, both TLS and SSL are essential technologies for ensuring the security and confidentiality of e-payment transactions, protecting sensitive financial information from unauthorized access and interception. It's important for e-payment service providers to implement these protocols correctly and keep them up to date to maintain a high level of security.

11.2 EMV (Europay, MasterCard, and Visa) Protocol

EMV is a widely adopted global standard for secure payment card transactions. EMV cards, also known as chip cards or smart cards, contain embedded microprocessors that provide stronger security than traditional magnetic stripe cards. EMV transactions involve a dynamic cryptogram generated for each transaction, making it more challenging for attackers to clone or counterfeit payment cards. It works in an e-payment system by establishing a secure and standardized process for conducting payment transactions using chip-based cards (smart cards) and EMV-compliant point-of-sale (POS) terminals. Here's a step-by-step overview of how the EMV protocol works in an e-payment system [80]:

11.2.1 Step 1 Card Authentication and Initialization

The payment card contains a microprocessor chip that stores sensitive payment data, including cardholder information, account details, and cryptographic keys. When the card is inserted into an EMV-compliant POS terminal, the terminal establishes communication with the card's chip. The terminal and the card initiate an authentication process to ensure the card's authenticity and integrity.

11.2.2 Step 2 Application Selection

The terminal requests a list of applications (e.g., credit, debit, loyalty) available on the card. The card responds with the available applications, and the user selects the desired payment method.

11.2.3 Step 3 Cardholder Verification

The terminal prompts the cardholder to enter a Personal Identification Number (PIN) for verification. Alternatively, other verification methods like signature or no verification may be used, depending on the transaction and card configuration.

11.2.4 Step 4 Terminal Generates Transaction Data

The terminal generates a unique transaction-specific data (often referred to as the "Application Transaction Counter" or ATC) and other necessary transaction data.

11.2.5 Step 5 Dynamic Cryptographic Authentication

The card generates a dynamic cryptogram (a unique transaction code) using the transaction data, the card's private key, and other cryptographic parameters. The card sends the dynamic cryptogram to the terminal.

11.2.6 Step 6 Terminal Verification and Online Processing

The terminal validates the dynamic cryptogram using the card's public key, ensuring the transaction's authenticity and integrity. If necessary, the terminal may perform additional checks, such as checking the card's expiration date and checking for cardholder verification method (CVM) requirements. If the transaction requires online authorization (e.g., for high-value transactions), the terminal sends the transaction details to the card issuer for approval.

11.2.7 Step 7 Transaction Approval

The card issuer receives the transaction request, validates the transaction details, and performs fraud checks. The issuer approves or declines the transaction and sends a response back to the terminal.

11.2.8 Step 8 Transaction Completion

The terminal displays the transaction result to the cardholder (approved or declined). The cardholder may be prompted to remove the card.

11.2.9 Step 9 Transaction Logging and Updates

The terminal and card update various counters and logs to maintain proper records of transactions and ensure consistency in future transactions.

11.2.10 Key Benefits of the EMV Protocol in E-Payment Systems

Security

The dynamic nature of transaction data and the use of cryptographic methods significantly reduce the risk of card counterfeiting and data breaches.

Fraud Prevention

EMV technology makes it challenging for attackers to clone cards and use stolen data for unauthorized transactions.

Global Acceptance

EMV is a globally recognized standard, making it easier for users to use their chip-based cards internationally.

Reduced Liability

In cases of fraudulent transactions, liability is shifted to the party (issuer or merchant) with the least secure technology.

EMV plays a crucial role in enhancing the security of e-payment transactions, offering a robust framework for secure and reliable payment processing [81].

11.3 3-D Secure (3DS)

3-D Secure (3DS) is an authentication protocol designed to enhance the security of online e-payment transactions. It is primarily used to verify the identity of the cardholder during card-not-present (CNP) transactions, such as online purchases, to reduce the risk of fraud. 3DS is often associated with the "Verified by Visa," "Mastercard SecureCode," "American Express SafeKey," and similar programs offered by various card networks. The goal of 3DS is to provide an additional layer of security by adding an extra step to the online payment process, where the cardholder's identity is verified before the transaction is authorized [82, 83]. Here's how the 3DS process works:

11.3.1 Step 1 Initiation of the Transaction

The cardholder initiates an online transaction on a merchant's website or mobile app. The merchant's payment system detects that the transaction involves 3DS authentication.

11.3.2 Step 2 Redirect to the Issuing Bank

After the cardholder submits payment details, the transaction is temporarily paused. The cardholder is redirected to their card-issuing bank's authentication page or a 3DS authentication window.

11.3.3 Step 3 Authentication Process

The cardholder is prompted to provide additional authentication information, which could include a One-Time Password (OTP) sent to their registered mobile phone or email, a biometric scan (e.g., fingerprint), or a knowledge-based question. The authentication information is submitted to the card-issuing bank.

11.3.4 Step 4 Issuer's Decision

The card-issuing bank evaluates the authentication information and decides whether to approve or decline the transaction. If approved, the bank generates an authentication response, indicating that the cardholder's identity has been verified.

11.3.5 Step 5 Response to Merchant

The authentication response is sent back to the merchant's payment system.

11.3.6 Step 6 Transaction Authorization

The merchant's payment system receives the authentication response and proceeds to authorize the transaction based on the authentication result.

11.3.7 Benefits of 3-D Secure

Reduced Fraud

3DS helps prevent unauthorized use of payment cards for online transactions, reducing the risk of fraudulent activity.

Liability Shift

In case of fraud, if the merchant uses 3DS and the cardholder's identity is successfully verified, the liability for the transaction may shift from the merchant to the card issuer.

Consumer Confidence

3DS provides an added layer of security, increasing consumer confidence in making online purchases.

Regulatory Compliance

3DS can help merchants comply with regulations related to online payment security.

Customization

Merchants can customize the 3DS process to balance security and user experience based on their preferences.

It's important to note that while 3-D Secure enhances security, it can also introduce friction into the checkout process, potentially leading to cart abandonment if not implemented thoughtfully. Many modern implementations of 3DS aim to balance security with a streamlined user experience to minimize disruption while still ensuring transaction safety.

11.4 Secure Electronic Transaction (SET)

The SET (Secure Electronic Transaction) protocol was a security framework designed to enhance the security of electronic payment transactions over the internet. It was developed in the late 1990s by Visa and Mastercard in collaboration with other organizations to address concerns about the security of online credit card transactions. While SET was innovative for its time, it is important to note that it is no longer widely used or supported, having been largely replaced by newer and more streamlined security protocols. Here is an overview of how the SET protocol worked) [84, 85]:

11.4.1 Step 1. Participant Roles

Cardholder

The individual making the online purchase.

Merchant

The online store or entity receiving payment.

Issuer

The bank or financial institution that issued the credit card to the cardholder.

Acquirer

The bank or financial institution that handles the merchant's account.

11.4.2 Step 2 Digital Certificates

Each participant (cardholder, merchant, issuer, and acquirer) had a digital certificate issued by a trusted Certificate Authority (CA). Digital certificates were used to establish the identity and authenticity of the participants in the transaction.

11.4.3 Step 3 Transaction Steps

The cardholder initiated an online purchase on the merchant's website. The merchant's system generated a payment request and sent it to the acquirer.

11.4.4 Step 4 Payment Gateway

The acquirer forwarded the payment request to the issuer for verification. The issuer verified the cardholder's identity and credit card details.

11.4.5 Step 5 Generation of Dual Signature

Once the cardholder's identity was verified, a dual signature was generated for the transaction. The dual signature included the merchant's digital signature and a portion encrypted with the cardholder's private key.

11.4.6 Step 6 Authorization

The dual signature, along with other transaction details, was sent back to the merchant. The merchant's system forwarded the transaction to the acquirer for authorization.

11.4.7 Step 7 Transaction Settlement

If authorized, the acquirer settled the transaction with the merchant. The cardholder's account was charged, and the funds were transferred to the merchant's account.

11.4.8 Key Advantages of SET

Strong Security

SET aimed to provide strong authentication and encryption to protect sensitive payment data during online transactions.

Privacy

SET aimed to separate the payment card number from the merchant, reducing the risk of the cardholder's data being exposed to the merchant.

Liability Shift

Similar to modern authentication protocols, SET offered a liability shift in case of fraudulent transactions where the cardholder's identity was verified.

Despite its security features, SET faced challenges such as complexity, interoperability issues, and the emergence of newer and more streamlined security protocols like SSL/TLS and 3-D Secure. As a result, SET did not achieve widespread adoption and was eventually phased out in favor of more practical and efficient payment security solutions [86, 87].

11.5 Near Field Communication (NFC)

NFC is a short-range communication technology used in contactless payment methods, such as mobile wallets (e.g., Apple Pay, Google Pay) and contactless cards. NFC-enabled devices communicate securely over short distances, making tap-and-go payments possible. NFC (Near Field Communication) is a technology that enables short-range wireless communication between devices, typically within a few centimeters. NFC is commonly used for a variety of applications, including e-payments. In the context of e-payment, NFC allows for contactless transactions where payment information is exchanged between a customer's device (like a smartphone or contactless card) and a payment terminal [88, 89]. Here's how NFC works in e-payment:

11.5.1 Step 1 Device Interaction

The customer's NFC-enabled device (such as a smartphone or contactless card) and the payment terminal (merchant's device) have NFC chips embedded in them. These devices have both a sender and a receiver, allowing them to communicate when brought close together.

11.5.2 Step 2 Initiating the Transaction

The customer initiates an e-payment transaction by tapping or bringing their NFC-enabled device near the payment terminal. This establishes a connection between the two devices.

11.5.3 Step 3 Data Exchange

Once the devices are in close proximity, they exchange data using radio-frequency signals. The customer's device sends payment information, which may include the card details, token, or other relevant information required for the transaction.

11.5.4 Step 4 Authentication and Authorization

The payment terminal receives the payment information and forwards it to the payment network or issuer for authentication and authorization. The issuer verifies the transaction details and determines whether to approve or decline the payment.

11.5.5 Step 5 Transaction Confirmation

If the transaction is approved, the payment terminal communicates the approval to the customer's device.

11.5.6 Step 6 Completion of Transaction

The customer receives confirmation on their device that the payment was successful. The payment terminal may also provide a receipt or confirmation for the customer.

11.5.7 Key Benefits of NFC in E-Payment [90]

Convenience

NFC-enabled payments are quick and convenient. Customers can complete transactions by simply tapping their devices on the payment terminal.

Security

NFC transactions use encryption and other security measures to protect payment information during transmission. This reduces the risk of interception or unauthorized access.

Reduced Contact

NFC-based contactless payments minimize physical contact between the customer and the payment terminal, which can be especially advantageous for hygiene reasons, as seen during the COVID-19 pandemic.

Tokenization

Many NFC-based e-payment systems use tokenization, where a unique token is used for each transaction instead of transmitting the actual card information. This adds an extra layer of security by ensuring the actual card data isn't exposed.

Versatility

NFC technology can be used for various payment methods, such as credit and debit cards, mobile wallets (like Apple Pay, Google Pay), and even wearable devices.

It's important to note that both the customer's device and the payment terminal must support NFC technology for this type of contactless payment to work. Additionally, security measures like PIN entry or biometric authentication (e.g., fingerprint or facial recognition) may be required for certain transactions to further enhance security [91, 92].

References

1. André, P. (2014). *Network security*. Wiley. Retrieved December 17, 2023, from https://www.iste.co.uk/book.php?id=814
2. *What is a firewall? Definition and explanation*. kaspersky. Retrieved June 20, 2023, from https://www.kaspersky.com/resource-center/definitions/firewall
3. The evolution of firewalls: from packet filtering to machine learning-powered NGFWs. paloaltonetworks. Retrieved https://www.paloaltonetworks.com/cyberpedia/the-evolution-of-firewalls-from-packet-filtering-to-machine-learning-powered-ngfws.
4. W. Ahmed et al., Security in next generation mobile payment systems: A comprehensive survey. May 2021, doi: https://doi.org/10.1109/ACCESS.2021.3105450.
5. Z. Wang. Research on feature and architecture design of AI firewall. *Proceedings* - 2021 *5th international conference on data science and business analytics,* ICDSBA 2021, pp. 75–78, 2021, doi: https://doi.org/10.1109/ICDSBA53075.2021.00024.
6. Rathi, M., & Lohia, M. (2014). Research paper on firewalls system. *IJIRT, 1*(5), 185–187.
7. *What is a software firewall?* Nordlayer. Retrieved https://nordlayer.com/learn/firewall/software/
8. Sangfor Technologies. *What is software firewall? Difference between hardware firewall and software firewall*. Sangfor. Retrieved https://www.sangfor.com/blog/cybersecurity/what-software-firewall-difference-between-hardware-firewall-and-software.

9. Collins Ayuya. *7 best firewall software solutions: 2023 firewall comparison*. Enterprise Networking Planet. Retrieved https://www.enterprisenetworkingplanet.com/guides/best-firewall-software/.
10. *What is a hardware firewall? Hardware vs. software firewalls*. Fortinet. Retrieved https://www.fortinet.com/resources/cyberglossary/hardware-firewalls-better-than-software. A hardware firewall is a physical device much like a, plugged into the firewall first.
11. Nitishkumarpatel191. *Difference between hardware firewall and software firewall*. GeeksforGeeks. Retrieved https://www.geeksforgeeks.org/difference-between-hardware-firewall-and-software-firewall/.
12. Chiradeep BasuMallick. *Top 10 firewall hardware device in 2022*. Spiceworks. Retrieved https://www.spiceworks.com/it-security/network-security/articles/top-10-firewall-hardware-devices/
13. *Hybrid firewall advantages and disadvantages*. Fortinet. Retrieved https://www.fortinet.com/resources/cyberglossary/hybrid-firewall-advantages-disadvantages. What Is a Hybrid Firewall, firewall acts as a proxy.
14. Wang, Q., Li, L., & Hu, S. (2021). Computer network information security protection faced by digital art museums based on the internet of things. *Wireless Communications and Mobile Computing, 2021*, 1–10. https://doi.org/10.1155/2021/2297733
15. Kaplesh, P., & Goel, A. (2019). Firewalls: A study on techniques, security and threats. *Pramana Research Journal, 9*(June), 201941–201952.
16. *What is a circuit level firewall?* NordLayer. Retrieved December 17, 2023, from https://nordlayer.com/learn/firewall/circuit-level/
17. Sophos Ltd. *Sophos XG Firewall*. pp. 4–377, 2017., Retrieved https://www.sophos.com/en-us/medialibrary/pdfs/factsheets/sophos-xg-series-appliances-brna.pdf .
18. J. Zhu. Research on E-payment model. In *Proceedings of the International Conference on E-Business and E-Government, ICEE 2010*, 2010, pp. 339–341. doi: https://doi.org/10.1109/ICEE.2010.93.
19. Keith Murphy. *Benefits of E-payment*. Planergy. Retrieved https://planergy.com/blog/electronic-payments/
20. Zahid Halim. *Current-issues-in-the-acceptability-of-e-payment-systems-world-wide*. researchgate. Retrieved December 18, 2023, from https://www.researchgate.net/figure/Current-issues-in-the-acceptability-of-e-payment-systems-world-wide_fig1_267687068.
21. Nguyen, T. D., & Huynh, P. A. (2018). The roles of perceived risk and trust on E–payment adoption. In *Studies in Computational Intelligence* (Vol. 760, pp. 926–940). Springer Verlag. https://doi.org/10.1007/978-3-319-73150-6_68
22. Rachna, & Singh, P. (2013). Issues and challenges of electronic payment systems. *IJRMP, 2*(9), 25–30.
23. S. Solat, *Security of electronic payment systems: A comprehensive survey*. Paris, Jan. 2017. Retrieved https://hal.science/hal-01433025
24. Universal Academy. *Security Requirements in E-Patment systems*. EduRev. Retrieved https://edurev.in/t/115447/Security-Requirements-in-Electronic-Payment-System
25. Kaur, J., & Singh, S. (2021). An overview of E-payment systems. *International Journal of Research in Engineering and Science, 9*(8), 06–09. www.ijres.org
26. *What is a smart card*. GoCardless. Retrieved https://gocardless.com/guides/posts/what-is-a-smart-card/.
27. Laxmi Ashrit, *What is smart card—How it works, specifications, types and applications*. Electricalfundablog. Retrieved December 18, 2023, from https://electricalfundablog.com/smart-card-works-specifications-types/
28. *Prepaid Cards 101- A definitive guide for growth and profitability*. M2P. Retrieved https://m2pfintech.com/blog/prepaid-cards-101-a-definitive-guide-for-growth-and-profitability/
29. *Prepaid cards*. Federal Trade Commission. Retrieved https://consumer.gov/managing-your-money/prepaid-cards#what-it-is.

30. Justin Pritchard. *Get to know the parts of a debit or credit card.* the balance. Retrieved https://www.thebalancemoney.com/parts-of-a-debit-or-credit-card-front-and-back-315489
31. Matthew D'angelo. *How do credit card payments work?* Investopedia. Retrieved https://www.investopedia.com/how-do-credit-card-payments-work-5069924
32. Marilyn Starkenberg. *What are mobile payments?* kevin. Retrieved https://www.kevin.eu/blog/what-are-mobile-payments/
33. Pillai, V. N. (2021). Present status of E-banking in India. *International Journal of Creative Research Thoughts, 9*(9), 556–561.
34. Jake Frankenfield. *What is online banking? Definition and how it works.* Investopedia. Retrieved https://www.investopedia.com/terms/o/onlinebanking.asp
35. C. Kinza Yasar. Lauren Horwitz. *Digital wallet.* WhatIs.com. Retrieved https://www.techtarget.com/whatis/definition/digital-wallet
36. *Digital wallets-features, advantages, popularity and more.* paytm blog. Retrieved https://paytm.com/blog/payments/mobile-wallet/digital-wallets-features-advantages-popularity-and-more/.
37. *What is cryptocurrency?* javaTpoint. Retrieved https://www.javatpoint.com/what-is-cryptocurrency.
38. Nikunj Gundaniya. *Security and privacy concerns in degital payments.* customer THINK. Retrieved https://customerthink.com/security-and-privacy-concerns-in-digital-payments/. Hacking and cyber-attacks are,as the theft of funds.
39. Arshad, A., Rehman, A. U., Javaid, S., Ali, T. M., Sheikh, J. A., & Azeem, M. (2021). A systematic literature review on phishing and anti-phishing techniques. *arXiv preprint arXiv*, 163–168.
40. Bhavsar, V., Kadlak, A., & Sharma, S. (2018). Study on phishing attacks. *International Journal of Computers and Applications, 182*(33), 27–29. https://doi.org/10.5120/ijca2018918286
41. *Malware attacks: Definition and best practices.* Rapid7. Retrieved https://www.rapid7.com/fundamentals/malware-attacks/
42. *What is ransomware.* Trellx. Retrieved https://www.trellix.com/en-in/security-awareness/ransomware/what-is-ransomware.html.
43. *Understanding, recognizing and preventing malware.* Nationwise. Available: https://www.nationwide.com/lc/resources/personal-finance/articles/how-malware-works
44. Javeed, D., & MohammedBadamasi, U. (2020). Man in the middle attacks: Analysis, motivation and prevention. *International Journal of Computer Networks and Communications Security, 8*(7), 52–58. https://doi.org/10.47277/ijcncs/8(7)1
45. Conti, M., Dragoni, N., & Lesyk, V. (2016). A survey of man in the middle attacks. *IEEE Communications Surveys and Tutorials, 18*(3), 2027–2051. https://doi.org/10.1109/COMST.2016.2548426
46. *What is a data breach?* IBM. Retrieved July 18, 2023, from https://www.ibm.com/topics/data-breach
47. Patrick Smyth. *How do data breaches happen?* PAYSIMPLE. Retrieved July 18, 2023, from https://paysimple.com/blog/how-to-prevent-data-breach/
48. M. Abbas and Robin Harmon. *What is card skimming.* SEATTLE CREDIT UNION. Retrieved July 20, 2023, from https://www.seattlecu.com/askus/what-is-card-skimming#
49. Kiki Sheldon. *Card skimming: What it is and how to prevent it.* ithemes. Retrieved July 20, 2023, from https://ithemes.com/blog/card-skimming-what-it-is-and-how-to-prevent-it/
50. Michael Hill. *The 12 biggest data breach fines, penalties, and settlements so far.* CSOonline. Retrieved July 21, 2023, from https://www.csoonline.com/article/567531/the-biggest-data-breach-fines-penalties-and-settlements-so-far.html.
51. S. M. K. Saheed Oladimeji. *SolarWinds hack explained: Everything you need to know.* techtarget. Retrieved July 20, 2023, from https://www.techtarget.com/whatis/feature/SolarWinds-hack-explained-Everything-you-need-to-know.

52. Dan Avery. *Capital One $190 million data breach settlement: Today is the last day to claim money*. CNET. Retrieved July 20, 2023, from https://www.cnet.com/personal-finance/capital-one-190-million-data-breach-settlement-today-is-deadline-to-file-claim/#:~:text=A 2019 data breach exposed, 100 million Capital One customers. Today is the final day,100 million people's personal information.
53. Sara Merken. Wawa to pay $8 million in data breach settlement with state AGs. REUTERS. Retrieved July 20, 2023, from https://www.reuters.com/legal/litigation/wawa-pay-8-mln-data-breach-settlement-with-state-ags-2022-07-26/.
54. SecureWorld News Team. *3 ways ticketmaster failed at cybersecurity*. SECUREWORLD. Retrieved July 20, 2023, from https://www.secureworld.io/industry-news/3-ways-ticketmaster-failed-at-cybersecurity#:~:text=The Ticketmaster U.K. data breach,security codes, and expiration dates.
55. *British Airways data breach*. WIKIPEDIA. Retrieved https://en.wikipedia.org/wiki/British_Airways_data_breach#:~:text=British Airways said the attack, 380,000 total customers being compromised.
56. R. A. Vindu Goel. *Card data stolen from 5 million Saks and Lord & Taylor Customers*. The New York Times. Retrieved July 20, 2023, from https://www.nytimes.com/2018/04/01/technology/saks-lord-taylor-credit-cards.html
57. Khan Danish. *OnePlus data breach: India's cybersecurity agency issues advisory for users; reveals details*. Telecom.com. Retrieved December 18, 2023, from https://telecom.economictimes.indiatimes.com/news/oneplus-data-breach-indias-cybersecurity-agency-says-around-3000-oneplus-users-data-exposed/72246434# OnePlus which is still in, exposure of bank card details.
58. Catalin Cimpanu. *Dunkin' donuts accounts compromised in second credential stuffing attack in three months*. ZDNET. Retrieved July 20, 2023, from https://www.zdnet.com/article/dunkin-donuts-accounts-compromised-in-second-credential-stuffing-attack-in-three-months/.
59. David Lukic. *Before considering your next dining experience, remember the chipotle data breach*. IDStrong. Retrieved July 20, 2023, from https://www.idstrong.com/sentinel/chipotle-credit-card-hacks/. The chipotle breach of 2017, locations for goods and services.
60. Carly Page. *Hotel giant IHG blames cyberattack for booking systems outage*. techcrunch. Retrieved July 20, 2023, from https://techcrunch.com/2022/09/07/ihg-hotels-outage-cyberattack/
61. *2017 equifax data breach*. WIKIPEDIA. Retrieved July 20, 2023, from https://en.wikipedia.org/wiki/2017_Equifax_data_breach
62. Jim Ducharme. *Payment authentication | A complete explanation*. OUTSEER. Retrieved July 20, 2023, from https://www.outseer.com/payment-security/payment-security-payment-authentication/. One-timepasscodes challenge questions, or other accounts are legitimate.
63. *Top 5 most reliable and user-friendly authenticatn methods in online payments*. CyberSecurity. Retrieved July 20, 2023, from https://cybersecurity.asee.co/blog/top-authentication-methods-in-online-payments/.
64. *What is a dynamic CVV*. clover. Retrieved July 20, 2023, from https://blog.clover.com/what-is-a-dynamic-cvv/
65. *Enable generation of dCVV2 codes to fight fraud*. corepay. Retrieved December 14, 2023, from https://corepay.net/articles/enable-generation-of-dcvv2-codes-to-fight-fraud/
66. *What is biometric payment and whatare the benefits*. ARATEK. Retrieved July 20, 2023, from https://www.aratek.co/news/what-is-biometric-payment. Biometric. payment is a form,time they make a purchase.
67. *What is payment Tekenization and how does it work*, EBANX. Retrieved July 21, 2023, from https://www.ebanx.com/en/resources/payments-explained/tokenization/
68. Vivian Tajeda. *Payment tokenization guide: Definition and uses*. Retrieved July 20, 2023, from https://www.shopify.com/retail/payment-tokenization
69. *Tokenization payment technology guide*. adyen. Retrieved July 20, 2023, from https://www.adyen.com/knowledge-hub/guides/tokenization-payment-technology-guide#how

70. *What is device fingerprinting*. Arkose Labs. Retrieved July 20, 2023, from https://www.arkoselabs.com/explained/device-fingerprinting/. Device fingerprinting is the process, order to uniquely identify it.
71. Tamas Kadar. *What is device fingerprinting and how exactly does it work?* SEON. Retrieved July 20, 2023, from https://seon.io/resources/device-fingerprinting/
72. *Detecting fraud with device fingerprinting*. Callsign. Retrieved July 21, 2023, from https://www.callsign.com/knowledge-insights/preventing-fraud-with-device-fingerprinting
73. *How is behavioral biometrics used for authentication?* INCOGNIA. Retrieved July 21, 2023, from https://www.incognia.com/the-authentication-reference/how-is-behavioral-biometrics-used-for-authentication.
74. Zhao, P., Ding, Z., Wang, M., & Cao, R. (2019). Behavior analysis for electronic commerce trading systems: A survey. *IEEE Access, 7*, 108703–108728. https://doi.org/10.1109/ACCESS.2019.2933247
75. *Top 5 payment gateway security protocols*. Binary Stream. Retrieved July 23, 2023, from https://binarystream.com/blog/top-5-payment-gateway-security-protocols. Secure sockets layer (SSL) and, collected by a payment gateway.
76. Domantas G. & Brian F. *What is SSL? Understanding secure sockets layer and how it works*, Hostinger. Retrieved December 14, 2023, from https://www.hostinger.in/tutorials/what-is-ssl
77. Satapathy, A., & Livingston, J. (2016). A comprehensive survey on SSL/ TLS and their vulnerabilities. *International Journal of Computers and Applications, 153*(5), 31–38. https://doi.org/10.5120/ijca2016912063
78. *What is SSL?* SSL.com. Rettrieved July 24, 2023, from https://www.ssl.com/faqs/faq-what-is-ssl/.
79. A. Aayush, Y. Aryan, and B. Muniyal. Understanding SSL protocol and its cryptographic weaknesses. *Proceedings of 3rd International Conference on Intelligent Engineering and Management*, ICIEM 2022, pp. 825–832, 2022, doi: https://doi.org/10.1109/ICIEM54221.2022.9853153.
80. S. Balfe and K. G. Paterson. E-EMV: Emulating EMV for internet payments with trusted computing technologies. *Proceedings of the ACM Conference on Computer and Communications Security*, no. May, pp. 81–92, 2008, doi: https://doi.org/10.1145/1456455.1456468.
81. N. El Madhoun et al. *The EMV payment system: Is it reliable? To cite this version: HAL Id: hal-02293847 the EMV payment system: Is It Reliable?*, no. CSNet, 2019.
82. D. Singer. Visa's 3-D Secure: Secure Online Payment Authentication. *GIAC Security Essentials Certification (GSEC) Practical Assignment Version 1.4c*, 2005.
83. Yungaicela-Naula, N. M., Vargas-Rosales, C., & Perez-Diaz, J. A. (2021). SDN-based architecture for transport and application layer DDoS attack detection by using machine and deep learning. *IEEE Access, 9*, 108495–108512. https://doi.org/10.1109/ACCESS.2021.3101650
84. A. Bouch. 3-D secure: A critical review of 3-D secure and its effectiveness in preventing card not present fraud by," University of London, no. March, pp. 1–104, 2011.
85. Swati Tawde. *Secure electronic transaction*. EDUCBA. Retrieved July 24, 2023, from https://www.educba.com/secure-electronic-transaction/
86. Karthryn Pomroy. *SET protocol: What is it and how does it work?*. Payment Cloud. Retrieved July 24, 2023, from https://paymentcloudinc.com/blog/set-protocol/
87. Jake Frankenfield. *Secure electronic transaction(SET): Definition and how it works*. INVESTOPEDIA. Retrieved July 24, 2023, from https://www.investopedia.com/terms/s/secure-electronic-transaction-set.asp
88. D. Clarisa and D. Marlena. Design of secure NFC e-Payment with ambient conditions-based solutions and Chaskey Algorithm. *Proceedings - IWBIS 2021: 6th International Workshop on Big Data and Information Security*, pp. 139–144, 2021, doi: https://doi.org/10.1109/IWBIS53353.2021.9631840.
89. Borowski-Beszta, M. (2018). Mobile payments using NFC technology in the light of empirical research. *Torun Business Review, 3*(17), 5–16. https://doi.org/10.19197/tbr.v17i3.311

90. Nikunj Gundaniya. *All you need to know about NFC payments*, digipay.guru. Retrieved July 26, 2023, from https://www.digipay.guru/blog/all-you-need-to-know-about-nfc-payments/
91. Olivia Deng. *How an NFC payment works & why more businesses are accepting them*, Hubspot. Retrieved July 25, 2023, from https://blog.hubspot.com/sales/nfc-payment. An NFC payment is a, Apple Pay and Google Pay.
92. Marilyn Starkenberg. *What are NFC payments, and how do they work?*, kevin. https://www.kevin.eu/blog/nfc-payments/

Smart and Secured Ways of E-Payment: Design of New Frameworks

M. Savithri, S. Vijayalakshmi, A. Arul SelvanGnanamonickam, and V. Shobana

1 Introduction

Electronic payment (e-payment) systems have completely changed how transactions are carried out in the modern digital era by providing unmatched efficiency and convenience. Convenience is great, but it also means that strong security measures are absolutely necessary to protect private financial data from fraud and online threats. In order to maintain user trust and guarantee the smooth operation of digital economies, new frameworks that give equal weight to intelligence and security must be designed as e-payment methods continue to develop. The state of financial technology is exemplified by clever and safe e-payment frameworks, which combine front-line technologies and encoding strategies to improve user experience while bolstering security against malevolent actors. These frameworks are designed to prevent new dangers in the digital realm and solve the inherent vulnerabilities of traditional payment systems.

From the introduction of the e-commerce various improvements are done to provide numerous benefits to the customers and the organisation making their business worldwide. The development of technologies like internet banking, mobile banking, and various other payment systems like GPay, Paytm, UPI, etc. made the e-commerce platform still more comfortable for the customers to buy their products by making

M. Savithri (✉) · S. Vijayalakshmi
Department of Data Science, Christ University, Bengaluru, India
e-mail: savithri.m@christuniversity.in; s.vijayalakshmi@christuniversity.in

A. A. SelvanGnanamonickam
Department of Software Systems, KG College of Arts and Science, Coimbatore, India

V. Shobana
Department of Computer Science with Cyber Security, Dr. N.G.P. Arts and Science College, Coimbatore, India
e-mail: shobana@drngpasc.ac.in

S. Vijayalakshmi et al. (eds.), *Smart Power Systems*, Power Systems,
https://doi.org/10.1007/978-3-031-89459-6_18

the payment gateway easy. The future of electronic commerce looks bright and it is becoming more and more popular over time, becoming a new growth point [1]. As a result, the need for electronic payments has emerged and it is now thought that these are essential for successful business and financial services. As such, it is time to re-evaluate some of the issues that are impeding the success of electronic commerce, one of which is the purchase of electronic contents and services also known as e-products.

Due to the increased value of e-transaction, the rate of security breaches is also increasing constantly which leads to high risk and uncertainty. This also leads to various consequences globally on the security and economic factor of the nations [2]. As per the report of verizon [3], there was a substantial increase in the various attacks while compared to the previous year 2022. As per the report, the breaches are tripled. The breaches are increased 180% when compared to the last year this is due to the effect of MOVEit and zero-day vulnerabilities.The use of e-payment is increasing and it is also believed that to improve the use of e-commerce trust on security and the perception about the trust on good security is required. Due to the various mobile payment system that are used, the transactions using the e-payment system is also increasing day by day which might also increase the potential risk on security issues that might also have an impact on the business. It is necessary that both the customers and the business organisations should be aware about the various threats that are associated with the e-payment system and should take necessary action to prevent the attacks against the organisations transaction and also to protect the information of their customers. A survey done by cybersecurity experts has shown that there are more breach expectations on mobile payment which might increase by 87% with in next 12 months. There are various types of vulnerabilities like data breach, cyberattack, etc. so it is required to provide a framework that will be a good and secured platform for e-payment system.

2 Roles of E-Payment System

In this digital era, most of the customers are purchasing the products that they need through online shopping. The e-payment system plays a major role in online shopping. After COVID-19, most of the payments are made through e-payment system like Paytm, GPay, UPI, etc. Cashless transactions are increasing since it is easy for the customers to make the payment, no need to carry cash, availability of e-payment service in all places. In today's business and financial activities, e-payment systems play a number of crucial functions. By acting as intermediaries, they make it possible for parties to transfer money quickly and easily without requiring actual currency or cheques. Additionally, by automating payment procedures, these systems increase efficiency by lowering the time and resources needed for financial transactions. E-payment systems use encryption and other security measures to safeguard-sensitive financial data, reducing the possibility of fraud and unauthorised access. Security is of the utmost importance. Additionally, these systems produce useful transaction data that companies may use to gain understanding of consumer behaviour, and governments are depending more and more on e-payment systems to optimise public services and distribute benefits. E-payment systems, which provide safe, effective, and convenient financial transactions in a variety of industries, are essentially essential to contemporary economies.

3 Money Payment System

3.1 Money, Payments, and the Current Payment System

Money plays a vital role in the modern economics, and it is a medium of exchange which is universally accepted deferred payment standard. It helps to facilitate the transaction and act as a common intermediary in the selling and buying of products. It also acts as a unit of account that will help us to compare the value of the goods and the services. Money also acts as a value which helps individuals to their wealth over a time as holding cash or deposit or wealth. It is also used for deferred payment like credit transaction [4], investment, etc. So, money plays an important role in the economy by facilitating the interaction and to enable the allocation of resource and it also helps to foster the economic growth.Diverse payment systems cover a broad range of procedures that make transactions between people, companies, and financial institutions easier. Cash, cheques, credit cards, debit cards, bank transfers, electronic funds transfers (EFT), mobile payments, contactless payments, cryptocurrency, and prepaid cards are a few of the most often used ways. For in-person transactions, cash is still frequently utilised since it provides privacy and quick settlement. Even though their use is dwindling, cheques are still used for some transactions, especially business-to-business payments, as paper-based authorization documents for money transfers between bank accounts.

Financial institutions issue credit cards, which let customers make purchases on credit. The transactions are paid for at the end of each month's payment cycle. In contrast, debit cards provide instantaneous money transfers at the point of sale since they are directly connected to the user's bank account [5]. Bank transfers can be made online, using mobile banking apps, or in person at bank offices. They entail the electronic transfer of money between bank accounts, either within the same bank or between separate banks. Automated clearinghouse (ACH) transactions, bank transfers, wire transfers, direct deposits, and other electronic payments are all included in electronic funds transfer (EFT), which makes safe and effective cash transfers possible.

Mobile wallets and payment apps, along with technologies like Near Field Communication (NFC), QR codes, and peer-to-peer (P2P) payment systems, can all be used for mobile payments. A subtype of mobile payments known as contactless payments allow customers to pay at the point of sale more quickly and easily by just touching or waving their cards or mobile devices close to a contactless-enabled terminal. Peer-to-peer transactions are made possible by cryptocurrencies like Bitcoin and Ethereum, which use cryptographic techniques to provide safe transactions and decentralised control, doing away with the need for middlemen like banks. Reloadable or non-reloadable prepaid cards are loaded with money and can be used for transactions up to the card's balance. For those without bank accounts or credit histories, prepaid cards provide a safe and easy substitute for cash.

These many payment methods offer choices for both online and offline transaction processing, satisfying a variety of consumer preferences. Payment system innovations are driven by ongoing technical innovation, which improves accessibility, security, and efficiency for users everywhere.

The present payment system is a sophisticated infrastructure that combines the efforts of numerous parties, such as banks, payment processors, and regulatory agencies, to enable the seamless transfer of payments between consumers, companies, and financial institutions. Below is a summary of every element:

Banks offer accounts for people and companies to keep money and by enabling transactions, banks play a crucial part in the payment system. They provide checking, savings, and merchant accounts, among other services. Additionally, banks serve as middlemen in transaction settlement, guaranteeing the safe movement of payments between accounts. Payment processors that help banks, merchants, and customers authorise and settle electronic payments are known as payment processors. They offer the infrastructure and technology required to handle credit cards, debit cards, and mobile payments, among other payment options. Payment processors control risk, validate the legitimacy of transactions, and guarantee safe and correct money transfers.

Regulatory Bodies keep an eye on the payment system to make sure it runs smoothly, safely, and in accordance with all relevant rules and laws. The regulation of payment systems and the supervision of the operations of banks and payment processors are under the purview of central banks or other government bodies in numerous nations. They set guidelines and standards for payment processing, keep an eye out for fraud and abuse, and enforce adherence to consumer protection and anti-money laundering laws.

Interbank Networks systems link several banks and financial organisations to enable the transfer of money and information. Payments may be handled swiftly and effectively, thanks to these networks, which let banks talk to one another and settle transactions. The ACH (Automated Clearing House) and SWIFT (Society for Worldwide Interbank Financial Telecommunication) networks are two instances of interbank networks.

Clearing and Settlement Systems are the systems that help the parties to a transaction transfer money to each other at the end. While settlement involves the actual transfer of money between accounts, clearing involves the verification, netting, and matching of transactions. By guaranteeing precise and safe payment completion, these technologies lower the possibility of fraud and mistakes.

The banks, payment processors, regulators, interbank networks, clearing and settlement systems, and other players make up the intricate ecosystem that is the current payment infrastructure. These parties collaborate to maintain financial stability and economic activity by ensuring that money may be moved between participants in the economy promptly, safely, and effectively.

3.2 Payment of Today's Consumers

Consumers of today have converged convenience, security, and technical innovation in their payment preferences. In order to promote trust and confidence in digital payment systems, stakeholders need to give equal weight to usability and security as cashless transactions continue to gain popularity. We can guarantee a seamless and safe payment experience for all parties involved by addressing customer concerns, embracing technology improvements, and fostering a culture of cybersecurity awareness. Cashless payment methods are becoming more and more popular among customers nowadays due to factors including ease, developments in technology, and changing consumer preferences. Let's examine each in detail:

Consumer Behaviour: An increasing propensity for cashless purchases characterises the current consumer landscape. Electronic payment options including credit and debit cards, mobile wallets, and digital payment platforms are steadily replacing more conventional payment methods like cash and cheques. This change is being driven by accessibility, speed, and convenience. Customers value the convenience of completing transactions with a single tap, swipe, or click, particularly in our fast-paced world. Furthermore, as individuals place a higher priority on safety and hygiene in the wake of the COVID-19 outbreak, contactless payment acceptance has increased.

Security Concerns: Although cashless payments are convenient, customers' top concern is still security. Concerns over the security of online transactions have increased in light of instances of fraud, identity theft, and data breaches. Customers are cautious about possible weaknesses in digital payment systems, such as fraudulent transactions and illegal access to their financial data. Payment service providers and financial institutions make significant investments in cybersecurity solutions like biometric verification, tokenization, encryption, and real-time fraud detection algorithms to allay these worries. Furthermore, principles for protecting payment data and consumers' financial interests are outlined in regulatory compliance standards such as the Payment Card Industry Data Security Standard (PCI DSS).

Convenience vs. Security: When creating efficient payment systems, striking the right balance between the two is crucial. Customers appreciate that cashless transactions are simple to use and provide a smooth experience, but they also worry about safeguarding their private financial information from misuse and unauthorised access. Collaboration between stakeholders is necessary to deploy strong security measures without sacrificing usability in order to balance these conflicting objectives. To improve digital payment systems' security posture while maintaining customer comfort, innovations in data encryption techniques, transaction monitoring systems, and authentication technologies are crucial. Building a robust and secure payment ecosystem also requires instilling confidence in electronic payment systems and educating consumers about recommended practices for protecting their payment information.

3.3 Innovation in Payment

Innovation is essential to the advancement and competitiveness of the payment industry. It enables companies to adjust to shifting consumer tastes, technical developments, and legal obligations. Cutting-edge payment options improve efficiency, security, and convenience while also improving the overall customer experience. Moreover, by meeting a range of needs and preferences, innovation stimulates market expansion, competitiveness, and financial inclusion. Through disruptive technologies, fintech startups are transforming the existing payment systems landscape. Fintech companies use cutting-edge technology like blockchain, AI, and machine learning to provide creative payment solutions that subvert the dominance of traditional financial institutions. Peer-to-peer payment networks, digital wallets, mobile payment apps, and alternative financing services are all included in these solutions. Fintech disruption democratises financial services for both individuals and enterprises by increasing competition, lowering transaction costs, and improving accessibility.

The payment industry is undergoing a revolution, thanks to blockchain technology and cryptocurrencies, especially in terms of security, transparency, and decentralisation. Blockchain minimises processing delays and transaction costs by enabling peer-to-peer transactions without the need for middlemen. Peer-to-peer and permissionless transactions are made possible by decentralised digital currencies such as Bitcoin, Ethereum, and Ripple. Blockchain technology and cryptocurrencies, albeit still in their infancy, have the potential to revolutionise the world of payments by offering safe, effective, and censorship-resistant payment options. In order to improve security and convenience, biometric verification techniques—like thumbprint and facial recognition—are being incorporated into payment systems more and more. By using distinct physiological traits for user authentication, biometric payments reduce the danger of lost or stolen cards and password theft. Businesses may provide secure and frictionless payment experiences that align with the demands of modern consumers for seamless transactions by integrating biometric authentication into payment terminals and mobile devices.

Thanks to Near Field Communication (NFC) technology, contactless payment methods have become more popular, which has changed consumer behaviour and payment practices. With contactless payments, customers only prerequisite is to trace their cards or mobile devices at payment terminals to rapidly and securely complete transactions. This technology is especially pertinent in light of the COVID-19 epidemic because it not only makes life more convenient but also encourages hygiene and lowers the risk of physical contact. Consequently, a wide range of industries, including retail, hospitality, and transportation, have adopted contactless payments. Innovation in the payment industry promotes competitiveness, advances the state of affairs, and improves customer satisfaction in general. From the use of blockchain technology and fintech disruption to biometric authentication and contactless payments, creative solutions are redefining the way that transactions are carried out and opening the door to a more efficient, safe, and accessible payment environment.

4 Need for Secured Payment System

Financial stability, economic prosperity, and customer trust are just a few of the reasons why a secure payment system is critical. A secure payment system is essential for Customer Trust to Gaining the confidence and trust of customers requires a secure payment method. People want their financial information to be protected from fraud, identity theft, and unauthorised access when they make payments, whether they do so online or in person. Businesses and financial institutions may see a decline in revenue as a result of a security breach that damages customer confidence. Thus, strong security protocols are required to safeguard confidential information and maintain customer confidence in the payment ecosystem. To protect against fraud a secured payment system is crucial for reducing the risks of fraud given the sophistication of cyber threats, which include malware attacks, phishing schemes, and data breaches. Through the utilisation of multi-layered security protocols, encryption strategies, and real-time fraud detection systems, enterprises may effectively prevent fraudulent actions and protect their own as well as their clients' interests. Tokenization and biometric authentication are examples of proactive procedures that provide an additional layer of safety, making it extra tough for hackers to corrupt financial transactions. By reducing the possibility of financial crimes including money laundering, financing of terrorism, and illicit transactions, secured payment systems help to promote financial stability. Financial organisations can stop illegal money from entering the financial system and preserve the integrity of the payment ecosystem by abiding by regulatory compliance standards and anti-money laundering (AML) laws. Strong security measures also promote stability and resilience in the face of outside shocks and uncertainty by strengthening investor trust in the financial markets. Guaranteeing the security of payment systems is essential for businesses to preserve their reputation and the ability to operate. A security breach can cause financial losses, interfere with corporate operations, and damage a brand's reputation. To effectively reduce risks and address possible attacks, firms must invest in cybersecurity infrastructure, personnel training, and incident response capabilities. Businesses may preserve their reputation in the marketplace, protect their assets, and protect their consumers by putting security first. Another strong argument in favour of keeping a protected payment method is regulatory compliance. To guarantee the accuracy, openness, and safety of financial transactions, financial authorities set strict rules and regulations. Businesses who intrude upon these rules risk harsh fines, legal implications, and harm to their repute.

Therefore, in order to achieve compliance and reduce related risks, firms need to keep up with changing regulatory requirements and build strong controls.

5 Smart and Secured Ways of E-Payment

E-payment solutions that are clever and secure are essential for guaranteeing seamless transactions and protecting private financial data. Some of the safe e-payment methods that we are following at present are encryption to protect sensitive data during transmission, use end-to-end encryption. This makes certain that the data cannot be retrieved by unauthorised individuals, even if it is interrupted. Tokenization is used to distinct tokens in place of private cardholder data. In this manner, the token cannot be used to carry out illegal operations without access to the original data, even in the event that it is intercepted. Two-factor authentication: Use two-factor authentication techniques, such as authenticator applications, biometric authentication, or SMS verification. By forcing consumers to present two pieces of identification before completing a transaction, this increases security even more. Make sure your website or application makes use of Secure Sockets Layer (SSL) certificates in order to create a secure connection between the user's browser and your server. In order to prevent unwanted access, this encrypts the data being transferred between the two parties. PCI compliance is used to safeguard cardholder data, fulfil with the Payment Card Industry Data Security Standard (PCI DSS). This entails keeping an information security policy up to date, establishing strict access rules, monitoring and testing networks on a regular basis, and maintaining a secure network. Biometric authentication helps us to confirm users' identities, utilise biometric authentication techniques like facial recognition and fingerprint scanning. Because biometric data is hard to fake, it adds an extra degree of protection. Blockchain technology provides decentralised and immutable ledgers, and hackers find it more difficult to manipulate transaction records. Fraud Detection Systems are used in fraud detection that are able to instantly identify suspicious activity and analyse transaction trends. These systems can aid in stopping fraudulent transactions in their tracks. Frequent Security updates all systems and software with the most recent security patches and updates. This aids in addressing any weaknesses that hackers might try to take advantage of. Inform users of the value of security precautions including using strong passwords, avoiding important transactions on public Wi-Fi, and being on the lookout for phishing scams. User awareness is essential to preserving security.

6 Framework for Smart and Secured E-Payment

Security plays the most important role in the e-payment system. Due to the increasing use of e-commerce platform, there is an increase in security attacks. It is most challenging issue for all e-commerce system to face the security requisites. There are two categories of security requirement: they are authentic users who use the system in safe way and the second category is lacking of malicious user who may succeed by attacking the implemented security system. The framework for designing the secured e-payment might be based on the known security attacks while during the development phase of the system countermeasures will be taken to prevent unexpected attacks and the developed system will be tested after the post-release [6]. It is more expensive to fix the defects based on the e-commerce design model [7]. Building a strong foundation for intelligent and safe electronic payments requires a multifaceted strategy that includes risk analysis, legal observance, and technology infrastructure. It entails putting strict security measures in place to protect sensitive data and lessen fraudulent activity, such as multi-factor authentication, encryption techniques, and real-time transaction monitoring [8]. Encryption of data in transit and storage, along with data minimization techniques and routine audits, assure data privacy and protection. While user education and awareness programmes help to create a culture that is security-conscious, incident response protocols and continuous monitoring are crucial for quickly addressing security incidents. The e-payment system's resistance to changing threats and legal requirements is further strengthened by efficient vendor management and a dedication to ongoing development. Figure 1 shows the security service model design by NIST for e-payment. It shows the various services and elements that are used in the implementation of security with the help of technology and the relationship among the various

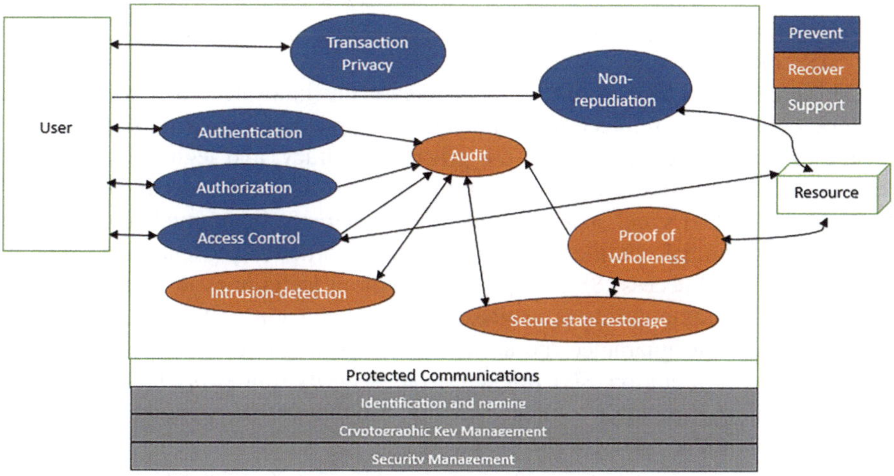

Fig. 1 NIST Model for integrating security service

elements involved in it. Based on the primary purpose, it is classified into three parts as Prevent, Recover, and Support. As the name implies, Prevent is used to prevent the attacks, Recover is used to identify the attack and recover the system from the breach, and the last is Support which is generic and it contains most of the security capabilities related to information technology. An organised framework to improve e-payment system security is provided by the NIST Security Services Model. It outlines several security service tiers designed to strengthen various system components. Establishing safe routes for data exchange and using encryption algorithms to protect data from interception and manipulation are priorities at the Communication Security Layer. To control user access to important functions and data, the Access Control Layer focuses on strict authentication procedures and fine-grained permission restrictions. The Data Security Layer places a strong emphasis on data encryption and integrity maintenance techniques to protect stored information from unwanted access and alteration. The System Integrity Layer promotes ongoing surveillance and intrusion detection in order to quickly detect and eliminate security risks. The Resiliency Layer emphasises how crucial redundancy, backup, and recovery plans are to maintaining system availability and continuity in the face of interruptions. By following this strategy, e-payment systems can strengthen their defences against a variety of security threats, promoting dependability and trust among stakeholders and users alike.

The NIST authentication model outlines a comprehensive strategy for safely confirming users' identities. It highlights how crucial it is to use a variety of authentication methods and elements that are appropriate for the transaction's context and degree of risk. In order to improve security, the model recommends using a variety of elements, including smart cards, passwords, biometrics, and one-time passwords. Moreover, it emphasises the necessity of adaptive authentication, in which the authentication procedure dynamically modifies according to contextual elements including user behaviour, location, and device specifications. The model also emphasises the significance of usability, promoting the deployment of user-friendly authentication techniques while guaranteeing strong security. Figure 2 represents the NIST model for authentication.

Mastercard and Visa collaborated to create the innovative Secure Electronic Transactions (SET) architecture, which aims to improve the security of online electronic payments. Concerns about the privacy, accuracy, and legitimacy of online transactions are the focus of this paradigm. SET utilises digital certificates to confirm the cardholder's and merchant's identities, guaranteeing mutual authentication and building mutual confidence. Furthermore, in order to prevent eavesdropping and unauthorised access, SET employs encryption techniques to secure sensitive payment information during transmission. Digital signatures are also incorporated into the system to guarantee the accuracy of transaction data and guard against manipulation or alteration. SET offers a strong security framework for online transactions, making it possible to make safe and reliable online payments that inspire confidence in both customers and businesses.

Robust security measures are incorporated into an e-commerce system that integrates with Secure Electronic Transactions (SET) to guarantee secure and

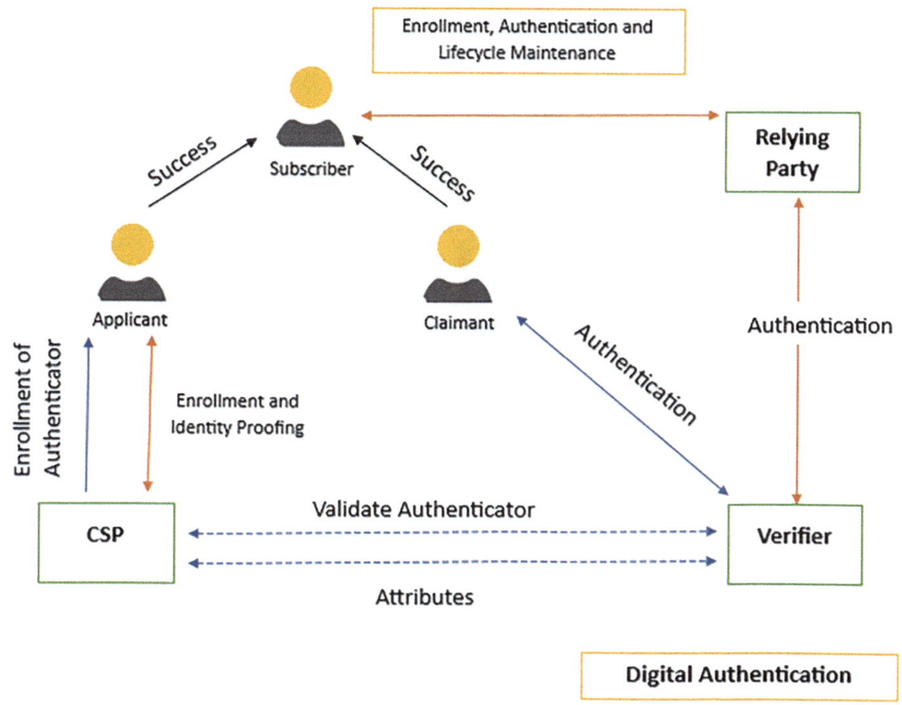

Fig. 2 NIST model for authentication

dependable online transactions. This system, which establishes confidence in the transaction process, uses digital certificates for mutual authentication between buyers and sellers by leveraging the SET architecture. Sensitive payment information is protected during transmission via encryption techniques, preventing possible interception by unauthorised parties. Digital signatures are also used to confirm the accuracy of transaction data, reducing the possibility of fraud or tampering. This solution offers a safe environment for making online transactions by incorporating SET into the e-commerce infrastructure, boosting customer confidence and promoting a reliable marketplace experience. Figure 3 represents the e-commerce platform with SET integration.

7 Internet Payment Systems

The platforms and technologies that make it easier to send money or complete financial transactions online are referred to as internet payment systems. With the rise in e-commerce and online transactions, these systems are becoming more and more common. Internet payment systems come in a variety of forms, with varying features, security protocols, and ways of working [9]. Credit Card Payments is a

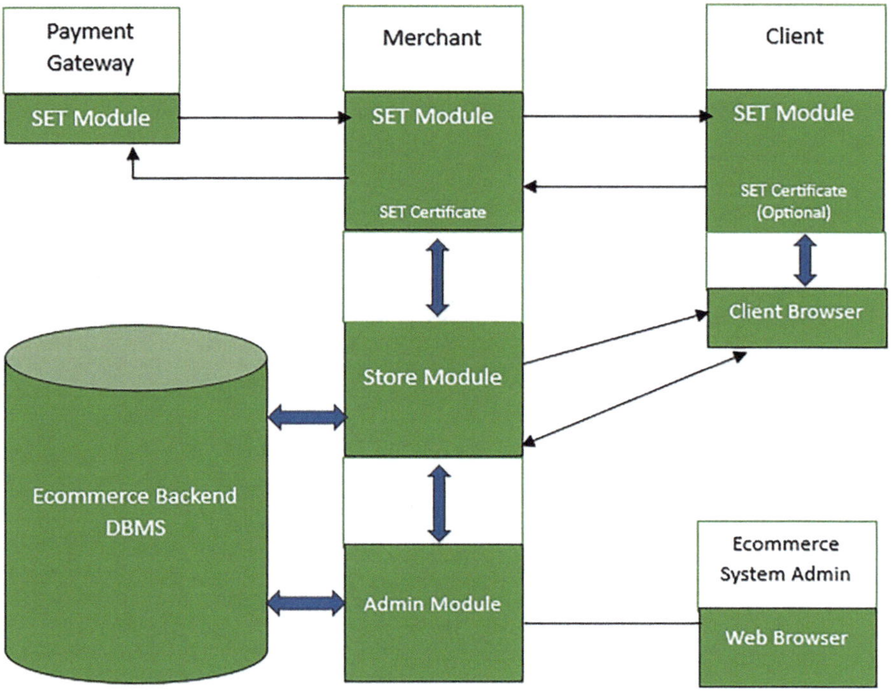

Fig. 3 E-commerce platform with SET integration

widely used payment option that enables consumers to make online purchases by submitting their credit card details over a secure payment gateway. Credit card networks handle transactions, and sensitive data is usually protected using security protocols like SSL/TLS encryption. Payments with a debit card are just like credit card payments; debit card purchases take money straight out of the user's bank account. The infrastructure and security protocols used for credit card payments are frequently also used for debit card purchases. Digital wallets, sometimes referred to as e-wallets, are online payment systems that save users' credit card details and facilitate rapid, easy transactions. PayPal, Apple Pay, Google Pay, and Samsung Pay are a few examples. Tokenization and encryption are frequently used by digital wallets to protect transactions. Bank Transfers, customers can electronically move money from their bank account to the merchant's account. This option may take longer to process than card payments because it usually includes authentication using online banking credentials. Payments with cryptocurrencies are like Bitcoin, Ethereum, and others, several retailers now take payments using digital currencies. Pseudonymity and cryptographic security features are provided by cryptocurrency payments; however, they also have volatility and legal issues. Mobile Payments use applications or near-field communication (NFC) technology, mobile payment systems allow customers to make transactions using smartphones and other mobile devices. Tokenization and biometric authentication (such as fingerprint or facial recognition) are examples of security features. Peer-to-peer (P2P) Payment systems enable people to send money to one another directly, frequently using internet or mobile apps. Zelle, Cash App, and Venmo are a few examples. To safeguard transactions, these systems usually use authentication and encryption.

8 Security Challenges in E-Payment

Sensitive financial information's availability, confidentiality, and integrity are all at risk from a variety of security flaws in e-payment systems. Data breaches are a major worry because they allow hackers to obtain credit card numbers, passwords, and personal information by taking advantage of holes in payment systems. Phishing attacks, which fool people into disclosing their login passwords or financial information, are another serious concern. They target users with misleading emails or websites [10]. Furthermore, e-payment systems can be compromised by malware and ransomware, which can infect machines and either steal or hold sensitive data ransom. Another common problem is identity theft, where thieves use stolen personal data to pose as authorised users and carry out illegal activities. Additionally, because cyber threats are constantly changing, it's important to be vigilant and take proactive steps to reduce risks. Some of these proactive measures include putting strong authentication mechanisms, encryption protocols, fraud detection systems, and frequent security audits in place. To protect e-payment systems and uphold confidence in online transactions, tackling these security issues necessitates a comprehensive strategy encompassing technology, regulations, user education, and cooperation among stakeholders.

9 Steps Taken to Overcome the Challenges

There are various preventive steps that can be taken in order to get around the security issues that come with e-payment systems. First off, implementing strong encryption techniques reduces the possibility of unauthorised parties intercepting confidential data during transmission by guaranteeing its confidentiality [11]. By adding multi-factor authentication, you may increase user verification and lower the risk of unauthorised access using credential theft. Early detection of suspicious activity is made possible by real-time fraud detection systems and continuous monitoring, which enables prompt intervention to stop fraudulent transactions. Additionally, frequent penetration tests and security audits aid in locating holes and flaws in the system, allowing for quick fixes and boosting resilience all around. Furthermore, funding user education and staff training initiatives increases knowledge of security best practices and equips users to identify and steer clear of possible dangers like phishing scams. The security posture of e-payment systems is further strengthened by working together with industry partners and regulatory agencies to share threat intelligence and uphold established security requirements. Through the implementation of a comprehensive strategy that includes procedural, technological, and educational activities, organisations can efficiently reduce security risks and foster trust in the dependability and integrity of electronic payment transactions.

10 Conclusion

Ensuring the integrity of online transactions and tackling the ever-changing cyber threat scenario need the development of innovative frameworks for intelligent and secure electronic payment systems. Strong encryption, multi-factor authentication, and real-time fraud detection systems are just a few of the ways these frameworks help e-payment platforms strengthen their security posture, protecting sensitive financial data and upholding user confidence. Moreover, staying ahead of developing threats and vulnerabilities requires constant monitoring, frequent security audits, and cooperation with industry partners. Additionally, user education and awareness campaigns are essential for encouraging security-conscious behaviour and reducing the danger of social engineering frauds like phishing assaults. The creation of new frameworks needs to be flexible and robust in order to meet new difficulties and comply with changing regulatory requirements as technology develops. Organisations may promote a more secure and reliable online transaction environment and stimulate innovation and growth in the digital economy by giving security first priority when designing e-payment frameworks.

References

1. Kim, C., Tao, W., Shin, N., & Kim, K. S. (2010). An empirical study of customers' perceptions of security and trust in e-payment system. *Electronic Commerce Research and Applications, 9*(1), 84–95.
2. Emmanuel Chinanu, O. A. U. (2018). Comparative survey of cyber-threat and attack trends and prediction of future cyber-attack patterns. *International Journal of Innovative Research in Computer and Communication Engineering, 6*(4).
3. *Verizon business*. (2024). Retrieved April 4, 2024, from https://www.verizon.com/business/resources/reports/dbir/2024/summary-of-findings/.
4. J. Wilkinson. (2023). *Takepayments*. takepayments.com. Retrieved April 2, 2024, from https://www.takepayments.com/blog/product-information/different-payment-methods/.
5. Badak, S., Kolte, V., Agrawal, M., & Gupta, S. (2023). Revolution of digital payment in India. *Journal of Mobile Computing, Communications & Mobile Networks, 10*(3).
6. Hassan, M. A., Shukur, Z., Hasan, M. K., & Al-Khaleefa, A. S. (2020). A review on electronic payments security. *MDPI, 12*(8).
7. Andeh, C., Oluyemi, A., Omeiza, A., Hammawa, M. B., & Scholar, R. (2019). Designing a new framework for securing electronic commerce Systems at Design Phase: With a special reference to an emerging economy. *International Journal of Avances in Scientific Research and Engineering, 5*(9), 92–106.

8. Surya, S., Jagtap, S. R., Ramnarayan, R., Priyadarshini, M., Ibrahim, R. K., & Alazzam, M. B. (2023). Protecting online transactions: A cybersecurity. In *2023 3rd International Conference on Advance Computing and Innovative Technologies in Engineering (ICACITE).*
9. Khando, K., Islam, M. S., & Gao, S. (2022). The emerging technologies of digital payments and associated challenges: A systematic literature review. *MDPI, 15.*
10. Chinanu, U., & Amujo, O. (2018). Comparative survey of cyber-threat and attack trends and prediction of future cyber-attack patterns. *International Journal of Innovative Research in Computer and Communication Engineering, 6*(4).
11. Khan, B. U. I., Olanrewaju, R. F., Baba, A. M., Langoo, A. A., & Assad, S. (2017). A compendious study of online payment systems: Past developments, present impact, and future considerations. *International Journal of Advanced Computer Science and Applications, 8*(5).

Index

S. Vijayalakshmi et al. (eds.), *Smart Power Systems*, Power Systems,
https://doi.org/10.1007/978-3-031-89459-6

T

V

Z

GPSR Compliance

The European Union's (EU) General Product Safety Regulation (GPSR) is a set of rules that requires consumer products to be safe and our obligations to ensure this.

If you have any concerns about our products, you can contact us on ProductSafety@springernature.com

In case Publisher is established outside the EU, the EU authorized representative is:

Springer Nature Customer Service Center GmbH
Europaplatz 3
69115 Heidelberg, Germany

Batch number: 10755329

Printed by Printforce, the Netherlands